# Practical Electron Microscopy

# Practical Electron Microscopy

Edited by
Nickolas Holloman

Willford Press
www.willfordpress.com

Published by Willford Press,
118-35 Queens Blvd., Suite 400,
Forest Hills, NY 11375, USA



ISBN: 978-1-64728-453-4

**Cataloging-in-publication Data**

Practical electron microscopy / edited by Nickolas Holloman.
p. cm.
Includes bibliographical references and index.
ISBN 978-1-64728-453-4
1. Electron microscopy. 2. Electron microscopy--Technique.
3. Electron microscopes. I. Holloman, Nickolas.
QD906.7.E37 P73 2023
548.83--dc23

For information on all Willford Press publications
visit our website at www.willfordpress.com

Willford Press

# Contents

# Preface

I am honored to present to you this unique book which encompasses the most up-to-date data in the field. I was extremely pleased to get this opportunity of editing the work of experts from across the globe. I have also written papers in this field and researched the various aspects revolving around the progress of the discipline. I have tried to unify my knowledge along with that of stalwarts from every corner of the world, to produce a text which not only benefits the readers but also facilitates the growth of the field.

Electron microscopy (EM) is a technique through which high resolution images of biological and non-biological specimens are obtained. Electrons are the primary source of illuminating radiation that provide high resolution images. This technique is applied in biomedical research to examine the detailed structure of tissues, cells, organelles and macromolecular complexes. The two main types of electron microscopes include transmission electron microscope (TEM) and scanning electron microscope (SEM). The TEM is used to view thin specimens such as tissue sections and molecules through which electrons can pass, generating a projection image. Common applications of TEM include viewing the interior of cells (in thin sections), the structure of protein molecules (contrasted by metal shadowing), and the arrangement of protein molecules in cell membranes (by freeze-fracture). In SEM, the image is formed by scanning a focused electron beam onto the surface of the specimen in a raster pattern. SEMs with appropriately equipped devices can be used to study the topography and atomic composition of specimens. The book studies, analyzes and presents the recent developments related to electron microscopy and its significance in modern times. Students, researchers, and experts of electron microscopy will likely benefit from this book.

Finally, I would like to thank all the contributing authors for their valuable time and contributions. This book would not have been possible without their efforts. I would also like to thank my friends and family for their constant support.

**Editor**

1

# Scanning Electron Microscopy with a Retarded Primary Beam

Luděk Frank

**Abstract**

The general trend for reducing the energies of primary electrons in electron microscopy has been faced with a gradual deterioration of the image resolution. Biasing the sample to a high negative voltage and making the electrons arbitrarily slow solely on and inside the sample has shown itself to be far more feasible than originally expected. The fundamental aberration coefficients (spherical and chromatic) of a combination of an objective lens and an immersion electrostatic lens formed by the biased sample decrease with the decreasing landing energy of the electrons. As a result, the spot size in scanning systems may become nearly independent of the landing energy of the electrons. The requirements placed on samples are strict but feasible, and detection of signal electrons is greatly facilitated by the acceleration of both reflected and transmitted electrons in the field of the biased sample and their collimation toward the optical axis. The interaction of slow electrons is not only more intensive than that at standard energies but even scattering phenomena appear which are not otherwise observed. Several application examples are presented. The benefits of very low energy EM are still being uncovered after its having been in routine use for several years.

**Keywords:** Scanning electron microscopy, scanning transmission electron microscopy, slow electrons, electron microscopy of materials, biomedical electron microscopy

## 1. Introduction

Historically, progress in electron microscopy and its applications in virtually all branches of science and technology, including health care, was "instrument driven" in the initial decades until around the end of the previous millennium. During this time, efforts in developing instrumentation and methodology were concentrated primarily on improving the image resolution up to its physical limitations. If we restrict ourselves to scanning microscopy techniques, the goal was to minimize the dimensions of the spot of the primary beam incident on a sample. A second aim was to seek possibilities of combining the basic electron optical imaging tool with various analytical attachments. Tailoring the ultimate solution of the instrument to a specific microstructure- or nanostructure-bound task was not unfamiliar but certainly did not predominate. Progressively enhanced tools for the computer-aided simulation of the properties of electron optical elements and systems up to the complete simulation of the imaging process enabled the community of instrumentation scientists to turn to the design of a device according to the quantities and processes that are to be monitored and measured. A concentration on the deeper understanding of the collection of the image signal via identification of those parts of the energy and angular distributions of the emitted species that are acquired by detectors goes hand-in-hand with this. To put it more simply, understanding and interpretation of the image begins to overshadow the problems of generating and shaping the exciting primary electron beam. This kind of activity is surely facilitated if the energy of the electrons incident on the sample can be freely chosen in such a way that this degree of freedom can be fully utilized.

The range of energies of the primary electron beam available for the formation of sufficiently sharp beams in the scanning electron microscope (SEM) was for a long time limited by the practical obstacles that emerge when lowering the beam energy. These include the decreasing efficiency of extraction of electrons from the gun cathode, the increasing undesired manifestation of external spurious influences in proportion to the time of flight of the beam, and, most importantly, the enlargement of the size of the disks of confusion due to chromatic and diffraction aberrations making the primary spot dimension steeply extending toward low energies. However, since the very beginning of the electron microscopy era, the possibility of having an electron energy very low at the sample and sufficiently high in the column was known from devices using immersion objective lenses [1,2]. These traditionally appear in emission electron microscopes, where the sample itself is the source of electrons, mostly excited by incident photons, which are usually emitted at units or tens of electronvolts and strongly accelerated immediately above the sample surface biased to a high negative potential. Moreover, the idea of reverting the ray path and using the sample bias to retard the impinging primary beam generated and formed at a high energy in a standard electron gun and column is similarly old [3]. This idea appeared in the literature several times in the following years, though without serious attempts at implementation, with the exception of a study presenting the very first experimental results but nevertheless not at a convincing quality [4]. Surprisingly, the first successful implementation of the idea of retarding the electron beam just on the sample was not conditioned by preliminary simulation studies or the assembly of a dedicated device; enough courage to take on a task generally considered to have little prospects was sufficient [5,6]. The crucial objection met at the very beginning was the anticipated existence of various lateral fields close to the sample surface that would destroy the image by smearing both the primary and signal fluxes when they are so slow. Practical experience proved much less critical, and the first series of micrographs of a quality consistent through the full energy range down to 1 eV was published in 1993 [7].

The first decade of collecting experience with this innovative SEM mode was summarized in a review [8], and shorter reviews were then published at intervals of a few years [9,10]. More specific reviews concerning materials science and nanotechnology applications have appeared recently [11,12]. Here, we are to extend the series of reviews presenting in brief the instrumental and methodological fundamentals and listing the recent successful application examples.

## 2. Motivation

When operating the scanning electron microscope (SEM) between, say, 5 and 10 keV, we enjoy an optimum compromise between conditions governing the primary beam formation and those characterizing the beam-sample interaction. We are able to extract sufficient electron current from all main types of sources and transport the beam over a column of tens of centimeters in length at a constant energy with a reasonable impact of spurious influences. Similarly, the corresponding wavelength of electrons and the energy spread of the beam in tenths of electronvolts, together with main aberration coefficients in millimeters, enable us to get a spot size of 1 nm or even slightly less. At the same time, the interaction volume of electrons of this speed in solids varies between units and hundreds of nanometers in size according to the sample material, although the true image resolution might be nearer to the spot size, provided sufficiently sharp structure details are present on the sample surface. Total yields of secondary and backscattered electrons are more or less comparable and can be easily separated as image signals. Conventional detectors acquire a part of the secondary electron (SE) emission attracted to a side-attached assembly with a front bias or a hollow cone of straight trajectories of backscattered electrons (BSE) with a ring-shaped diode or scintillator. The relief contrast of SE and material and channeling contrasts of BSE dominate routine SEM practice.

The observation of thin samples in a scanning transmission electron microscope (STEM) is traditionally performed at around 200 keV in dedicated instruments and around 30 keV in SEM devices equipped with a STEM attachment. While tissue sections for the former microscopes are prepared at hundreds of nanometers in thickness and samples containing heavier elements are correspondingly thinner, at tens of kiloelectronvolts, we have to go down to 100 nm and below. Detectors of transmitted electrons are split into several concentric rings so that the signal species can be sorted according to the scattering angle from the optical axis. This helps in obtaining sufficient contrasts from samples providing larger differences in local thickness or in atomic numbers of constituting elements. Of utmost importance here are the challenges of sample preparation, mainly as regards the aims of highlighting certain structure details considered interesting. Thanks to the concentrated illumination of primary electrons, samples for STEM can be somewhat thicker than those for TEM, but otherwise the experimentation issues are similar.

The traditional arrangement characterized above suffers from several drawbacks. Fast primary electrons interact mainly with atom cores, elastic forward scattering dominates, and inelastic scattering is weak. Scattering phenomena can be mostly explained within the laws of classical physics and the emitted signals do not respond to the quantum mechanical interior of the samples. The interaction volume of the beam in solid targets is too large, for which reason embedded structures such as precipitates are imaged fuzzy and both thin surface films and a great many relief details are invisible. The majority of signal species generated in depth do not penetrate to the surface to be emitted. Nonconductive materials charge up negatively to a surface potential in the kilovolt range. The edge effect, i.e. overbrightening of steeply inclined surface facets, impedes the measurement of distances and dimensions in SEM images. Fast BSE move along straight trajectories and the majority of them escape detection and impinge on the chamber walls. The main problems of STEM at standard energies include low contrasts, particularly with samples of living matter that need to be prepared with salts of heavy metals that highlight some but not all structure details, and the averaging of structure details across the sample thickness, which may hide some of them completely.

When decreasing the primary beam energy well below 5 keV or even to hundreds, tens, or units of electronvolts, we enter a different world of imaging conditions. Scattering events are increasingly dependent on the complete 3D potential distribution inside the target, i.e. they sense truncation of atomic potential by electrons and electron-electron interactions in general. Inelastic scattering and elastic backscattering become more important, so the crystallinity contribution to image contrast is enhanced. Below about 30 to 40 eV, electrons entering the target start behaving as Bloch waves, so their penetration inside is conditioned by the presence of unoccupied electron states in the particular direction and slow electron reflectivity can serve as a measure for the density of states and, therefore, can serve to fingerprint the orientation of grains in polycrystals, for example [13–15]. Interaction volume diminishes and information carried by emitted electrons is better localized laterally, while the surface sensitivity also increases, enabling one to observe all surface details and coverage. The edge effect disappears when the penetration depth stops exceeding the escape depth of signal species. Surface charging is reduced with increasing yield of SE, and at the "critical energy", at which the fluxes of incident and outgoing electrons are identical, no charge is dissipated and noncharging microscopy is possible under any vacuum conditions. The total yield of signal electrons reaches its maximum somewhere in the range of hundreds of electronvolts, exceeding the unit level with minor exceptions. In this range, nonconductive samples charge positively, although charge balance is again achieved at the even lower first critical energy. The positive charging is moderate only, thanks to the partial retraction of the slowest emitted electrons by the field of the surface potential. The wavelength of electrons extends, and when it becomes similar to interatomic distances at units of electronvolts, interference of waves reflected on both sides of surface atomic steps (divided wavefront interference) or on both surfaces of ultrathin surface coatings (divided amplitude interference) becomes a new contrast mechanism. As in LEED, backscattering in the range of tens of electronvolts concentrates BSE to diffracted spots that may reveal the surface crystallinity when selectively detected. Moreover, electrons leaving the sample close above the surface barrier are partially reflected, so the height and shape of the barrier contribute to the imaging signal.

When reducing the primary beam energy in a STEM, we obtain, first and foremost, higher contrast. The differential cross-sections governing both the elastic and the inelastic scattering mechanisms are inversely proportional to the energy of electrons, mostly to energy squared.

For this reason, the spatial density of the generation of the image information steeply increases so that, below 1 keV, we can get, for example, on tissue sections containing no agents with heavy metal salts not only very high contrast but also all structure details visualized, including those normally not highlighted with postfixation or staining media [16]. Naturally, the sections have to be thinner. For the range of hundreds of electronvolts, thicknesses below 10 nm are desirable. The production of such sections is, however, already feasible [17]. Moreover, thanks to reduced in-depth averaging of image layers, we get the chance of revealing tiny details unknown to date. With true 2D crystals of a single atom thickness, such as graphene, we can get nonnegligible electron penetration down to 1 eV [18] and obtain data that is extremely valuable for applications in nanoelectronics, for example.

The reasons for reducing the energy of electron incidence on targets in the scanning devices down to the lowest values, including the principle of implementation described below, have existed in the historical literature from the very beginning of the electron microscopy era and have been permanently demonstrated and supported more recently with indisputable results from the beginning of the 1990s. Starting from the 12th EUREM in Brno in 2000, the method was discussed at all large microscopy conferences and congresses in dedicated sessions. Still, the first commercially available SEMs allowing operation at tens of electronvolts did not appear until 2006/2007, and even then, the volume of application results in the literature remained extremely limited. Only quite recently, in the last 3 or 4 years, the "cathode lens" method has become a widely used standard tool in SEM practice. This circumstance starts erasing reasons for compiling review texts addressing just the method and transfers the observation material to reviews devoted to families of specimens. Nevertheless, the aim of this chapter is to summarize the fundamentals of the method, although it also features a list of successful applications in which very low energy SEM provides an important added value.

## 3. Implementation

The problem of having very slow electrons on the sample surface and fast electrons in the column outside the sample was solved as early as 1932 with the immersion objective lens [1]. The first considerations counted on the sample as the source of electrons, but since 1942 a design has existed for an electron microscope delivering fast electrons toward the sample biased to a high negative potential in order to retard the electrons immediately before their impact on the sample surface [3]. The leading idea at that time was to reach energies of impact that produce the maximum yield of secondary electrons. For these reasons, there was really nothing new in the attempts to revitalize the principle that appeared sporadically over the following 50 years, though without convincing experimental results. With hindsight, we can now say that past attempts were probably not initiated with sufficient seriousness because of doubts connected with the necessity of heavy biasing of the specimen, which indicated conditions that were later shown to be not so strict, such as the necessity of having the sample surface very finely polished, similarly to the electrodes of electrostatic lenses.

The principle of implementation of very low energy SEM and STEM is simply as follows: the sample is biased to a high negative potential similar to the negative potential of the cathode in the electron gun of the microscope. The energy of electron impact on the sample is then given by the difference between the two potentials. An adverse factor here is the fact that, while the potential values are subtracted, their fluctuations are added, which amplifies the noise at the lowest impact energies quite substantially. It is much better to implement a "low voltage booster", i.e. a positively biased axial tube insulated from the surrounding column that accelerates electrons just behind the gun anode and returns them to their original energy at the end of the objective lens [19,20]. The impact energy is then simply provided by another supply connected between the gun cathode and the sample. If we close the booster field at the end of the objective lens, we get a compound lens consisting of a (usually magnetic) focusing lens and an electrostatic retarding lens, an assembly long known [21] and incorporated into many commercially available electron microscopes under the name the Gemini lens.

The core concern is to know the electron optical parameters of the combination of the objective lens of the SEM with the electrostatic field above the sample. This field is, in fact, generated in a two-electrode immersion electrostatic lens with the sample serving as the cathode; this assembly is often called the "cathode lens" (CL). The anode of this lens has to have a central bore of some kind passing the primary beam so its field can be seen to be composed of a homogeneous retarding field and a field penetrating through the anode bore. On a three-electrode arrangement (cathode/grid/anode), Recknagel [2] derived an image formation theory employing expansion in $\varepsilon/U$, where $\varepsilon$ is the electron energy in the surface plane of the sample and $U$ is the acceleration voltage inside the lens. The analysis by Lenc and Müllerová [22,23], considering the anode field approximated with a second-order polynomial within a thin transition region in the anode plane, is more transparent. The main quantities, i.e. the spherical $C_S$ and chromatic $C_C$ aberration coefficients, were written (for large ratios $E_P/E$) as [23]

$$C_S \cong \frac{l}{E_P} E + \left( \frac{l^2}{D} + \frac{81}{16} C_S{}^f \right) \frac{1}{E_P{}^{3/2}} E^{3/2} + \ldots, \quad C_C \cong -\frac{l}{E_P} E + \frac{9}{4} \frac{C_C{}^f}{E_P{}^{3/2}} E^{3/2} + \ldots \tag{1}$$

where $E$ is the landing energy of electrons, $E_P$ is the primary energy before retardation in the cathode lens, $l$ is the length of the cathode lens field, $D$ is the diameter of the anode bore, and $C_S{}^f$ and $C_C{}^f$ are the relevant aberration coefficients of the focusing objective lens. Obviously, the lower the landing energy of the electrons, the smaller the aberration coefficients, which is a relationship opposite to that valid for standard configurations without immersion, i.e. with energy independent aberrations. When decreasing the landing energy down to units of electronvolts, the aberration coefficients tend to fall in the micrometer range, which compensates very efficiently for other influences adverse from the point of view of image resolution. In Figure 1, we see typical energy dependences of the spot size for the sample bias switched on and off, in both cases with the beam aperture tailored separately to each landing energy, together with a particular case of the beam aperture before retardation fixed to a value optimum for a certain landing energy. At very low energies, the spot size deteriorates proportionally to $E^{-3/4}$ in the standard SEM, while the cathode lens reduces this slope to $E^{-1/4}$. For larger aberrations of the focusing lens, we can even get an energy-independent spot size [8].

We should mention here that, in the case of the Gemini compound lens, the working distance of the immersion electrostatic lens is nonzero (in contrast to the CL), so the basic aberration coefficients do not keep decreasing proportionally to the landing energy $E$ without limitations, but the relevant equations contain an "absolute" member proportional to the working distance [24] so the spot size remains acceptable only down to a certain energy threshold. A comparison of the Gemini and cathode lenses is discussed in great detail in Ref. [25].

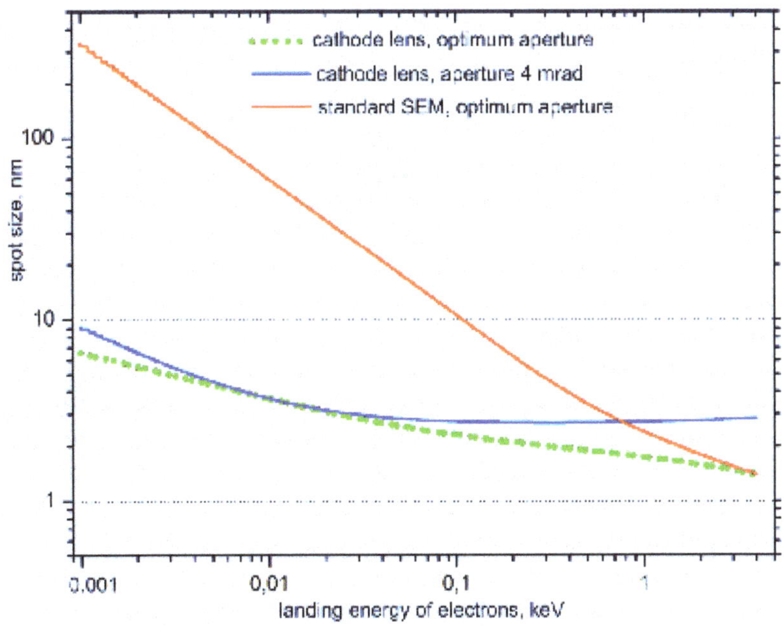


**Figure 1.** Energy dependences of the spot size for a typical magnetic focusing lens combined with a cathode lens.

The advantages of an assembly containing a cathode lens include the landing energy of electrons easily adjustable by the sample bias with the alignment of the microscope column untouched. Nevertheless, by altering the bias, we also vary the optical parameters of the electrostatic lens, first and foremost, the object distance of the sharp spot size and the dimensions of the field of view currently adjusted with scanning coils or electrodes. Approximate analytical calculations have shown these variations to be only moderate: at very low landing energies, the final magnification ranges between 1/2 and 2/3 of that with the CL off, in dependence on the position of the pivot point of the scanning system [7]. As regards the underfocusing necessary to compensate the cathode lens action, this amounts to units of micrometers per electronvolt at very low energies [7]. Algorithms suitable for correction of magnification and focusing in dependence on the cathode lens excitation can be found in Ref. [26]. Variations in the beam aperture and the beam impact angle with the landing energy are more important. In well-adjusted CL, we decrease only the axial component of the velocity of electrons, so the beam aperture enlarges, roughly in proportion to $(E_P/E)^{1/2}$. Similarly, amplification always occurs to the angle of impact of the primary beam on the sample surface when the rocking style of scanning is employed with a pivot point above the sample. Although the impact angle normally grows toward the margin of the field of view quite negligibly, after this "amplification", it has to be considered, particularly when imaging crystalline samples exhibiting channeling contrast. At very low energies, we also have the field of view restricted by the total reflection of electrons taking place at a distance from the optical axis at which the glancing impact of electrons is achieved.

The cathode lens is aligned when the above-sample field is homogeneous and the nondeflected primary beam impacts normally on the sample. If any sample tilt is required, we have to count on some smearing of the primary spot due to the lateral field generated by the tilt. Moreover, the inclination of the beam by the mechanical tilt angle is again multiplied by the ratio of axial velocities before and after retardation. In this case, multiplication of the impact angle might prove advantageous. Practical experience has shown that all these apparent drawbacks of an assembly containing a cathode lens are manageable in practice without placing a great burden on the operator.

The alignment of the cathode lens consists of placing the anode bore on the optical axis and adjusting the sample by means of slight tilts to a parallel position with respect to the anode. Both these conditions are easily controllable with an anode made from a scintillator plate, preferably single crystal, sensitive on both sides. We then "see" the upper surface of the anode/detector and, at low magnification, can align its bore laterally to the screen center. At very low energies, we begin observing the above-sample assembly by means of electrons backscattered from the sample surface near the total reflection. Electrons emitted from individual sites on the sample are collimated in the strong field into narrow bundles so those emitted around the center of the field of view mostly escape through the anode bore. For this reason, we see the same anode bore dark from the bottom, and when both these circles are made concentric, the CL field is homogeneous. Reliable adjustment requires a sample stage equipped with two independent, mutually perpendicular sample tilts. A combination of sample tilt and rotation may also be helpful. The alignment is illustrated in Figure 2. The same approach may also be successful for other versions of the anode if the anode bore can be made visible around the view-field center.

The adverse external influences typical of the SEM include stray fields, primarily an external magnetic field. The deflection of the primary beam due to such a field depends on the time of flight of the beam, which decreases when reducing the range of flying at low energy. This is minimized in the case of zero working distance of the immersion lens, i.e. the cathode lens. When comparing the deflection at 20 keV and 200 eV, we get a ratio of 1:10 without the cathode lens but just 1:1.5 if the landing energy of 200 eV is obtained by retardation from 20 keV in the CL [25].

A strong electric field around the sample affects the primary electrons, although it affects the signal species even more because of their widely varying energy and directions of motion. Now, let us consider the negatively biased sample inserted between two grounded detectors, of which the upper one is either the CL anode or any detector above the anode to which electrons are passing through the anode bore and the lower detector is of an arrangement typical for the scanning transmission electron microscope (STEM) instrument. The homoge-

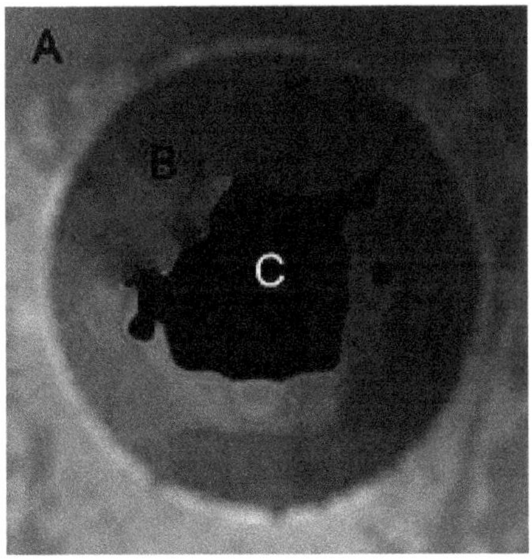


**Figure 2.** Illustration of a well-aligned cathode lens: the bored scintillator disc of the BSE detector also serving as the anode (A), sample surface biased to a mirroring voltage (exceeding the acceleration voltage in the gun) (B), and mirror image of detector bore (the irregular contour is due to the unevenness of the sample consisting of a foil deposited on a mesh) (C).

neous field of the CL accelerates the backscattered or transmitted electrons off the sample surface and collimates them into a bundle of a width decreasing with increasing field strength. The complete angular distribution for the full polar angle range (0, π/2) is concentrated in a circle of diameter [8].

$$d = \frac{4l}{\sqrt{k_e - 1}} \tag{2}$$

where $k_e$ is the ratio of the final and initial energies of the emitted electrons. Generally, electrons are emitted at energies ranging between zero and the landing energy $E$, and although this interval may be quite short in comparison to the final energy, the factor $k_e$ may vary widely. We then get the signal electrons in the anode plane sorted according to their emission energy, which enables a certain energy filtering. In particular, secondary electrons (SE), if released at a low landing energy, are usually collimated to within a diameter of tenths of millimeters so they mostly escape through the anode bore, and if the anode also serves as the detector, we get an image signal composed almost exclusively of backscattered electrons (BSE). It is important that the complete BSE emission is usually collimated to a diameter in units of millimeters so we also acquire BSE emitted at large angles with respect to the surface normal that are usually omitted in conventional SEMs. As we will see later, SE also contribute to the transmitted electron (TE) signal acquired with a below-sample situated detector because they are accelerated similarly as TE passing the sample with various energy losses. Having the TE detector composed of traditional concentric sensitive detector rings for acquisition of bright-field (BF), dark-field (DF) and high-angle annular dark-field (HAADF) signal components, we should get SE in the BF channel only.

In modern SEMs, the sample is often immersed in a strong magnetic field of an open objective lens in order to improve the electron optical parameters of the column. When combining the open magnetic focusing lens with the CL, we face the question of the trajectories of signal electrons, particularly as regards the interpretation of their initial angular distribution on the basis of currents obtained in detector channels. As Figure 3 shows, the mismatch in the angular distribution normally appearing with the sample in a magnetic field is nearly fully eliminated by means of the electric field for $k_e = 11$.

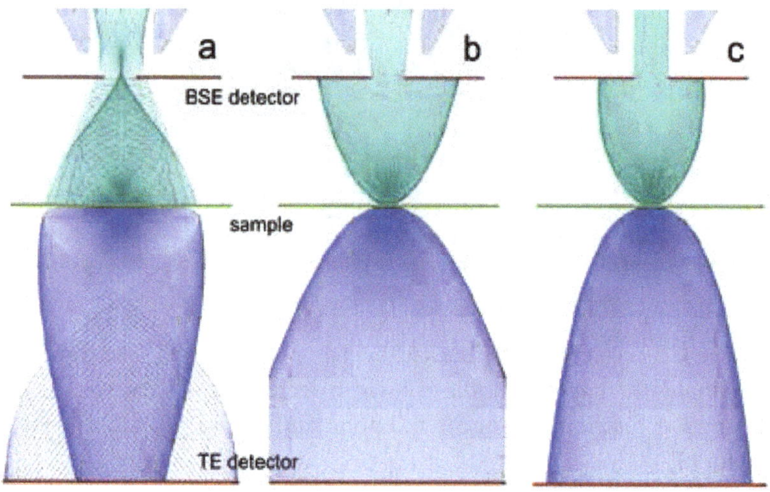


**Figure 3.** Trajectories of elastically backscattered and transmitted electrons from a sample immersed in the magnetic field of an open magnetic lens (a), from a sample to which the primary electrons are retarded 11 times (b), and from a sample surrounded by both fields (c).

As regards the detectors themselves, they are impacted by electrons accelerated approximately to the energy of electrons acquired in the microscope gun. The standard BSE detectors of SEMs, whether scintillator-based or semiconductor, are tailored to this energy, so no special precautions regarding detection are needed; every BSE detector installed in a particular SEM will work when we bias the sample and create a CL in its surroundings. Alternatives are listed in Ref. [8].

Slower electrons penetrate more shallowly into targets so the information depth of low-energy electron imaging shortens. Increased surface sensitivity provides enhanced information about the topography and possible coating of the surface, naturally including any contamination that may have concealed the surface to be examined. At hundreds of electronvolts and below, we enter the branch of instrumentation normally employed by surface physics methods such as low-energy electron diffraction, photoelectron, or Auger electron spectroscopy, in which ultrahigh vacuum (UHV) conditions are used and any sources of hydrocarbons have to be avoided. Experience has confirmed this assumption, although there is still the chance of improving surface cleanliness even under standard high (though dry) vacuum conditions where we have to remove primarily the adsorbed hydrocarbons loaded with the sample.

Electrons at energies below 50 eV have proven themselves in removing the hydrocarbon molecules instead of decomposing them and creating the well-known carbonaceous contamination marking previous fields of view in the SEM [18,27,28]. This kind of cleaning caused, for example, the transmissivity of single-layer graphene to increase 2.5 times after bombardment with a dose of 1.3 $Ccm^{-2}$ of 40 eV electrons [18].

Here we should provide answers to two frequently asked questions: How flat/smooth should the sample surface be? The CL field strength is usually about 2 or 3 kV/mm, i.e. 2 or 3 V/μm. Similarly strong lateral fields are generated with steeply inclined facets of surface unevenness details, with protrusions being more critical than dips. The tolerable relief height then depends on the landing energy we want to use. Protrusions in units of micrometers in height are well acceptable down to tens of electronvolts, whereas, at units of electronvolts, we can observe certain image deformations around surface steps around 1 μm in height. However, gently undulating surface structures with a p-p distance in units of micrometers are usually not apparent at all as disruptive damage to the image but are visible as surface topography details.

Can we image nonconductive samples? When placing a slab made from an insulating material on the conductive base of the sample holder and biasing the holder with respect to the surrounding ground, we get a situation equivalent to a capacitor with multilayer dielectric, composed here of a vacuum and a nonconductive sample slab. The potential drop then splits to both media according to their dielectric constants, so we get a somewhat lower negative bias on the sample surface than that led to the holder. However, when applying "good practice" in calibrating the landing energy scale, which consists of identifying its zero according to the disappearance of all topography details on a sample area as flat as available, we do not need to know the potential drop over the sample. What remains is to consider the landing energy scale modified according to the division between the sample and vacuum of every volt added to the holder bias, thereby giving the sample surface potential changed by less than 1 V. The potential drop across the sample can be detected when also checking the zero landing energy on the holder aside from the sample and the difference in both holder biases so determined can be used to correct the landing energy scale.

Since around a decade ago, the main producers of electron microscopes have been offering a cathode lens mode in their new instrument types, calling it beam deceleration, gentle beam, or decelerating optics. However, advances in the collection of new application results have been slow and seem to have been accelerating only quite recently as the Microscopy and Microanalysis Meeting 2015 has indicated. Owners of older SEMs are not completely excluded from enjoying the advantages of the CL mode in their instruments: simpler microscopes, more feasible adaptation; see Ref. [7]. The detection issue is solved by any functional BSE detector. All that is necessary is to insulate the sample from the stage and to connect it to a high-voltage feedthrough into the sample chamber. It is usually sufficient to insert the sample into a capsule made from a good insulating material and then load this capsule onto the stage. What remains is to connect the sample to the feedthrough in a way that does not cause havoc in the sample movements. More obstacles appear, of course, if the sample is loaded via an airlock. Nevertheless, plenty of positive experience was gathered with such adaptations in the 1990s.

## 4. Applications

Since the CL mode was first used for the practical examination of SEM samples, we have continued to map various families of samples in order to identify possible added value introduced by this mode. If the results of such a survey are to be efficiently demonstrated, it would seem to be natural to compare side-by-side micrographs taken at conventional energies at several kiloelectronvolts or more, with CL mode images differing not only in terms of the significantly reduced landing energy of electrons but usually also in terms of the collection of a broader or even complete angular distribution of backscattered electrons. In this chapter, we will follow this style of presentation for a selection of sample types often appearing in SEM practice.

### 4.1. Surfaces

The most straightforward expectation connected with decreasing the landing energy of primary electrons on the sample is their reduced penetration into the sample. Shortened information depth, along with reduced lateral diffusion, produces enhanced surface sensitivity, i.e. improved visibility of topographic details such as tiny dips, protrusions, and ridges, and also the sudden appearance of very thin surface coverage that is fully transparent and invisible at conventional energies. When comparing the two frames in Figure 4, we can identify examples of both these types of differences. Here, the penetration depth of primary electrons is the main factor; however, as shown below, the contribution of signal electrons from a broad range of polar angles of emission also plays a role.

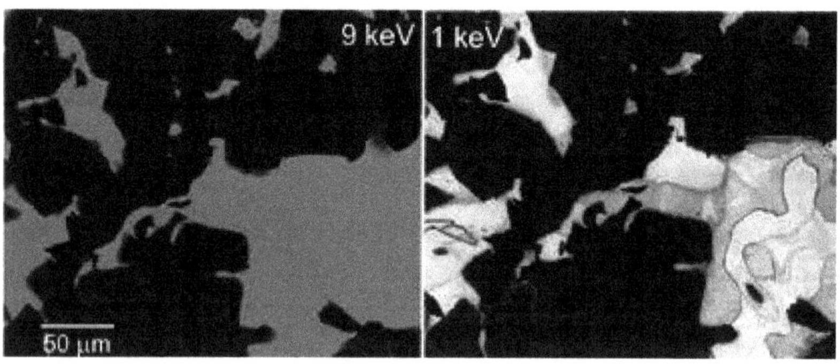


**Figure 4.** Carbon nitride film 200 nm in thickness deposited on a silicon substrate covered by around 5-nm-thick native $SiO_x$, delaminated due to compressive stress, CL mode, primary energy 9 keV.

The example shown in Figure 5 addresses a typical three-dimensional (3D) sample in which imaging of a highly complex surface structure requires the dimensions of the interaction volume of primary electrons in the material not exceeding the dimensions of the 3D details to be observed. The averaging of information over the volume from which BSE are capable of reaching the surface to be emitted smears out details smaller than, for example, hundreds of nanometers at 10 keV. Nevertheless, the interfaces between "bubbles" and vacuum are reproduced relatively sharply because the signal in pixels situated closely outside these bubbles does not include electrons diffusing through the material, so that the averaging is abruptly terminated at the margin.

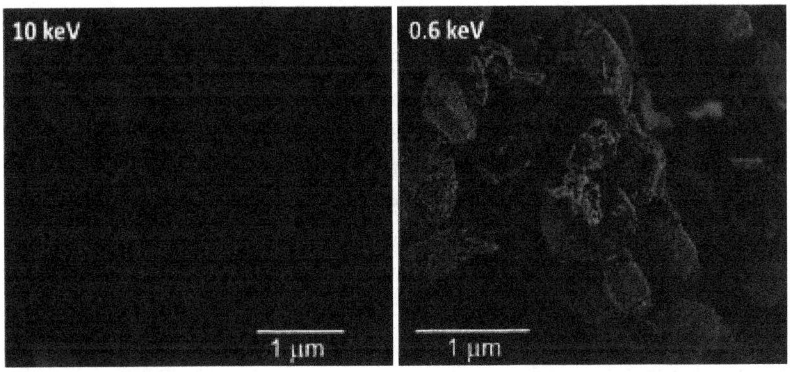


**Figure 5.** Mesoporous carbon nitride foam as a carrier for catalytic gold nanoparticles, CL mode, primary energy 10 keV.

We might complete the paragraph concerning surface imaging with one problematic issue often mentioned where low electron energies are concerned. It has been said many times that decreasing the energy of incident electrons leads to reduced radiation damage of the sample. On one hand, every electron brings less energy, so certain inelastic events may no longer take place. On the other hand, the penetration depth shortens with decreasing energy faster than linearly, so dissipation of the delivered energy takes place in significantly shallower subsurface layer at growing spatial density of the dissipated power. From this point of view, the radiation damage connected with mechanisms active down to any particular low energy grows downward this energy. A well-known example is the creation of "black rectangles" marking the fields of view bombarded for a longer time with illuminating electrons. This carbonaceous contamination comes from the decomposition of adsorbed hydrocarbons under the impact of electrons. The more intense frames of these rectangles are caused by diffusion of the hydrocarbon molecules from the surrounding, so the phenomenon might be partly suppressed if we first immobilize the hydrocarbons around the next field of observation by electron impact. It is easy to verify that the intensity of this kind of contamination increases down to 100÷200 eV, i.e. to the fuzzy threshold of the shortest inelastic mean free path of electrons in solids, and only then fades out. Below some 50 eV, the radiation damage may usually be neglected.

### 4.2. Immersed objects

In the previous paragraph, we presented samples with a heterogeneous or buckled or discontinuous surface coating and with a ragged spatial structure containing tiny protrusions and depressions. However, even when the sample is ideally flat and smooth and not covered with any thin layer, we may be confronted with imaging issues in cases in which very small objects are immersed just below the surface with the object tops lying on the same level as the neighboring surface. Good examples are precipitates in alloys prepared with an overall flat and smooth surface as shown in Figure 6. Here, we compare standard SEM micrographs at 10 keV in the BSE and SE signals. Precipitates are apparent in both frames, but most appear fuzzy. In the BSE image, the main contrast contribution comes from the atomic number difference, i.e. "material contrast". The precipitates are composed of either $Mg_2Si$ or $Mg_3Si$ [30,31], so the difference of the mean atomic number with respect to that of Al is only 0.33 or 0.5, which is obviously sufficient for good contrast. The relatively sharp edges of the precipitates indicate an in-depth thickness much smaller than the information depth of the imaging, which reveals the margin information borne only by the BSE scattered in the narrow upper part of the pear-shaped interaction volume. In contrast to this, the SE appearance of the precipitates is fuzzier due to the contribution of the SE2 species released by BSE returning from the sample depth in a broad flux. The CL mode frame, thanks to much smaller interaction volume, not only shows quite sharp edges of the precipitates but also reveals their internal structure with a bright frame and dark core. The explanation of the internal structure obviously also has to incorporate the crystal structure of the precipitates, providing a specific contrast contribution when a sufficiently broad angular range of BSE is acquired, as we will show below.

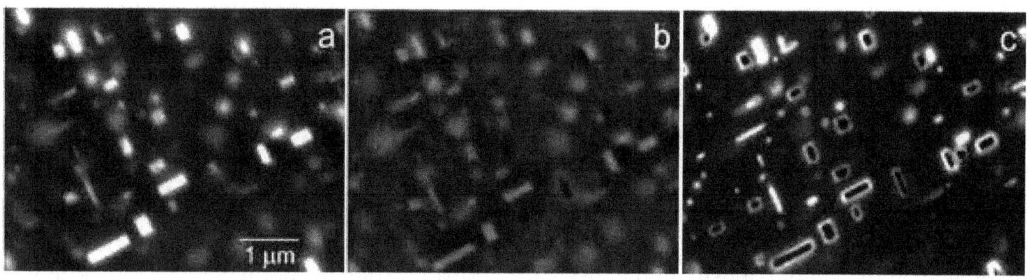


**Figure 6.** Precipitates in Al-1.0 mass% $Mg_2Si$ with 0.4% excess Mg alloy, annealed, quenched, and age hardened: standard BSE image at 10 keV taken with a coaxial detector (a), SE image at 10 keV taken with a side-attached Everhart-Thornley detector (b), and the cathode lens image at 1,500 eV for 10 keV primary energy (c).

### 4.3. Local crystallinity

At high or medium electron energies, i.e. down to hundreds of electronvolts, the local crystallinity of the sample manifests itself in the channeling contrast. Incoming electrons are scattered at a rate proportional to the density of atomic planes they face, and if dense planes are inclined or even perpendicular to the surface, electrons penetrate into depth along interatomic channels and the probability of their backscattering decreases. For this reason, we get a BSE signal dependent on the local crystallinity. At hundreds of electronvolts, the scattering on atom cores begins to be combined with interaction with electrons in the target [32], and at even lower energies, below some 30 or 40 eV, the scattering is on a pure electron basis. If the incident electrons are of an energy close above the vacuum level, after gaining the inner potential, they appear within unoccupied energy bands that are already modified with the internal 3D potential distribution from the parabolic bands of free electrons and hence acquire a dispersion characteristic to the crystal system and its spatial orientation. Effectively, the impinging electrons convert into Bloch waves and move preferentially in directions of a high density of states. The reflectivity of electrons this slow then depends on the local density of states in the direction of impact of electrons on the crystal, i.e. electron states coupled to the incident electron wave. These have a surface-parallel wave vector component equal to that of the incident wave or differing in any surface reciprocal lattice vector [13,15,33].

We have already mentioned the enhanced crystal information contained in the BSE signal acquired at angles further from the surface normal. As we see in Figure 7, the polar-angle-sorted BSE imaging is composed of multiple contributions. The sample shown in this figure does not exhibit any local material contrast of details distinguishable at the given magnification, so we can compare contrasts of the grain orientation, grain boundaries, and the surface topography. Visibility of the grain boundaries dominates nearest to the optical axis; at higher angles and at angles near to 90°, we get the topography, whereas, between these angular intervals, the signal dependence on the grain orientation is most pronounced. Possible material contrast would appear at angles near to the optical axis. When assessing this figure, we should take into account that SE are also accelerated in the cathode lens field and then appear detectable very near to the optical axis. Their contribution is responsible for the edge effect visible in Figure 7a. Thus, as follows from Figure 7, the crystal contrast is best acquired at emission angles above 40° or 50° that are abandoned in conventional configurations of BSE detectors.

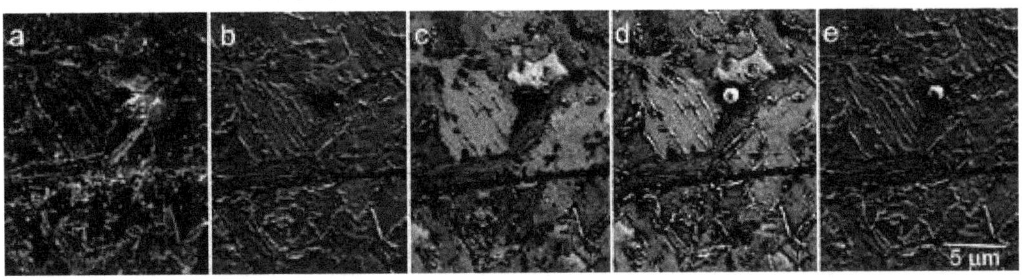


**Figure 7.** TRIP-aided bainitic ferrite (TBF) steel, imaged in the CL mode at 500 eV with primary energy of 4.5 keV, micrographs taken within the polar angle ranges of backscattered electrons: 0°–15° (a), 17°–26° (b), 28°–42° (c), 44°–61° (d), and 63°–90° (e).

Now, let us turn to the electron energy dependence of the crystal contrast. As Figure 8 shows, the acicular martensite structure is best visible around 500 eV, whereas, at kiloelectronvolt energies and also at tens of electronvolts, this kind of contrast is much weaker. The proposed explanation is as follows: at units of kiloelectronvolts, the image information is averaged within the depth range exceeding 100 nm, so the thin martensitic whiskers variously rotated with respect to the electron impact direction are averaged as regards their channeling ability, which reduces the resulting contrast. Moreover, this micrograph was taken with the cathode lens off, so only the BSE moving along straight trajectories within a rather narrow angular interval were acquired and the high-angle BSE providing more crystal contrast were abandoned. An optimum balance between the information depth and the thickness of the contrast producing structure details appears somewhere around 500 eV, and here, the full emission of the BSE is also detected. At 50 eV, the information depth falls below 1 nm, to which subsurface layer the structure features responsible for the contrast obviously do not raise. Having the information depth incorporating only two to three atomic layers, we may see only the surface reconstructed crystal that tends to convert differently oriented grains to rather unified structure. Another view might be based on a general fading of the channeling contrast at these energies.

**Figure 8.** Martensitic steel imaged in the CL mode with a primary energy of 6 keV.

Figure 9 refers to the above-mentioned contrast mechanism connected with the local density of electron states. We have to restrict ourselves to energies at which the absorption of hot electrons, proportional to the imaginary part of the crystal potential, is sufficiently low in order not to overwhelm the expected phenomenon [34]. As we see in the EBSD map, two of the Al grains in Figure 9, grains A and B, are near the orientation (111). Their reflectivity curves in Figure 9d are really quite similar, at least as regards the positions of the maxima and minima. The orientation of grain C is nearer to the middle of the color-coding stereographic triangle and the corresponding reflectivity dependence differs mainly below 20 eV. Grain D with an orientation near to (100) has a very different reflectivity. Obviously, having available reference curves for fundamental orientations of a particular crystal, we may be able to identify orientations of grains in a polycrystal of the same material [15]. As a potential alternative to the commonly used EBSD method, this approach would require an investment of decades of effort in data processing algorithms and tools, as was made during EBSD development. In principle, the low-energy reflectivity method offers better spatial resolution available at lower energy with a nontilted sample and faster data acquisition, thanks to only one-dimensional data collected for every pixel.

### 4.4. Internal stress

In heavily deformed materials, some residual internal stress remains even after any deforming action is terminated. Consequently, elastic or plastic deformation of the crystal or of the grains in a polycrystal exists and should be observable in view of the importance of this parameter for material development and diagnostics. In relaxed grains, the BSE signal level revealing the penetration of primary electrons along the interatomic channels is constant over the grain area

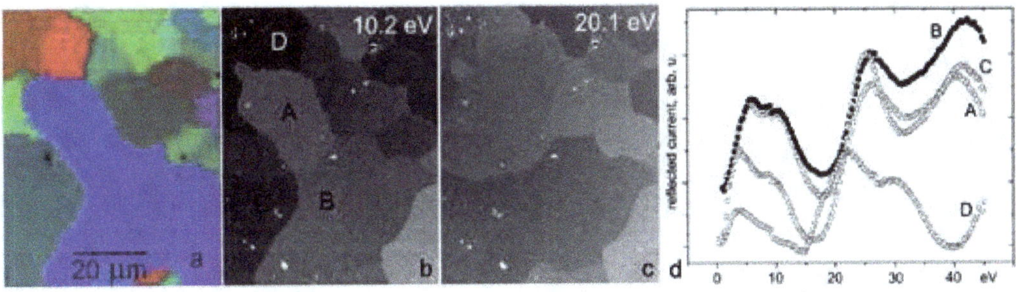


**Figure 9.** Identification of crystal grains in Al on the basis of reflectivity of very slow electrons: EBSD map (a), CL mode micrographs (b and c), and energy dependence of the reflectivity of selected grains (d) (reproduced from Ref. [29]).

up to the grain boundary where dislocations and other defects are normally clustered. If microscopic deformations exist inside a grain, the conditions for the channeling of electrons vary even inside the grain, which leads to signal variations proportional to the extent of deformation and correspondingly distributed. Gradual signal variations without abrupt changeovers are characteristic, as Figure 10 demonstrates on a margin of the Vickers indent. On the inclined wall of the indent, we may observe plastically deformed grains of internally varying brightness owing to changes in crystallographic orientation, whereas, outside the indent, the grains are of a constant signal over their full area.

**Figure 10.** Margin of a Vickers indent in an annealed polycrystalline Cu sample, CL mode at 500 eV showing the residual stress distribution inside the indent (left) and relaxed grains (right).

Figure 11 demonstrates the dependence of the visibility of grain deformation on the energy of electrons. Similarly as in Figure 8, we have the contrast culminating around 500 eV and nearly invisible at the nonreduced primary beam energy or at 50 eV. The reasons for this phenomenon

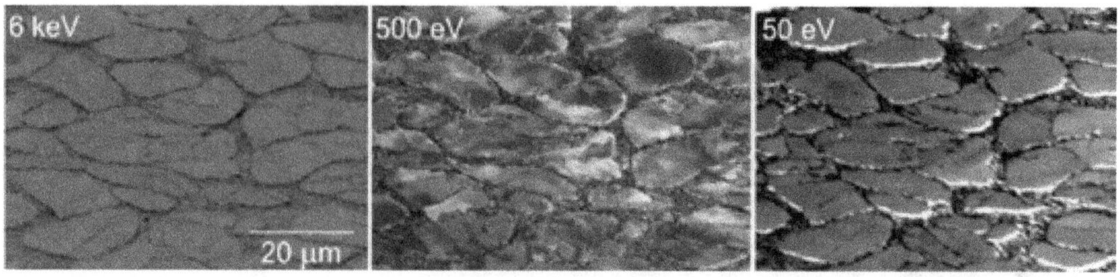


**Figure 11.** X210Cr12 ledeburitic steel heated to a semisolid state, heavily deformed and cooled, CL mode micrographs for the primary energy of 6 keV.

are very probably identical to those regarding contrasts between relaxed grains, as discussed in the previous paragraph.

### 4.5. Semiconductors

The measurement of the local density of active dopants in semiconductors is of crucial importance for semiconductor technology in its all versions and phases, particularly as regards integrated circuits. While silicon wafers recently began achieving 45 cm in diameter, the characteristic dimensions of circuit structure details have been stepwise diminished from 45 to 32 nm and then to 22 nm, allegedly with 16 nm as the future prospect. The gigantic number of doped patterns excludes any true interoperational checks during production, but the possibility of measuring both the critical dimensions and true local density of a dopant still poses a difficult task for instrumentation designers. Thanks to its nondestructive application, the flexible size of the field of observation, resolution below 1 nm, and possible response to all sample characteristics including topography, chemistry, crystallinity, and electronics, SEM is traditionally considered the most suitable diagnostic method. However, important obstacles have hindered the development of this SEM application. The measurement of the critical dimensions of patterns faces the edge effect, i.e. overbrightening of the margins of surface steps imaged with secondary electrons that are also emitted from the sidewall as an extended surface. The edge effect disappears when the penetration depth of primary electrons drops beneath the escape depth of SE; with silicon sample, this happens at a landing energy of around 300 eV [35], which requires the use of the CL mode. Although the visibility of doped patterns was verified long ago [36], the explanation of dopant contrast has fluctuated between a great many alternatives. A review of this topic up to 2000 [37] considered the governing factor to be the difference in ionization energy between p- and n-types and its balancing by above-surface electric "patch fields". Later works incorporated below-surface fields on the junction between semiconductor and carbonaceous contamination [38] and the filtering action of the surface potential barrier [39]. The first observations of doped semiconductors in the CL mode [40] provided data about the contrast dependence on the external electric field of the CL and on vacuum conditions and the configuration of the detector. An explanation of these observations relied on the influence of surface passivation and the CL field causing the dopant contrast enhanced with respect to conventional SEM imaging.

This chapter does not aim to make a decision about the proposed models of the dopant contrast mechanism but merely to gather the most important experimental data. In order to avoid the low repeatability of data recorded in the past on series of samples with only one density of dopant per sample, which was to a certain extent the result of nonidentical surface treatment causing differences in the potential barrier, our samples were prepared with patterns doped by four different dopant densities. Figure 12 confirms the main relation known for a long time, namely, the higher imaging signal from p-type patterns compared to n-type patterns. As we see, the dopant contrast increases from 4 keV to 1 keV. At 4 keV (and primary energy 7 keV), SE are accelerated to 3 keV before detection, which is not sufficient to obtain a high signal from a scintillator covered with a metallic layer. This change of landing energy also causes acquisition of a broader flux of BSE, which are collected completely at 1 keV. Simultaneously, the slowest SE escape detection through the central detector bore, although faster SE still contribute to the image. However, this finding indicating an increase in contrast due to the loss of the slowest SE somewhat contradicts the measurement of the p/n contrast made with energy filtering of the SE in which the contrast carried by SE up to 10 eV was higher than that for the full SE emission [41]. In our case, SE below 3 eV escape detection fully, whereas, at higher energies, the central parts of the collimated SE flux are not detected. An important factor is that, at around 1 keV, the image contrast is reliably proportional to the dopant density and can be quantified for the sake of density measurement.

When going down with landing energy of electrons, the dopant contrast in Figure 12 diminishes and at tens of electronvolts inverts, giving the p-type patterns darker than the n-type background. No explanation has yet been proposed for this phenomenon, connected solely with the BSE emission because SE completely escape through the detector bore. Down to units of electronvolts, this inversed contrast is enhanced, but its dependence on the dopant density is much weaker. Near to zero energy, one more contrast inversion appears and we again get brighter p-type silicon. This effect is discussed below.

**Figure 12.** p-type doped patterns of dopant concentrations: $10^{19}$ cm$^{-3}$ (B), $10^{18}$ cm$^{-3}$ (C), $10^{17}$ cm$^{-3}$ (D), and $10^{16}$ cm$^{-3}$ (E) on an n-type substrate ($10^{15}$ cm$^{-3}$, A), etched in the Buffered Oxide Etch, CL mode, primary beam energy 7 keV, current 500 pA.

It is no great surprise that the "opposite" structure of n-type patterns on a p-type substrate in Figure 13 exhibits an opposite brightness relation with darker strips. Again, we have the contrast increasing toward 1 keV, now with a less pronounced proportionality between the contrast and the dopant density. At tens of electronvolts, the contrast inverts again and becomes more dependent on the dopant density. No important change takes place near the mirror image.

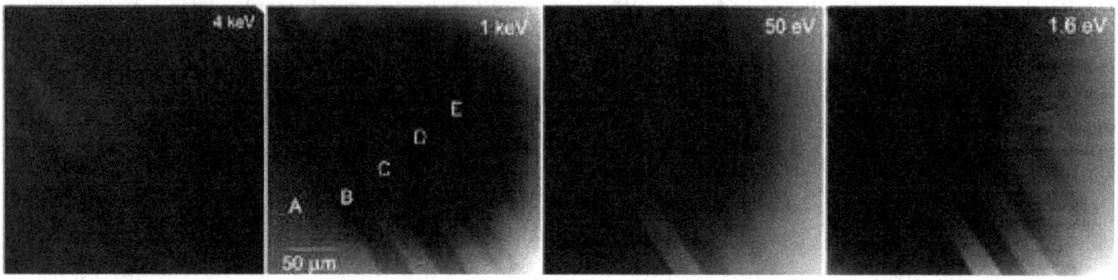


**Figure 13.** n-type doped patterns of dopant concentrations: $10^{19}$ cm$^{-3}$ (B), $10^{18}$ cm$^{-3}$ (C), $10^{17}$ cm$^{-3}$ (D), and $10^{16}$ cm$^{-3}$ (E) on a p-type substrate ($10^{15}$ cm$^{-3}$, A), etched in the Buffered Oxide Etch, CL mode, primary beam energy 6 keV, current 500 pA.

Important findings concern the influence of the surface status on the observed contrasts. Particularly, under standard vacuum conditions, when carbonaceous contamination is always present, we observe contrast dependence on the rate of contamination, impact angle of primary electrons, electron dose, detector geometry and position, etc. [42]. However, even the mere storage of an originally etched sample in air for several weeks leads to inversion of the contrast in the UHV microscope (see Figure 14).

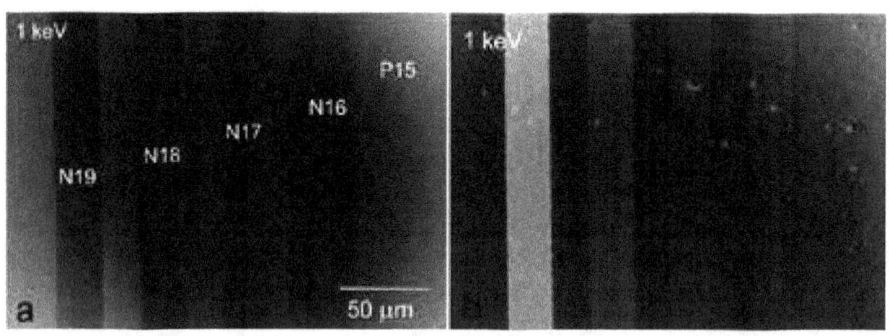


**Figure 14.** n-type patterns on a p-type substrate (see Figure 13) after etching in the Buffered Oxide Etch (a) and after several weeks in air (b).

Now, let us return to the ultimate inversion of the p/n contrast on a p-type pattern/n-type substrate sample near zero landing energy. Here, we obtain an extremely high signal from p-type patterns though with variously shaped black dots (Figure 15). The measurement shows that we are getting a "full" contrast here between the total beam current and no current. The explanation was based on the injection of electrons in the p-type patterns and their recombi-

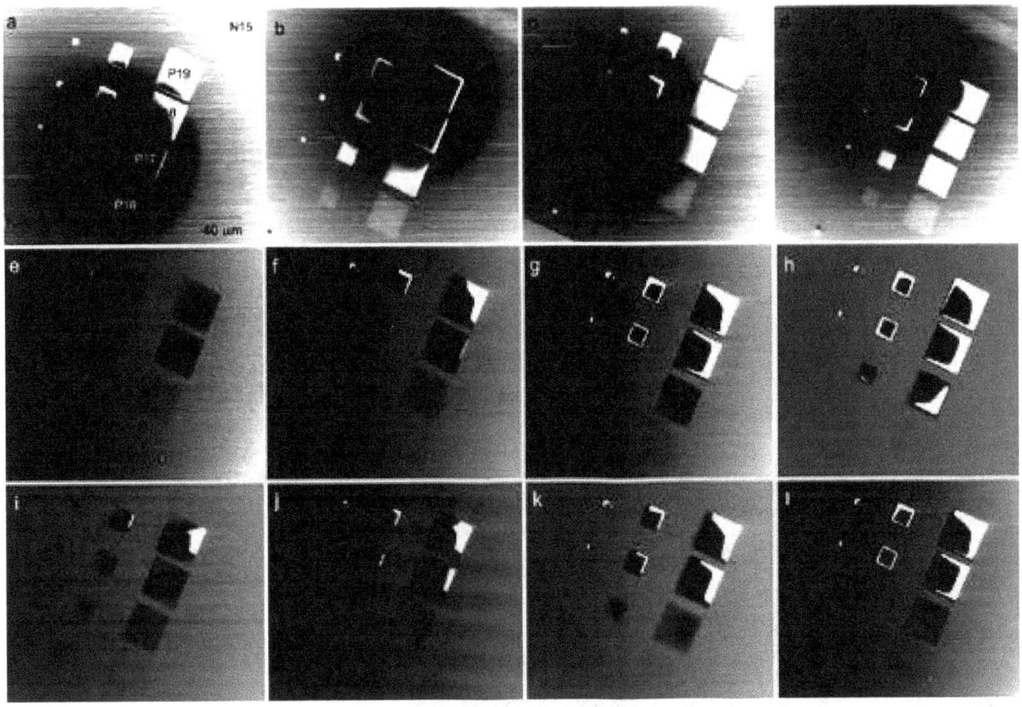


**Figure 15.** Total reflection phenomena on p-type doped patterns: lateral shifts of the pattern group influencing which pixels illuminate the active detector area around the central bore (a–d); dependence of pattern charging on the landing beam current: 30 pA (e); 100 pA (f); 600 pA (g); and 1.5 nA (h); and dependence on the electron dose proportional to the pixel dwell time: 560 ns (i); 1.76 μs (j); 5.36 μs (k); and 48.56 μs (l).

nation and on the ability of the small surface charge around 1 V thereby created to decrease the very low landing energy of incident electrons near enough conditions of total reflection [43]. The total reflection of the primary electron beam is directed toward the above-sample detector based on the bored scintillator disc, so we get white pixels where the reflected ray hits the scintillator surface and a black area where it hits the bore.

In Figure 15, we see the phenomenon dependent on the mutual position of the group of patterns and the detector and also on the beam current and dose. Obviously, the emergence of this phenomenon depends on a certain relationship between the electron dose sufficient to cause it and dopant density, which may offer another tool for measuring this density.

### 4.6. Thin sections

An extremely important family of microscopic samples consists of thin sections of tissues or more generally of live matter and various organic materials. These samples are usually amorphous and composed of light elements. For microscopic observation, they are cut into layers of submicrometer thicknesses and observed in TEM or STEM, traditionally using electrons in a range of hundreds of kiloelectronvolts and more recently of tens of kiloelectronvolts. An exception to this is low-voltage TEM operated at around 5 keV [44,45]; the success of this instrument is based on partially overcoming difficulties with the achievement of a sufficient image contrast in structures composed of light elements. In order to get good image contrast, sophisticated preparation procedures have been elaborated, particularly for tissue sections. These include fixation in various agents, immersion in resins, postfixation with osmium tetroxide, and staining with agents such as uranyl acetate or lead citrate. High contrast is then obtained where the heavy metal species from the chemicals used in preparation are located. Unfortunately, only some details of the structure are highlighted in this way, mainly with staining. When using STEM equipped with a CL mode and decreasing the landing energy of electrons, we observe a dramatic increase in contrast even when no heavy metal-containing substances are used in preparation [16] (see Figure 16).

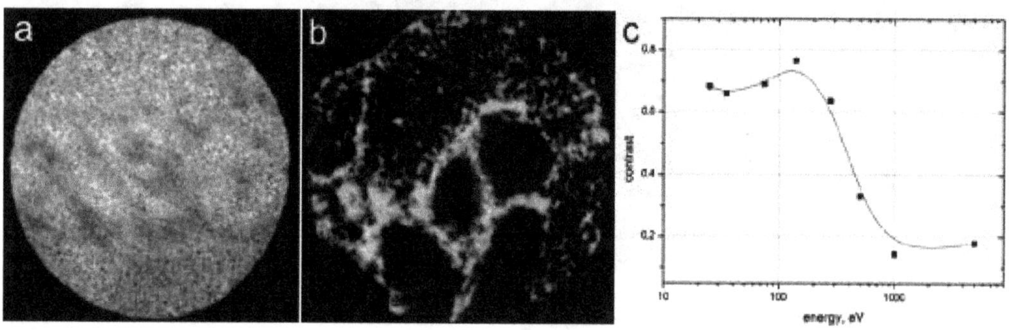


**Figure 16.** Section of mouse heart muscle, not fixed with osmium tetroxide and not stained: 10 nm section imaged by conventional TEM at 80 keV (a), CL mode micrograph taken at 500 eV with a primary energy of 4.5 keV (b), and energy dependence of the contrast of this sample in the CL mode (c).

An important finding made in a pioneering study [16] concerns changes in the embedding resin. It has been known that the resin is partially "radiation damaged" under electron bombardment with the consequence of the increased electron transmissivity of sections, but this increase was moderate only. The rate of this effect was compared at various electron energies, and whereas 5 keV STEM increases the transmitted flux two times after a dose of about $5\times10^{-3}$ $Ccm^{-2}$, at 500 eV, the increase was 20× at a doubled dose [16]. The idea is that, under electron impact, the resin is partly depolymerized and monomers are then released from the surface; no losses in the observable structure details have been discovered after this "ultimate" preparation step performed in situ.

Sections for ultralow-energy STEM have to be very thin, preferably below 10 nm [17]. These are transparent down to 500 eV if "electron bombardment thinning" is utilized and micrographs have even been taken at tens of electronvolts. The advantages here include averaging of image details over a much shorter trajectory across the layer, which may reveal some features, usually bright spots, not observable in thicker sections at any electron energy.

The application of ultralow-energy STEM in polymers, polymer blends, and composites forms a branch to be examined next.

### 4.7. 2D crystals

Intuitively, crystalline layers should be more transparent for slow electrons than amorphous sections because of the expected possibility of channeling among atomic planes or columns. For this reason, ultralow-energy STEM with the cathode lens mode has been applied to 2D crystals, firstly to graphene. The first results were achieved when comparing the Raman spectroscopy identification of flakes of a certain thickness with STEM observation [46]. Even flakes exhibiting a Raman spectrum corresponding to single-layer graphene were found to be composed of tiny flakes of various thicknesses and small holes. This finding argues in favor of introducing CL-mode STEM as an acknowledged method for the diagnostics of graphene and other 2D crystals.

Because the available graphene samples are generally composed of flakes only rarely exceeding micrometer sizes and usually overlapping each other at least partly, we need first and foremost to be able to obtain sufficient contrast between sites differing in thickness by a single layer of carbon atoms. As we see in Figure 17, this demand is met at about 100 eV in the transmission mode, whereas this kind of contrast is not available in the BSE signal.

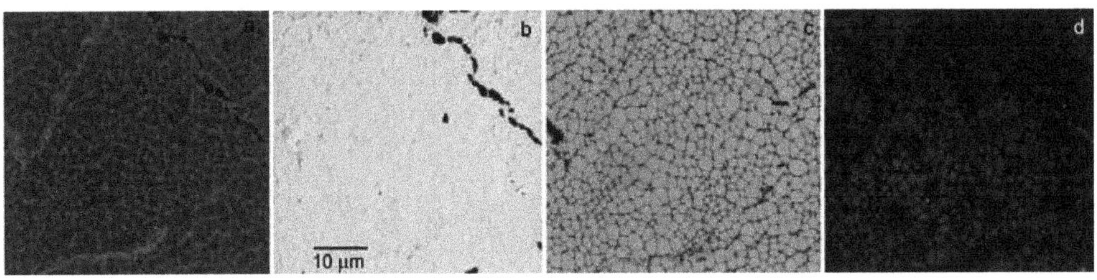


**Figure 17.** CVD graphene samples deposited on lacey carbon lying on a copper mesh, commercially available sample declared as three- to five-layer graphene: reflected signal for 1 keV (a) and 100 eV (b) landing energies and transmitted signal for 1 keV (c) and 100 eV (d).

One sort of commercially available graphene is a polycrystalline, CVD prepared material deposited on lacey carbon with eyes of up to 1 or 2 μm in diameter, which was used in our study. In Figure 18, we see the layer-by-layer contrast at higher magnification in a standard vacuum microscope, together with a detailed frame showing the polycrystalline structure of individual domains, though of continuous single-layer graphene (1LG hereinafter). These samples have been used for the measurement of the electron transmissivity through various numbers of graphene layers, assuming that the total area of nucleation centers visible in Figure 18a is negligible when averaging the transmitted signal over the lacey eye, and similarly, the existence of the domains does not influence the result.

We can notice in Figure 18d the electron penetration through the sample even at 4 eV. In fact, transmitted current was measurable down to or even below 1 eV. This enabled us to measure the transmissivity across the full energy scale from kiloelectronvolts to 1 eV. The result in Figure 19 was surprising in view of the expected increase in the inelastic mean free path of electrons below about 50 eV, which is not only generally observed [47] but also results from

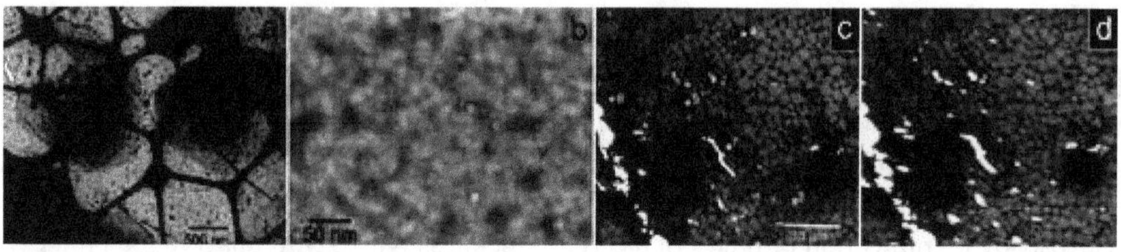


**Figure 18.** CVD graphene (see Figure 17) in transmitted electrons at higher magnification, 220 eV (a) and 1 keV (b), and at lowest energies, 40 eV (c) and 4 eV (d).

calculations [48]. The explanation should be sought in the fact that graphene in flakes sufficiently exceeding units of nanometers in size does not have an energy gap between the valence and conductance bands meeting at the Dirac point [49]. This causes interband transitions as the inelastic scattering mechanism working down to arbitrarily low energy losses in scattering events.

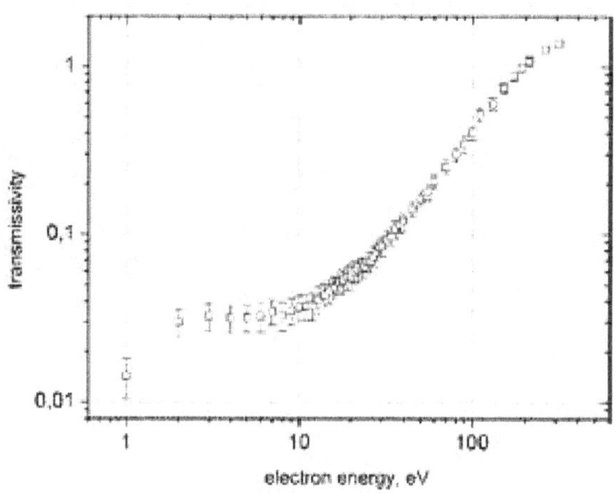


**Figure 19.** Energy dependence of the transmissivity of slow electrons through single-layer graphene.

In Figure 19, we see the transmissivity apparently exceeding 100% above 200 eV. This effect is caused by SE released inside the sample with original direction of movement toward the bottom surface where they are emitted and accelerated toward the detector together with the transmitted primary electrons. The excess current is then balanced from the earth. A detailed measurement of the electron transmissivity has been performed for up to 7LG. The ratio of transmissivities 1LG/7LG was found to be largest at 40 eV (over 6), so this energy is recommended for the reliable counting of graphene layers [18].

Electron microscopy of graphene below 10 eV was also examined in the reflection mode in a low-energy electron microscope (LEEM) [50]. The BSE signal was found oscillating in such a way that $n$ layers of graphene (grown on various surfaces or free standing) produce $n$-1 minima in the reflectivity between 0 and 8 eV [50,51]. We have also verified this relation in STEM and

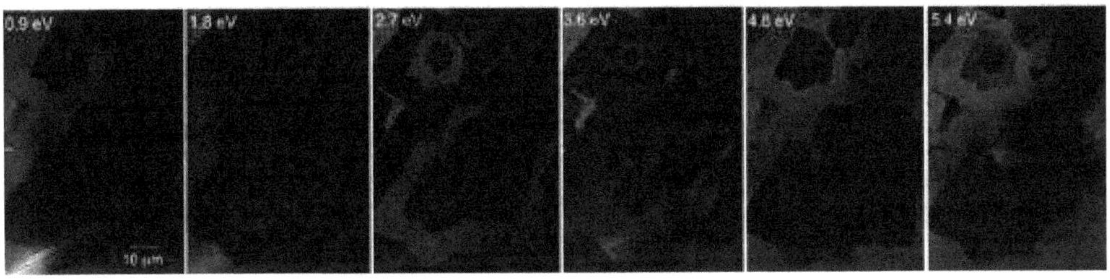


**Figure 20.** Multilayer graphene deposited by the CVD technique on a Cu foil, CL mode micrographs taken for the primary energy 6 keV.

were even able to confirm experimentally the second band of oscillations predicted by calculations in the range between 13 and 20 eV [27]. Oscillations are demonstrated in Figure 20 on a stack of multiple layers of graphene.

Finally, let us return to the "radiation damage" created by very slow electrons, mentioned above in connection with the decomposition of the resin in which the tissue section was embedded. Electrons below 40 or 50 eV were found to effectively clean the graphene samples increasing their transmissivity: a dose of 1.5 $Ccm^{-2}$ of 40 eV electrons increased the transmissivity of 1LG 2.5 times [18]. With graphene as a single atomic layer, there are no possible doubts about the mechanism of this radiation damage — if the sample is not bored, the only alternative is the removal of adsorbed gases, which was observed even under standard vacuum conditions. We suppose the removal of the hydrocarbon molecules instead of their decomposition normally generating the carbonaceous contamination, as happens immediately when increasing energy to, say, 200 eV. This kind of in situ surface cleaning promises a method for some surface studies performed outside UHV.

In Figure 21, the cleaning of graphene with slow electrons is shown on decreased reflectivity and increased transmissivity of longer bombarded parts of the fields of view.

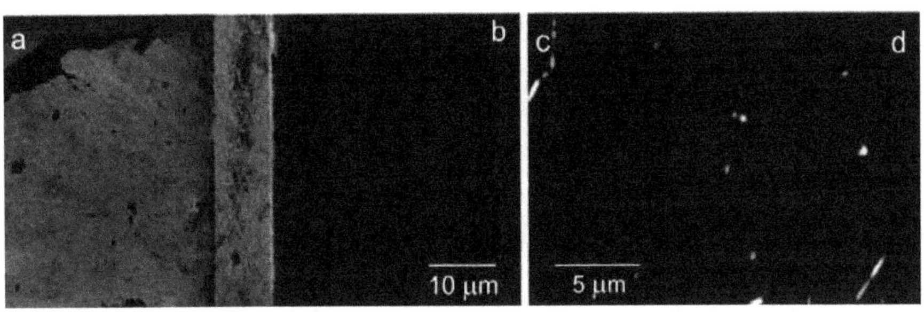


**Figure 21.** Demonstration of the cleaning effect of bombardment with very slow electrons, single-layer graphene, CL mode at 30 eV: original state shown in reflected (a) and transmitted (d) signal, decreased reflectivity (b), and increased transmissivity (c) after impact of about 1 $Ccm^{-2}$ of 30 eV electrons.

An observation of graphene samples grown on substrates at very low energies also makes it possible to distinguish between overlayer and underlayer growth of the second layer. A decision is made according to the comparison of micrographs taken at few hundreds of electronvolts, when all the "wedding cake" of stacked layers is visible, with a frame at units of electronvolts when we see only the topmost layer [27].

## 5. Conclusions

During a little more than 20 years of laboratory existence and nearly a full decade since the introduction of a commercial device, the cathode lens mode has slowly ceased to be considered a methodological novelty. If any future review of the method is to be written, its content will probably be limited to a particular type of samples and place the emphasis on models of contrast mechanisms rather than the mere comparison of the traditional and new appearance of micrographs. However, this time the older approach has been chosen. We have seen above that not all effects observed even when examining quite common samples have already been satisfactorily explained. This chapter aims to encourage colleagues capable of creating the desired physical models.

## Acknowledgements

This chapter mostly summarizes the results of research performed by the Group of Microscopy and Spectroscopy of Surfaces at the Institute of Scientific Instruments of the CAS in Brno, headed by Dr. Ilona Müllerová, who began CL mode implementation in an SEM and continuously manages the project. The presented data and micrographs have mostly been collected by former Ph.D. students of the group: Dr. Miloš Hovorka, Dr. Šárka Mikmeková, Dr. Eliška Mikmeková, Dr. Filip Mika, Dr. Zuzana Pokorná, and Dr. Ivo Konvalina. Thanks are also due to Dr. Aleš Paták, Mr. Jiří Sýkora, and Mr. Pavel Klein. Some of the samples have been provided by Dr. Jana Nebesářová (Biology Centre, České Budějovice), Professor Kenji Matsuda (University of Toyama, Japan), Professor Bohuslav Mašek (University of Western Bohemia, Plzeň), and Dr. Baowen Liu (UNIST, Ulsan, South Korea). In recent years, the program has been supported by the project TE01020118 of the Technology Agency of the Czech Republic (Competence Centre: Electron Microscopy) and, in part, by MEYS CR (LO1212).

## Author details

Luděk Frank*

Address all correspondence to: ludek@isibrno.cz

Institute of Scientific Instruments, Czech Academy of Sciences, Brno, Czech Republic

## References

[1] Brüche E, Johannson H. Elektronenoptik und Elektronenmikroskop. Naturwissenschaften. 1932;20:353–358. DOI: 10.1007/BF01504926

[2] Recknagel A. Theorie des elektrischen Elektronenmikroskops für Selbststrahler. Zeitschrift für Physik. 1941;117:689–708. DOI:10.1007/BF01668875

[3] Zworykin VK, Hillier J, Snyder RL. A scanning electron microscope. ASTM Bulletin. August 1942:15–23. DOI: 10.1520/STP10082B

[4] Paden RS, Nixon WC. Retarding field scanning electron microscopy. Journal of Physics E: Scientific Instruments. 1968;1:1073–1080. DOI: 10.1088/0022-3735/1/11/305.

[5] Müllerová I, Lenc M. The scanning very-low-energy electron microscopy (SVLEEM). In: Proceedings of the 2nd Workshop of the European Microbeam Analysis Society, Electron Microbeam Analysis, Mikrochimica Acta Supplement 12; May 1991; Dubrovnik. Wien: Springer; 1992. p. 173–177.

[6] Müllerová I, Frank L. Use of the cathode lens in scanning electron microscope for low voltage applications. Mikrochimica Acta. 1994;114/115:389–396. DOI: 10.1007/BF01244565

[7] Müllerová I, Frank L. Very low energy microscopy in commercial SEMs. Scanning. 1993;15:193–201. DOI: 10.1002/sca.4950150403

[8] Müllerová I, Frank L. Scanning low energy electron microscopy. In: Hawkes PW, editor. Advances in Imaging and Electron Physics, Vol. 128. Amsterdam: Elsevier; 2003. p. 309–443. DOI: 10.1016/S1076-5670(03)80066-6

[9] Müllerová I, Frank L. Very low energy scanning electron microscopy. In: Méndez-Vilas A, Diaz J, editors. Modern Research and Educational Topics in Microscopy. Badajoz: Formatex; 2007. p. 795–804.

[10] Frank L, Hovorka M, Konvalina I, Mikmeková Š, Müllerová I. Very low energy scanning electron microscopy. Nuclear Instruments and Methods in Physics Research Section A. 2011;645:46–54. DOI: 10.1016/j.nima.2010.12.214

[11] Müllerová I, Hovorka M, Mika F, Mikmeková E, Mikmeková Š, Pokorná Z, Frank L. Very low energy scanning electron microscopy in nanotechnology. International Journal of Nanotechnology. 2012;9:695–716. DOI: 10.1504/IJNT.2012.046749

[12] Frank L, Hovorka M, Mikmeková Š, Mikmeková E, Müllerová I, Pokorná Z. Scanning electron microscopy with samples in an electric field. Materials. 2012;5:2731–2756. DOI: 10.3390/ma5122731

[13] Jaklevic RC, Davis LC. Band signatures in the low-energy-electron reflectance spectra of fcc metals. Physical Review B. 1982;26:5391–5397. DOI: 10.1103/PhysRevB.26.5391

[14] Pokorná Z, Frank L. Mapping the local density of states by very-low-energy scanning electron microscope. Materials Transactions. 2010;51:214–218. DOI: 10.2320/matertrans.MC200921

[15] Pokorná Z, Mikmeková Š, Müllerová I, Frank L. Characterization of the local crystallinity via reflectance of very slow electrons. Applied Physics Letters. 2012;100:261602. DOI: 10.1063/1.4729879

[16] Frank L, Nebesářová J, Vancová M, Paták A, Müllerová I. Imaging of tissue sections with very slow electrons. Ultramicroscopy. 2015;148:146–150. DOI: 10.1016/j.ultramic.2014.10.009

[17] Nebesářová J, Vancová M. Preparation of different biological objects for low voltage electron microscope. In: Proceedings of the 16th International Microscopy Congress, Vol. 1; 3–8 September 2006; Sapporo. Sapporo: Publication Committee of IMC16; 2006. p. 500.

[18] Frank L, Mikmeková E, Müllerová I, Lejeune M. Counting graphene layers with very slow electrons. Applied Physics Letters. 2015;106:013117. DOI: 10.1063/1.4905221

[19] Beck S, Plies E, Schiebel B. Low-voltage probe forming columns for electrons. Nuclear Instruments and Methods in Physics Research Section A. 1995;363:31–42. DOI: 10.1016/0168-9002(95)00254-5

[20] Plies E, Degel B, Hayn A, Knell G, Neumann J, Schiebel B. Experimental results using a "low-voltage booster" in a conventional SEM. Nuclear Instruments and Methods in Physics Research Section A. 1999;427:126–130. DOI: 10.1016/S0168-9002(98)01560-5

[21] Frosien J, Plies E, Anger K. Compound magnetic and electrostatic lenses for low-voltage applications. Journal of Vacuum Science and Technology B. 1989;7:1874–1877. DOI: 10.1116/1.584683

[22] Lenc M, Müllerová I. Electron optical properties of a cathode lens. Ultramicroscopy. 1992;41:411–417. DOI: 10.1016/0304-3991(92)90220-E

[23] Lenc M, Müllerová I. Optical properties and axial aberration coefficients of the cathode lens in combination with a focusing lens. Ultramicroscopy. 1992;45:159–162. DOI: 10.1016/0304-3991(92)90047-N

[24] Lencová B. Electrostatic lenses. In: Orloff J, editor. Handbook of Charged Particle Optics. 1st ed. Boca Raton: CRC Press; 1997. p. 177–221.

[25] Frank L, Müllerová I. Strategies for low- and very-low-energy SEM. Journal of Electron Microscopy. 1999;48:205–219.

[26] Káňová J, Zobač M, Oral M, Müllerová I, Frank L. Corrections of magnification and focusing in a cathode lens-equipped scanning electron microscope. Scanning. 2006;28:155–163. DOI: 10.1002/sca.4950280304

[27] Mikmeková E, Frank L, Müllerová I, Li BW, Ruoff RS, Lejeune M. Study of multi-layered graphene by ultra low energy SEM/STEM. Diamond and Related Materials, in press. DOI: 10.1016/j.diamond.2015.12.012.

[28] Mikmeková E, Frank L. Examination of graphene with very slow electrons. In: Proceedings of the 6th Nanocon International Conference 2014; 5–7 November 2014; Brno. Brno: Tanger; 2015. p. 741–746.

[29] Frank L, Mikmeková Š, Pokorná Z, Müllerová I. Scanning electron microscopy with slow electrons. Microscopy and Microanalysis. 2013;19[S2]:372–373. DOI: 10.1017/S1431927613003851

[30] Matsuda K, Ishida Y, Müllerová I, Frank L, Ikeno S. Cube-phase in excess Mg-type Al-Mg-Si alloy studied by EFTEM. Journal of Materials Science. 2006;41:2605–2610. DOI: 10.1007/s10853-006-7819-6

[31] Matsuda K, Ishida Y, Müllerová I, Frank L, Ikeno S. Chemical analysis of the cube phase in Al-Mg-Si alloy by EFTEM. In: Proceedings of an International Conference on Solid-to-Solid Phase Transformations in Inorganic Materials 2005, Vol. 1; 29 May–3 June 2005; Phoenix. Warrendale: TMS; 2005. p. 371–376.

[32] Bauer E. Low energy electron microscopy. Reports on Progress in Physics. 1994;57:895–938. DOI: 10.1088/0034-4885/57/9/002

[33] Strocov VN, Starnberg HI. Absolute band-structure determination by target current spectroscopy: Application to Cu(100). Physical Review B. 1995;52:8759–8766. DOI: 10.1103/PhysRevB.52.8759

[34] Bartoš I, van Hove MA, Altman MS. Cu(111) electron band structure and channeling by VLEED. Surface Science. 1996;352–354:660–664. DOI: 10.1016/0039-6028(95)01204-4

[35] Pejchl D, Müllerová I, Frank L. Unconventional imaging of surface relief. Czechoslovak Journal of Physics. 1993;43:983–992. DOI: 10.1007/BF01595290

[36] Chang THP, Nixon WC. Electron beam induced potential contrast on unbiased planar transistors. Solid State Electronics. 1967;10:701–704. DOI: 10.1016/0038-1101(67)90099-8

[37] Sealy CP, Castell MR, Wilshaw PR. Mechanism for secondary electron dopant contrast in the SEM. Journal of Electron Microscopy. 2000;49:311–321.

[38] El Gomati MM, Wells TCR, Müllerová I, Frank L, Jayakody H. Why is it that differently doped regions in semiconductors are visible in low voltage SEM? IEEE Transactions on Electron Devices. 2004;51:288–292. DOI: 1109/TED.2003.821884

[39] Schönjahn C, Broom RF, Humphreys CJ, Howie A, Mentink SAM. Optimizing and quantifying dopant mapping using a scanning electron microscope with a through-the-lens detector. Applied Physics Letters. 2003;83:293–295. DOI: 10.1063/1.1592302

[40] Müllerová I, El Gomati MM, Frank L. Imaging of the boron doping in silicon using low energy SEM. Ultramicroscopy. 2002;93:223–243. DOI: 10.1016/S0304-3991(02)00279-6

[41] Rodenburg C, Jepson MAE, Bosch EGT, Dapor M. Energy selective scanning electron microscopy to reduce the effect of contamination layers on scanning electron microscope dopant mapping. Ultramicroscopy. 2010;110:1185–1191. DOI: 10.1016/j.ultramic.2010.04.008

[42] Mika F, Frank L. Two-dimensional dopant profiling with low-energy SEM. Journal of Microscopy. 2008;230:76–83. DOI: 10.1111/j.1365-2818.2008.01957

[43] Frank L, Müllerová I. The injected-charge contrast mechanism in scanned imaging of doped semiconductors by very slow electrons. Ultramicroscopy. 2005;106:28–36. DOI: 10.1016/j.ultramic.2005.06.004

[44] Delong A, Kolařík V, Martin DC. Low voltage transmission electron microscope LVEM-5. In: Proceedings of the 14th International Congress on Electron Microscopy, Vol. 1; 31 August–4 September 1998; Cancun. Bristol: IOP Publishing; 1998. p. 463–464.

[45] Drummy LF, Yang J, Martin DC. Low-voltage electron microscopy of polymer and organic molecular thin films. Ultramicroscopy. 2004;99:247–256. DOI: 10.1016/j.ultramic.2004.01.011

[46] Mikmeková E, Bouyanfif H, Lejeune M, Müllerová I, Hovorka M, Unčovský M, Frank L. Very low energy electron microscopy of graphene flakes. Journal of Microscopy. 2013;251:123–127. DOI: 10.1111/jmi.12049

[47] Seah MP, Dench WA. Quantitative electron spectroscopy of surfaces: A standard data base for electron inelastic mean free paths in solids. Surface and Interface Analysis. 1979;1:2–11. DOI: 10.1002/sia.740010103

[48] Fitting HJ, Schreiber E, Kuhr JC, von Czarnowski A. Attenuation and escape depths of low-energy electron emission. Journal of Electron Spectroscopy and Related Phenomena. 2001;119:35–47. DOI: 10.1016/S0368-2048(01)00232-8

[49] Warner JH, Schäffel F, Bachmatiuk A, Rümmeli MH. Graphene, Fundamentals and Emergent Applications. 1st ed. Amsterdam: Elsevier; 2013. Chapter 3. DOI: 10.1016/B978-0-12-394593-8.00001-1

[50] Hibino H, Kageshima H, Maeda F, Nagase M, Kobayashi Y, Yamaguchi H. Microscopic thickness determination of thin graphite films formed on SiC from quantized oscillation in reflectivity of low-energy electrons. Physical Review B. 2008;77:075413. DOI: 10.1103/PhysRevB.77.075413

[51] Feenstra RM, Srivastava N, Gao Q, Widom M, Diaconescu B, Ohta T, Kellogg GL, Robinson JT, Vlassiouk IV. Low-energy electron reflectivity from graphene. Physical Review B. 2013;87:041406. DOI: 10.1103/PhysRevB.87.041406

# 2

# Sample Preparations for Scanning Electron Microscopy – Life Sciences

Mogana Das Murtey and Patchamuthu Ramasamy

**Abstract**

Sample preparations are essential in scanning electron microscopy. Flawed sample preparations can undermine the quality of results and lead to false conclusions. Thus, the aim of this chapter is to equip researchers, post graduate students and technicians with essential knowledge required to prepare samples for scanning electron microscopy (SEM) investigations in the life sciences.

**Keywords:** Scanning electron microscopy, sample preparation

## 1. Introduction

Obtaining acceptable SEM images with good ultrastructural preservation requires careful application of the SEM sample preparation methods. The image in Fig. 1a can be easily accepted as a reasonable image for publication. However, a better sample preparation technique would have produced an image as in Fig. 1b. The bacterial cells in Fig. 1a are flattened and shrunken. The non-removal of the mucus layer has obscured the surface structures, unlike the cells in Fig. 1b.

## 2. Primary consideration in SEM sample preparations

### 2.1. Collection of samples

In the laboratory, collection of samples for SEM preparations generally involves the dissection of an animal or incision of a plant material to remove tissues from the main

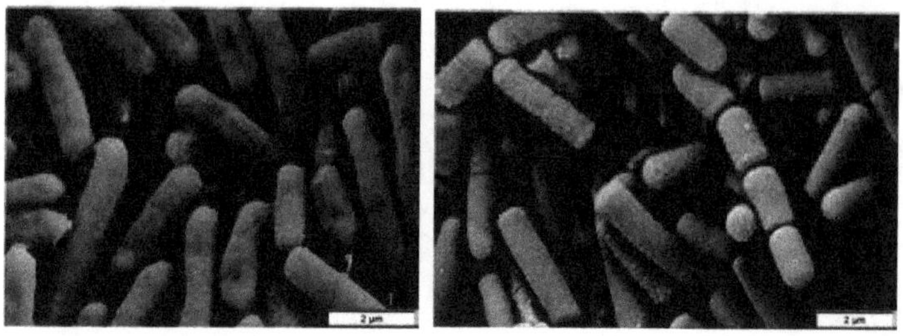

(a) (b)

Courtesy of EM Unit, Universiti Sains Malaysia

**Figure 1.** (a) Bacillus cereus. The cells were washed with distilled water and oven-dried at 40°C. The cells are flattened, shrunken and covered with a layer of dried mucus. (b) Bacillus cereus. The cells were air-dried using the HMDS protocol as described in 4.1.2b. The cells appear well rounded and fulsome while the dividing cells are well exposed.

body. Immediate fixing of the removed samples is very important to avoid autolysis, putrefaction and drying effects, which may destroy the ultrastructural integrity of the tissues. It is recommended that the organs are flooded with the primary fixative before incisions are made to remove the tissues. Once removed, the tissues should be immediately placed in a vial containing the fixative until the start of the processing protocol. If it is not possible for the samples to be placed in a fixative immediately, for example biopsies removed in a surgical theatre, then, the alternative could be to leave them in a suitable physiological saline and fixed soon after at the earliest possible time. All incisions must be performed with fresh sharp blades to avoid deformation of tissues from the undue physical forces needed with blunt blades. Once the samples have been placed in the vial containing the primary fixative, the same vial should be used throughout the sample preparation protocol until the 'specimen mounting' stage prior to viewing the sample in the SEM. Simply decant or pipette out the changes without any form of physical contact on the surface of the sample. If the fixed sample could not be processed on the same day, then it is advisable to leave the vial with the sample in a refrigerator (but never in a freezer or freezing compartments!). Every effort must be made to keep the tissue moist till the 'drying' processes at the end of all sample preparation protocols. For field collections, the researchers should have vials of fixative in hand for immediate immersion of the collected samples in the fixative. Immersion fixation may not be suitable for fungal samples growing on leaves and barks, due to the curling and collapsing of the fluffy hypha structures when they come in contact with liquids. Vapour fixation followed by freeze drying, as described in Section 4.3, gives better results. These field collected fungal samples must be carefully placed in a closed container space, kept moist by placing wet filter papers within, to minimise the drying of the fine structures, while being taken to the laboratory for vapour fixation. Students working with such fungal specimens in the author's laboratory have encountered these drying problems even for the samples that were brought to the laboratory from nearby trees within the campus and was minimised only by employing the method of transfer described above [1, 2, 3, 4].

## 2.2. Sample from cultures

Microbial cultures of bacteria and fungus need strict safety measures for the fixation process. Biosafety cabinets should be used wherever possible. The sample should be removed from the biosafety cabinets only after the fixation process. The fixatives can be added directly into the culture plates or broth cultures as needed. Generally, fresh microbial cultures give better results (Figs. 2a and 2b). Avoid the selection of organisms in their death phase of their growth curve or overgrown cultures except when doing comparative studies. Many students leave their cultures in a refrigerator for long period of times and produce disappointing results in the SEM [5, 6].

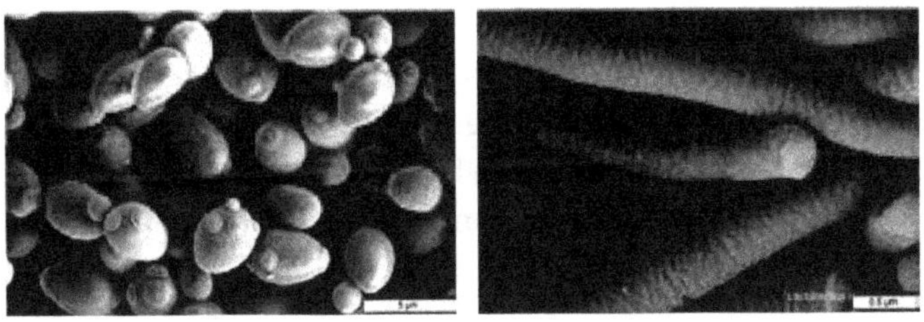

(a) (b)

Courtesy of EM Unit, Universiti Sains Malaysia

**Figure 2.** (a) A fresh culture of yeast Saccharomyces cerevisiae, prepared using the air drying method with HMDS as described in 4.1.2c. (b) A fresh culture of Lactobacillus acidophilus, prepared using the air drying method with HMDS as described in 4.1.2b.

## 2.3. Fresh water and marine samples

Organisms from fresh water habitat can be fixed in the routine EM fixatives. However, for marine samples, the fixatives should be prepared in filtered or artificial seawater with its osmolarity matching their natural environment (Figs. 3a and 3b) [1, 2, 4, 7]

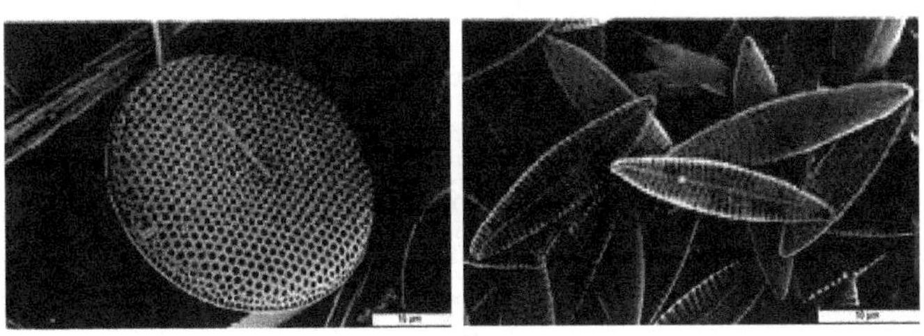

(a) (b)

Courtesy of EM Unit, Universiti Sains Malaysia

**Figure 3.** Marine diatoms fixed in 4% glutaraldehyde prepared in seawater, followed by the air drying protocols with HMDS as described in 4.1.2b, showing good structural integrity.

### 2.4. Samples with heavy mucous films

The presence of mucous films on sample surfaces can obstruct the clarity of the surface ultrastructures (Fig. 1a). This problem is often encountered when processing samples like mucous producing organisms and parasites removed from organs. Although clearing the mucous layer is sometimes achieved by using gentle centrifugation, alcohol, glycerol or enzymes, the application methods need to be worked out very carefully to avoid any consequential damaging effect on the fine structures of the sample. In the absence of the availability of suitable chemicals to remove these mucous layers, washing of the samples with a suitable physiological saline before fixation does help. However, this has to be done by employing numerous changes of the physiological saline solution accompanied with very gentle agitations for every change of the saline solution. If the organisms are small, performing these washings in an embryo dish, under a stereomicroscope can be useful. Students have reported to the author that parasites removed from guts needed about 50 times of washings before the fine structures on these parasites were rendered visible (Figs. 4a and 4b). It has to be stressed here that the washings have to be done before fixation. Once the organism or tissue has been fixed, the removal of the mucous can be extremely difficult [4, 8].

(a) (b)

Courtesy of EM Unit, Universiti Sains Malaysia

**Figure 4.** The removal of the mucous layer in these intestinal parasites was achieved by 25–50 gentle washings with physiological saline. The number of washings has to be experimentally worked out for the type and intensity of the mucous covering.

### 2.5. Avoiding stress in samples

Small organism such as worms, water insects and zooplanktons tend to show some forms of stress or even struggle when immersed into the fixative. These stress effects may result in a change or loss of some of the ultrastructural components of the sample. It may be advisable to narcotize or slow down the organism before fixation. Some researchers leave the samples in a fridge at about 5–8°C for a short period of time before fixation while others immerse the samples in magnesium chloride solutions or dilute alcohol solutions before fixation [4, 7, 9].

### 2.6. Dry samples

The sample processing techniques to be discussed here, namely air drying, critical point drying and freeze drying involves the drying of the samples to ensure maximum structural preservation. It should be realised that samples which are *already* dry cannot be reprocessed to bring it back to its original state. However, some temperature dried samples (Figs. 14a and 14b) do provide useful data, for which the techniques for particulate and bulk samples as described in Sections 7.0 and 8.0 can be employed.

### 2.7. Handling of samples

SEM, unlike TEM, allows the processing of larger organisms and tissues. However, if the samples are too large, it would be wise to cut them into smaller areas of interest, while ensuring that the surface areas of interest are not touched in the process. Even the slightest grazing of your tool on the sample surface of interest can introduce mechanical damage to your sample destroying the fine structures permanently. In the author's unit, these tasks are always performed under a stereomicroscope with the use of storkbill forceps (Fig. 5) and fresh sharp blades.

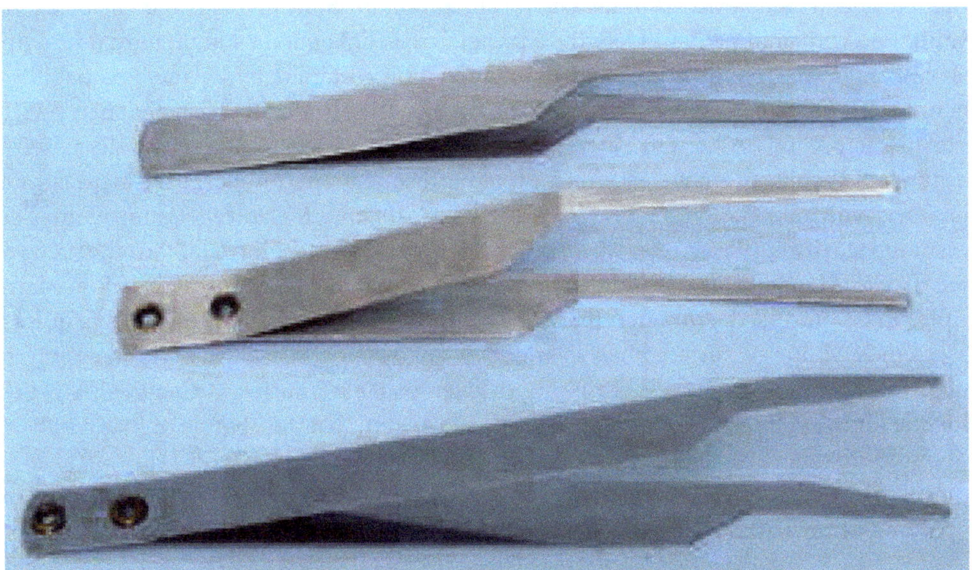

**Figure 5.** Storkbill (or Insect) forceps, which are made of flexible materials, aid in careful handing of samples.

### 2.8. Safety issues

All fixatives are volatile and harmful to living cells. Even their vapours can fix epithelial cells of the mouth, nose, hand and corneal membranes. Avoid any form of exposure by using gloves and fume hoods. Some workers prefer to use a double layer of gloves during the fixation process. HMDS and TMS are highly volatile and flammable liquid should be used only in well-ventilated areas or in a fume hood. The preparation and use of osmium tetroxide should be very strict in a fume hood. The vial containing the sample should only be taken out of the fume hood after the 'washings' with buffer or distilled water as the protocol prescribes. Please refer Section 9.0 for the protocols to prepare osmium tetroxide stock solutions [1, 2, 3 and 4].

## 3. Factors influencing SEM sample preparation

### 3.1. Fixation

Fixation of samples is probably the most crucial step in SEM sample preparation protocols, which could determine the eventual quality of the images obtained. Ideally, the fixation process should preserve and stabilize the structures of the cells, tissues or organisms keeping them structurally as faithful as possible to its living state. The process should also prevent autolysis and putrefaction in the cells. This is normally achieved by placing the sample in a chemical fixative formulation while also providing an optimal environmental condition in terms of pH, temperature and osmolarity. Routine chemical fixation for SEM involves the use of fixative formulations containing glutaraldehyde, a protein cross-linker and osmium tetroxide, a lipid cross-linker. Formaldehyde is used in combination with glutaraldehyde, due its capability to penetrate faster into the tissue, although it is not known to be a good fixing agent on its own for electron microscopy. Although, there are numerous fixative formulations discussed in literatures and books, one may start with the fixative formulations routinely used in the author's laboratory, which are 4% glutaraldehyde, Karnovsky's fixative, McDowell-Trump fixative and 1% osmium tetroxide. (Please refer 9.0 for the recipe). McDowell-Trump fixative is the preferred fixative for SEM in the author's unit for its better penetrative properties, which allows larger samples to be processed. Although there is a notion that penetration of fixatives is of lesser concern in SEM due its surface analysis mode, it must be realised that poorly fixed internal structures may result in inward shrinking or even collapsed surface structures. A general guideline for the volume of fixative to be used is to ensure that the volume of fixative should be about 15–20 times greater than the volume of the tissue [1, 2, 3, 4, 10 and 11].

### 3.2. pH and osmolarity

The choice of an appropriate buffer solution serves to keep the pH and osmolarity of the fixing solution within the physiological range as required for the sample, while acting as a vehicle for the fixing agent. The pH of buffers for the ultrastructure preservation is generally adjusted between pH 7.2 and 7.4 while the concentrations of the buffers are adjusted between 0.5 and 1.0 molar solution. For marine samples, osmolarity is achieved by using seawater instead of buffers (Figs. 3a and 3b)Hypertonic solutions give rise to cell shrinkage while hypotonic solutions result in cell swelling and poor fixation. The most commonly used buffers for electron microscopy are phosphate and cacodylate buffers. Phosphate buffers are safer to use and thought to be closer to cytoplasmic environments of most biological samples although it may produce electron dense precipitates in the presence of calcium ions. Precipitation is of lesser concern for cacodylate buffers but its formula contains arsenic, which is hazardous, and proper safety measures should be employed in the use and disposal of these solutions [1, 2, 3, 4].

### 3.3. Temperature of fixation

Fixation is routinely carried out at room temperature although it is believed that fixation at 0–4°C reduces the possibility of the extraction of cytoplasmic elements from cells [1, 2, 3 and 4].

### 3.4. Duration of fixation

The time of fixation is dependent on the size and density of the sample to be fixed. A general rule for the penetration of the fixative is '1hr per 1mm' of the sample for most fixatives. Since SEM is for surface analysis, the width of the sample may not be a limiting factor but largely dependent on the width of SEM sample holder available. However, the thickness of sample is best kept within about 3 mm to facilitate the penetration of fixatives. The shortest dimension of the sample determines the fixation time. If the sample is about $1mm^3$, a fixation period of 2 hours at room temperature or in a refrigerator may be sufficient. For larger samples, 6–24 hours may be necessary. Post-fixation of samples with osmium tetroxide is generally for 1–2 hours, but not any longer than 2 hours. The preparation and use of osmium tetroxide should be very strict in a fume hood. The vial containing the tissue should only be taken out of the fume hood after the 'washings' with buffer or distilled water as the protocol prescribes. Please refer Section 9.0 for the protocols to prepare osmium tetroxide. If the samples cannot be processed immediately, they can be kept in the glutaraldehyde-based fixative (McDowell-Trump Fixative, Karnovsky's Fixative or 4% Glutaraldehyde in buffer) in a refrigerator for a few weeks, with changes of the fixative periodically. Storage of the fixed samples in the refrigerator for very long periods should be avoided as degradation of the samples may occur [1, 2, 3 and 4].

## 4. SEM sample preparation methods

Almost all living organisms are composed of significant water components in their bodies. However, image formation in an electron microscope requires a high vacuum environment. Thus, the drying of samples becomes a prerequisite for the viewing and obtaining of good images in normal high vacuum SEM systems. Although there are low vacuum SEM systems, which allow the viewing of wet samples, these systems are generally not known for very high resolution and high magnification images. Thus, the challenge posed in SEM biological sample preparation is to dry the samples without any structural damage or changes. The main cause of these structural damage are the 'surface tension forces' of the water during the drying process when water transforms into its gaseous phase from its original liquid phase. Thus, the drying methods employed are primarily developed to circumvent the effect of these 'surface tension forces'. The three most common SEM sample preparation methods employed to minimise the structural changes associated with drying are 'air drying', 'critical point drying' and 'freeze drying'[1, 2, 4, 12, 13, 14, 15, 16, 17, 18, 19, 20, 21, 22].

Generally, the techniques involve the following pathways:

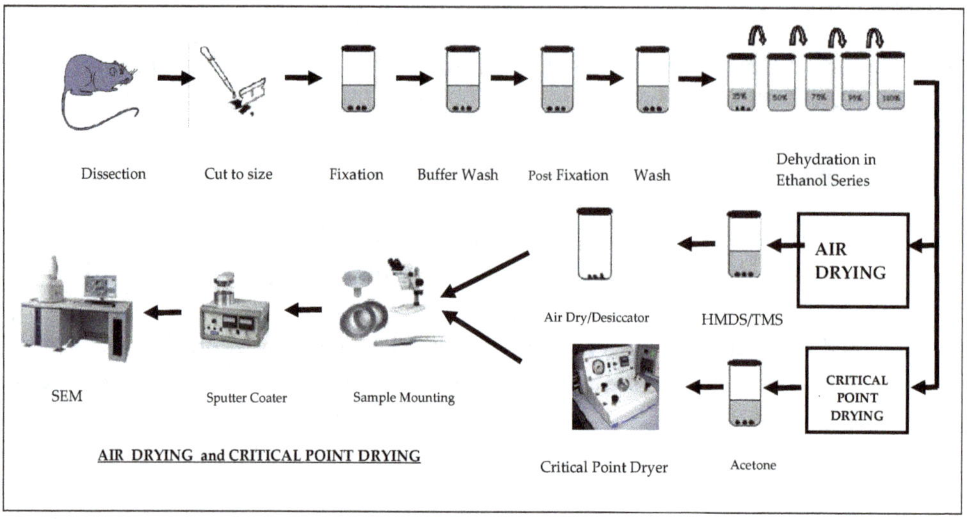


(a)

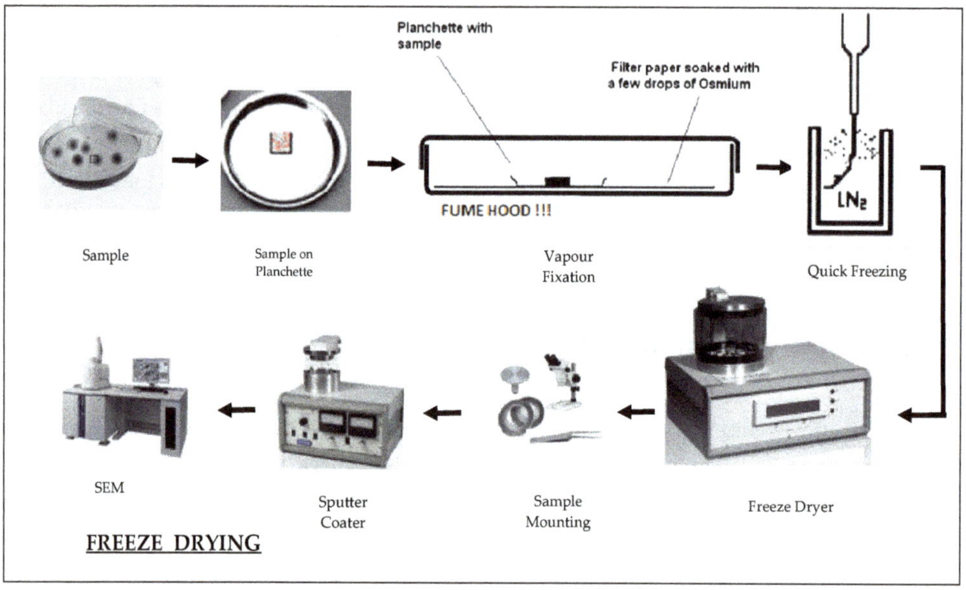


(b)

**Figure 6.** (a) Flowchart of the 'Air Drying' and ' Critical Point Drying' sample preparation methods. It could be noted that the protocols are the same for both the methods till the end of the dehydration process. (b) Flowchart of the 'Freeze Drying' sample preparation method.

## Air Drying, Critical Point Drying and Freeze Drying – A Comparison

All the three methods are known to be equally effective in ensuring good preservation of the structural integrity by minimising or eliminating the impact of surface tension effects on the samples.

Different workers have their own preferences after assessing the effectiveness of one method over the other in relation to their own samples. However, the following are some pertinent points for consideration to assist in making a choice:

**a.** Air drying with hexamethyldisilazane (HMDS) or tetramethylsilane (TMS) is the newest of the three methods and is gaining popularity as a reliable technique. It requires limited expertise and equipment. No liquid gases are required. It has been reported to be suitable for even very soft tissues. It also allows larger samples to be processed as compared to the critical point drying and freeze drying, both of which are limited by the size of sample holders provided for the equipment (Figs. 10b and 12b)

**b.** Critical point drying with the use of liquid carbon dioxide (and freon in the 70s) has been the method of choice for biological samples for over 50 years. Some workers have observed that critical point drying is a better choice for some of their botanical samples.

**c.** Fungal hypha and similar soft structures tend to lose their rigidity (and curl up) irreversibly when they come in contact with liquids. This reaction can be compared to wetting a cotton wool with water, which would make the cotton wool to lose its fluffiness irreversibly that even drying may not restore its original fluffiness. This problem effectively rules out the use of air drying and critical point dying for fungal hypha and similar samples due to the need for immersion of the samples in various liquids. The freeze drying method, which employs vapour fixation and quick freezing in liquid nitrogen, offers the best solution for these types of samples (Figs. 11a and 11b)

**d.** The freeze drying method do not involve the immersion of the sample in any water- or alcohol-based solutions, which ensures minimal physical contact of the sample with any media until the end of the drying process (with the exception of liquid nitrogen). However, the only disadvantage in this circumstance is that, for unclean samples, any presence of dirt, salts and mucous on the sample surface remains on the sample till the end of the processing protocol and realized only at the SEM viewing stage. Some of these extraneous materials may obscure surface details. The air drying and critical point drying has an advantage on this aspect by providing a washing effect on the samples while undergoing many changes in the fixatives, buffers and alcohols [12, 13, 14, 18, and 20].

## Standard Requirements for SEM Sample Preparation

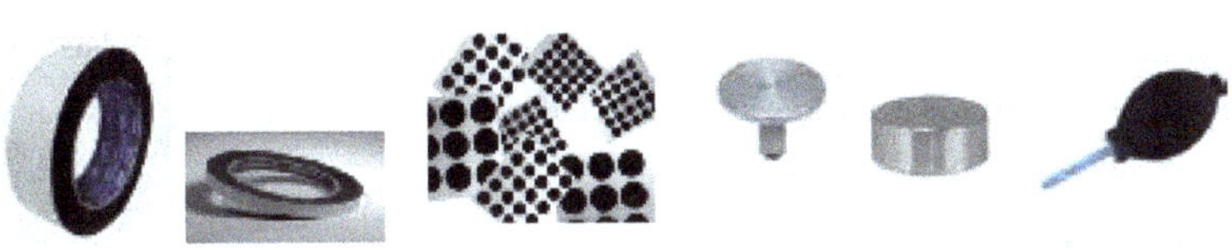

(a) (b) (c) (d)

**Figure 7.** (a) Double-sided carbon adhesive tape used extensively for sample attachment on sample stubs. (b) Double-sided carbon adhesive tabs also an excellent item for sample attachment. (c) **SEM** Sample stubs, shapes and sizes of which are dedicated to the SEM model in use. (d) Air dust blower, useful when preparing particulate samples, Section 7.

## 4.1. Air drying technique

### *4.1.1. General Concept*

The air drying technique is based on the use of highly volatile organic compounds such as hexamethyldisilazane (HMDS) and tetramethylsilane (TMS), in the drying process to minimise the effect of surface tension forces on cell ultrastructures.

However, the use of HMDS and TMS has to be preceded by the ethanol dehydration process of using increasing ethanol concentrations to gradually remove or rather dilute the water in the tissue, until the water component in the cells is completely replaced with 100% ethanol. It should be noted that the protocols for air drying and critical point drying techniques are same up to the 100% ethanol stage (Fig. 6a) [12, 13, 17, 18, 21].

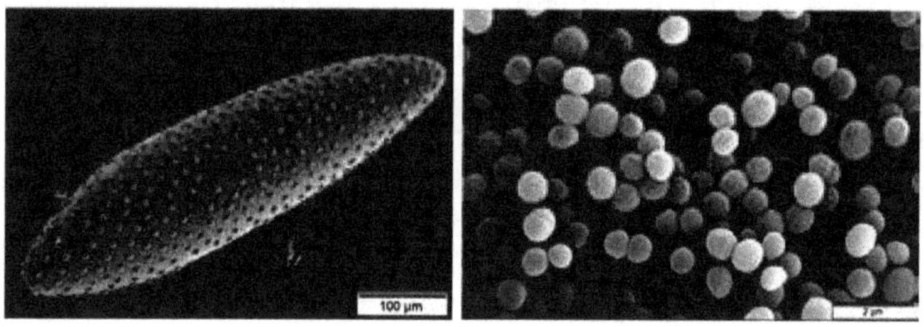

(a) (b)

Courtesy of EM Unit, Universiti Sains Malaysia

**Figure 8.** (a) Mosquito egg prepared using the air drying method with HMDS as described in 4.1.2a. (b) Staphylococcus aureus cells prepared using the air drying method with HMDS as described in 4.1.2b.

### *4.1.2. Standard protocols for the Air Drying Technique*

| | **a For tissues, insects and organisms that can be held or picked up with forceps** | |
|---|---|---|
| 1. | Fix in *McDowell-Trump* fixative (**alternatives**: *Karnovsky*'s fixative or 5% Glutaraldehyde) 2–24 hours prepared in 0.1M phosphate buffer, pH 7.2, at 4°C | |
| 2. | Wash in buffer (Use the same buffer as in step 1) | 3 x 10 minutes |
| 3. | Postfix in 1% Osmium tetroxide prepared in the same buffer as above at room temperature. | 1–2 hours |
| 4. | Wash in distilled water | 2 x 10 minutes |
| 5. | Dehydrate the sample as follows: | |
| | 35 % Ethanol | 1 x 15 minutes |
| | 50% Ethanol | 1 x 15 minutes |
| | 75% Ethanol | 1 x 15 minutes |

| | | |
|---|---|---|
| | 95% Ethanol | 2 x 15 minutes |
| | Absolute Ethanol | 3 x 20 minutes |
| 6. | Immerse the dehydrated samples in 1–2 ml of hexamethyldisilazane (HMDS) or tetramethylsilane (TMS) for 10 minutes | 2 x 10 minutes |
| 7. | Decant the HMDS from the sample vial. Leave the sample vial with the samples in a desiccator to air-dry at room temperature | overnight |
| 8. | The dried samples are then mounted on to a SEM sample stub (Fig. 7c) with a double-sided sticky tape | |
| 9. | Sputter the sample with gold and view in the SEM (please refer Section 6.0 on conductive coating of samples) | |
| 10. | *It should be noted that the protocols for Air Drying and Critical Point Drying techniques are the same up to the 100% Ethanol stage.* | |
| | **b Protocol for Cultured Micro-organisms (Loose or Loosened Cells)** | |
| 1. | **For liquid cultures**, centrifuge the cells (pellet formation) and discard the supernatant | 10 minutes |
| 2. | Resuspend the pellet in McDowell-*Trump* fixative (**alternatives**: *Karnovsky*'s fixative or 5% Glutaraldehyde) prepared in 0.1M phosphate buffer, pH 7.2<br>After 30 minutes, centrifuge and discard the supernatant | 30 minutes |
| 2a | **For solid agar cultures**, The fixatives can be added directly into the culture plates. After a fixing period of about 30 minutes, the cells adhering to the agar are suspended into the fixative solution by using a cell scrapper. The suspension can then be transferred into a vial. Centrifuge and discard the supernatant. | 30 minutes |
| 3. | Resuspend the pellet in 0.1M phosphate buffer. Centrifuge and discard the supernatant | 2 x 10 minutes |
| 4. | Resuspend the pellet in 1% Osmium tetroxide prepared in 0.1M phosphate buffer<br>After 1 hour, centrifuge and discard the supernatant. | 1 hour<br>**Fume Hood !** |
| 5. | Resuspend the pellet in distilled water.<br>Centrifuge and discard the supernatant. | 2 x 10 minutes<br>**Fume Hood !** |
| 6. | Dehydrate the sample through the ethanol series and HMDS as follows, centrifuging and discarding the supernatant for each change: | |
| | 35 % Ethanol | 10 minutes |
| | 50% Ethanol | 10 minutes |
| | 75% Ethanol | 10 minutes |
| | 95% Ethanol | 2 x 10 minutes |
| | Absolute Ethanol | 2 x 10 minutes |
| | Hexamethyldisilazane (HMDS)<br>*[Note: Centrifugations are not necessary if the cells settle easily at the bottom of the tube and the ethanol changes do not result in excessive loss of the cells.]not necessary* | |
| 7. | Discard the supernatant of the second HMDS and leave the sample vial with the cells in a desiccator to air-dry at room temperature. | Overnight, |

| | | |
|---|---|---|
| 8. | The dried cells are then mounted on to a SEM sample stub (Fig. 7c) with a double-sided sticky tape. (please refer Section 7.0 on *'Techniques for Particulate Samples'*) | |
| 9. | Sputter the sample with gold and view in the SEM (please refer Section 6.0 on conductive coating of samples). | |
| | **c Protocol for Cultured Micro-organisms (On Solid Agar and other Substrates)** | |
| 1. | Cut out about 5mm x 5mm piece of the growth colony with a sharp blade as carefully as possible without disturbing the cells on the surface. Perform this in a biosafety cabinet if necessary. | |
| 2. | Place the piece into a 6cm diameter glass petri dish. Add McDowell-Trump fixative into the petri dish to wet the agar without reaching the colony surface but enough to allow the fixatives to diffuse through the substrate to reach the colonies. Cover the petri dish. | 1–2 hours |
| 3. | Pipette out the fixative gently and replace it with 0.1M phosphate buffer and again wet the agar without reaching the colony surface. | 2 x 20 minutes |
| 4. | After pipetting out the phosphate buffer, place 1 or 2 pieces of dry filter paper beside the agar colony piece. Wet these filter papers with drops of 1% osmium tetroxide. Close the petri dish and seal it with parafilm. Use of fume hood is essential. Alternatively, a few drops of the 1% osmium tetroxide can be placed to wet the agar but without reaching the colony surface. | 1 hour |
| 5. | Remove the filter papers with osmium tetroxide and wash the sample with distilled water in the same manner as in Steps 2 and 3 | 2 x 20 minutes |
| 6. | Dehydrate the sample through the ethanol series and HMDS in the same manner as in Steps 2 and 3 : | |
| | 35 % Ethanol | 1 x 30 minutes |
| | 50% Ethanol | 1 x 30 minutes |
| | 75% Ethanol | 1 x 30 minutes |
| | 95% Ethanol | 2 x 30 minutes |
| | Absolute Ethanol | 3 x 30 minutes |
| | Hexamethyldisilazane (HMDS)<br>*Allow 30 minutes or more per step for thorough diffusion through the agar blocks and colonies.* | 2 x 30 minutes |
| 7. | Once the second change of HMDS has been pipetted out, leave the sample in a desiccator with the petri dish cover slightly open to air-dry at room temperature. The dried samples are then mounted on to a SEM sample stub (Fig. 7c) with a double-sided sticky tape. | |
| 8. | Sputter the sample with gold and view in the SEM (please refer Section 6.0 on conductive coating of samples). | |
| | **d Technique for a) preparations with few cells b) fragile cells that may not withstand centrifugation and c) if pelleting is to be avoided** | |
| 1. | Prepare a solution of 0.1% poly-L-lysine in phosphate-buffered saline (pH 7.2). | |
| 2. | Place a drop of the poly-L-lysine solution on a coverslip (or microscope slide) and place it in a covered petri dish lined with moist filter paper for about 1 hour. | |

| | |
|---|---|
| 3. | Rinse the poly-L-lysine coated coverslip with distilled water and place a drop of the cell suspension on the coated area of the coverslip (in the covered petri dish lined with moist filter paper) for about 30–60 minutes to allow the organisms to settle on the sticky material. The smaller the organism, the longer it takes to settle. |
| 4. | Carefully touch on the **side** of the drop with a fresh piece of cut filter paper to remove the liquid and immediately place a drop of EM fixative to replace the earlier fluid. (Do not let it to dry *completely* in between the step). Leave it alone for about 10 minutes |
| 5. | Repeat the steps with the standard SEM sample preparation protocol through buffer, OsO4, distilled water, graded alcohols and HMDS, **BUT** in the covered petri dish lined with **DRY** filter paper. (The times can be shortened to 5–10 minutes in each step.) |
| 6. | *Important: Ensure that the liquids DO NOT dry up completely in between the changesImportant: Ensure that the liquids DO NOT dry up completely in between the changes* |
| 7. | At the 2$^{nd}$ HMDS stage, once the HMDS has dried completely, the coverslip/slide can be attached to the SEM sample stub, sputtered and viewed in the SEM. |
| 8. | **Note:** This technique can also be used for cells grown on coverslips, if the cells are well adhered on to the coverslip. |

#### *4.1.3. General Precautions*

**a.** Once the samples have been placed in the vial containing the primary fixative, the same vial should be used throughout the sample preparation protocol until the 'sample mounting' stage prior to viewing the sample in the SEM. Simply decant or pipette out the changes without any form of physical contact on the surface of the sample.

**b.** The need for centrifugation is needed only to form a loose pellet and therefore low speed centrifugations of 500–1500 should suffice.

**c.** Never let the samples to dry at any stage (till the samples are immersed in the HMDS).

**d.** Strict safety measures should be observed for the fixation of pathogenic materials. Fixatives can be added directly into the culture plates or broth cultures as a safety measure.

**e.** Times recommended are for small pieces of samples (smaller than 1–2 mm$^3$). For larger pieces, times may have to be varied accordingly.

**f.** Buffers: Normally, 0.1M of pH 7.2, but requirements may vary for different samples. Refer to published papers.

### 4.2. Critical point drying technique

#### *4.2.1. General Concept*

Critical point drying method is based on the principle that by increasing the pressure and temperature of a wet sample, it is possible to dry the sample by achieving the 'critical point' of that liquid at which the phase boundary between its liquid and gaseous states are eliminated. At this state, the gas is released slowly thus minimising the effect of surface tension forces on cell ultrastructures.

However, the critical point for water is extremely high at 228.5 bar (3212 psi) and 374°C, which could instantly destroy the samples. Critical point of ethanol is 60.8 bar (882 psi) and 241°C. These properties of water and alcohol make them unsuitable for the critical point drying method.

$CO_2$ liquid with a critical point at 73, 8 bar (1072 psi) and 31°C is generally used. Since $CO_2$ is not miscible with water, acetone is used as a transitional (intermediate) fluid in the process.

Just as has been done in the air drying technique, the use of the critical point dryer has to be preceded by the ethanol dehydration process of using increasing ethanol concentrations to gradually remove or rather dilute the water in the sample tissue, until the water component in the cells is completely replaced with 100% ethanol. The size of the samples should be determined carefully in view of the sample holder sizes provided with critical point dryer (Fig. 10b). Some CPD manufacturers provide sample holders for cells grown on coverslips or wafer materials. However, good adherence of the cells on these substrates must be confirmed before attempting this drying technique. This should be done to avoid the excessive loss of cells during the drying cycle. The use of poly-l-lysine helps in the adherence of most of the cells on these substrates (refer 4.1.2d) [1, 14, 15, 16, 19].

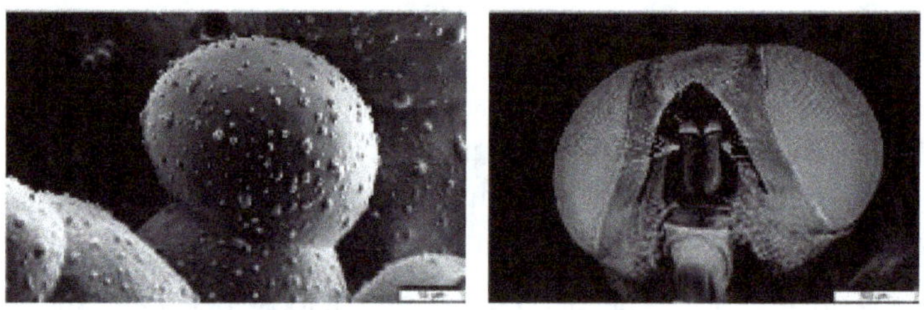

(a) (b)

(a) Courtesy of Dr. S. Sasidharan, Informm, Universiti Sains Malaysia. (b) Courtesy of EM Unit, Universiti Sains Malaysia

**Figure 9.** (a) Callus of Pomelo prepared using the critical drying method as described in 4.2.3. (b) Critical Point Dried House Fly prepared using the critical drying method as described in 4.2.3.

### 4.2.2. Materials

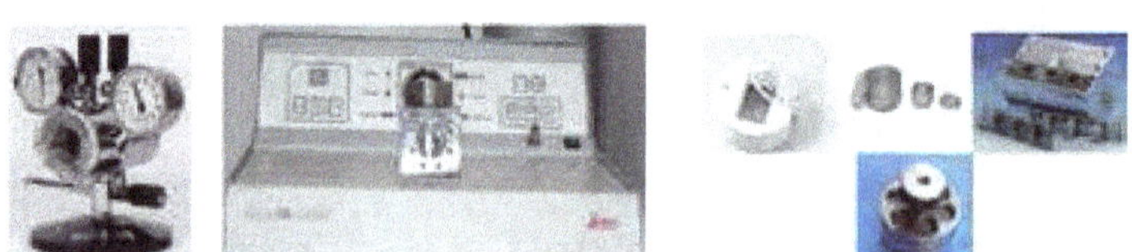

(a) (b)

**Figure 10.** (a) Critical Point Dryers (CPD). The $CO_2$ gas cylinders are not shown. It is advisable to use a $CO_2$ gas of at least 99.8% purity and the gas cylinder is fitted with a siphon tube to draw liquid $CO_2$ from the bottom of a cylinder. (b) CPD Sample Holders, which are supplied with the CPD apparatus.

#### *4.2.3. Standard protocol*

| General Protocol | | |
|---|---|---|
| 1. | Fix in *McDowell-Trump* fixative (**alternatives:** *Karnovsky*'s fixative or 5% Glutaraldehyde) prepared in 0.1M phosphate buffer, pH 7.2, at 4°C | 2 – 24 hours |
| 2. | Wash in buffer. (Use the same buffer as in step 1) | 3 x 10 minutes |
| 3. | Postfix in 1% Osmium tetroxide prepared in the same buffer as above at room temperature. | 1–2 hours |
| 4. | Wash in distilled water. | 2 x 10 minutes |
| 5. | Dehydrate the sample as follows: | |
| | 35 % Ethanol | 1 x 15 minutes |
| | 50% Ethanol | 1 x 15 minutes |
| | 75% Ethanol | 1 x 15 minutes |
| | 95% Ethanol | 2 x 15 minutes |
| | Absolute Ethanol | 2 x 15 minutes |
| | Acetone | 1 x 15 minutes |
| 6. | Transfer the samples into the CPD specimen holder or baskets (Fig. 10b) .The sample holder should contain (or immersed in) enough acetone to cover the samples. | |
| 7. | Perform the critical point drying as described in the instruction manual for the apparatus (Fig. 10a) | |
| 8. | The dried samples are then mounted on to a SEM sample stub (Fig. 7c) with a double-sided sticky tape. | |
| 9. | Sputter the sample with gold and view in the SEM (please refer Section 6.0 on conductive coating of samples). | |

#### *4.2.4. General Precaution*

The exchange and release of the $CO_2$ liquid and gas should be performed in a controlled manner to avoid turbulence within the CPD, which may be detrimental to the sample structures.

### **4.3. Freeze drying technique**

#### *4.3.1. General concept*

Freeze drying method is based on the principle of sublimation by shifting a solid phase directly into its gaseous phase in a vacuum, bypassing the liquid phase in the process, thus eliminating the problems associated with surface tension forces, which are dynamic only during 'liquid to gas' phase change [1, 14, 15, 16, 19].

(a) (b)

(a) Courtesy of Ms. Wardah Abd. Rahman, Universiti Sains Malaysia. (b) Courtesy of EM Unit, Universiti Sains Malaysia

**Figure 11.** (a) Freeze Dried Aspergillus niger prepared using the freeze drying method as described in 4.3.3. (b) Pollen of Orthosiphon stamineus prepared using the freeze drying method as described in 4.3.3.

### 4.3.2. *Materials*

(a) (b) (c)

**Figure 12.** (a) SEM Freeze Dryers are generally fitted with thermoelectric peltier cooled stages, adjustable to -60°C to enable careful sublimation of frozen samples under vacuum. The vacuum pump is not shown. (b) Aluminium Planchettes, useful to hold samples for freeze drying. After the freeze drying and sputtering processes, the planchette with the sample can be attached on the SEM stub in one piece for viewing. (c) Tissue-Tek, generally used in cryostat sectioning, aids in the firm attachment of frozen samples on the planchette when used together with double-sided carbon adhesive tapes or tabs.

### 4.3.3. *Standard protocol*

| **Freeze Drying** | |
|---|---|
| 1. | Prepare a planchette with double-sided sticky tape and a thin layer of 'Tissue-Tek' on the tape (Fig. 12b and 12c). Label the sample positions carefully. *(The sticky tape tends to detach from the planchette if used on its own. Similarly, 'Tissue-Tek' sometimes slide off with the sample once frozen when used on its own. In the author's unit, the sticky tape, preferably carbon, is firmly attached on to the planchette followed by a thin layer of 'Tissue-Tek' covering the sticky tape and beyond. The sample can then be placed on the 'Tissue-Tek' layer for vapour fixation and the liquid nitrogen plunge.)* |

| Freeze Drying | |
|---|---|
| 2. | Cut out a small piece of agar/substrate containing the bacterial/ fungal growth (about 5mm x 5mm) and quickly place it on the double-sided sticky tape with 'Tissue-Tek'. **[FUME HOOD]** |
| 3. | Place the planchette in a filter paper lined petri dish; wet the filter paper with a few drops of 2% osmium tetroxide, away from the planchette and close the petri dish immediately. Leave it alone in the fume hood for about 1–2 hours. This process is known as 'vapour fixation'. **[FUME HOOD]** |
| 4. | Once the sample has been 'vapour fixed', the planchette is plunged into 'slushy nitrogen (-210°C) and transferred on to the 'peltier-cooled' stage of the Freeze Dryer (Fig. 12a) and left to freeze dry for about 10 hours. (Refer to manual for the operation of the freeze dryer.) |
| 5. | After the freeze drying process, the planchette with the sample must be kept in a desiccator (if the sample cannot be viewed immediately). |
| 6. | The planchette is then mounted on to a SEM sample stub (Fig. 7c) with a double-sided sticky tape. Sputter the sample with about 5–10 nm of gold before viewing it in the SEM. |

#### *4.3.4. General Precaution*

Keep the sample in a moist environment at all times till the freezing process. If the sample is already dry or partially dry (i.e. if some removal of water from the sample has occurred prior to vapour fixation), structural alteration or deformation is to be expected.

## 5. Mounting of samples

Mounting of samples on the SEM stubs needs utmost care and the use of a low power stereomicroscope is highly recommended during the process. It can be heartbreaking to see the region of interest of a sample damaged by a simple unintentional graze with a tool, after all the hours spent on processing the sample. Storkbill forceps or insect forceps (Fig. 5), fine spatulas, double-sided carbon adhesives (Fig. 7a), Leit-C (Fig. 15a) and sharpened soft wood sticks are some of the tools which could come handy when performing the mounting process. Factors which need to be considered in the sample mounting process are sample orientation, sample stability and ensuring good adherence of the sample on the SEM stub. Leit-C can aid in placing the sample in the required orientation for larger samples. [1, 2, 4].

## 6. Conductive coating of samples for sem

Biological samples, being non-conductive, give rise to charging problems in the SEM as the bulk of primary electrons from the electron beam remain in the sample to form clouds of negative charges. This built-up charge *interferes with the primary beam* to bring about image distortions, loss of contrast with very bright and dark areas, known as 'charging effect' (Fig. 13b). To overcome these problems, *a thin layer of metal* is sputtered on the sample, thereby *increasing the conductance* on the sample to enable the absorbed electrons to find their way to ground. A sputter coater (Fig. 13a) is used for this purpose. The preferred metals for sputtering are gold, gold-palladium, platinum and chromium. However, platinum and chromium are the choice metals for high resolution imaging (more than 50, 000x) as gold sputtering results in visible graininess on the surface of the sample (Fig. 13c). Other materials used for sputtering are iridium and carbon. In the present state of art Field Emission SEMs, metal sputtering can be completely ignored with the use of low KV imaging techniques of less than 500 volts [1, 2, 4].

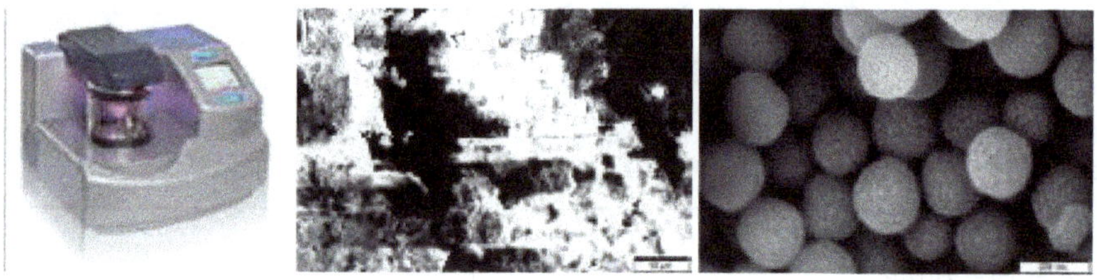

(a) (b) (c)

Courtesy of EM Unit, Universiti Sains Malaysia

**Figure 13.** (a) Sputter Coater – employs the physical vapour deposition principle by using argon ions to eject target material (Au, Au-Pd, Pt, Cr, etc.) from the target surface to sputter a thin film of the metal or alloy on the sample, in a vacuum. (b) Charging Effect on a wood sample in the SEM, thus metal sputtering is needed to enable the absorbed electrons to be dissipated. (c) Nano Silica Powder with visible graininess from gold sputtering. Chromium or Platinum would have been a better choice for this SEM magnification.

## 7. Technique for particulate samples

Particulate samples of dry powders (Fig. 14a), fine crystals (Fig. 14b), nanoparticles (Fig. 13c), dried bacterial cells and spores do not need any special sample preparation protocols. However, care must be taken to avoid the particles loosely 'stacking' one over the other. It should also be ensured that the particles are firmly stuck on the specimen stub.

Procedure:

**a.** Sprinkle a little of the sample evenly but lightly on a SEM sample stub with the double-sided sticky tape.

**b.** Use a hand blower (Fig. 7d) to blow away the loose particles. Always blow away from yourself into a bin or sink. Employ increased safety procedures if the particles are known to be harmful

**c.** The sample is now ready for viewing in the SEM. Sputter with Au, Pt or Cr, if necessary.

Stacking of particles is generally the main cause of image instability and charging problems. Sometimes, after the blowing away of the loose particles, the sticky tape may appear to be very clean without any particles left for SEM study. It will be noted that this type of preparation tends to give lesser agglomeration and lesser charging. It is possible for fine particles prepared on carbon sticky tape in this manner to be viewed employing the backscatter mode without the need for metal sputtering [1, 2, 4].

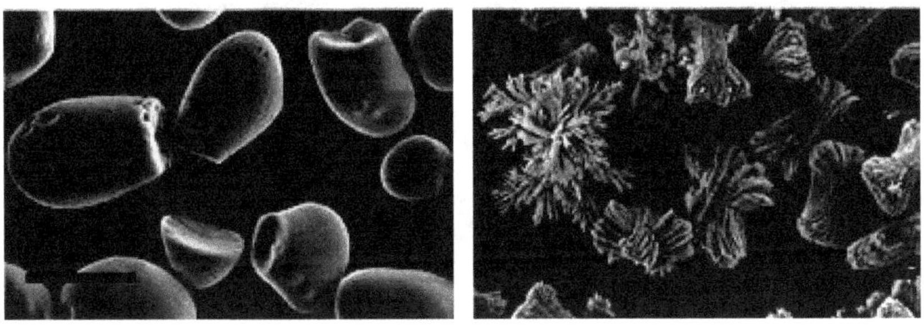

(a) (b)

(a) Courtesy of EM Unit, Universiti Sains Malaysia, (b) Courtesy of Prof. Zakaria Mohd. Amin, Universiti Sains Malaysia.

**Figure 14.** (a) Starch Powder, prepared by sprinkling the powder on a SEM sample stub with double-sided carbon adhesive tape followed by air blowing and the sputtering of gold. (b) Calcium Oxalate in dendritic forms prepared by sprinkling the crystals on a SEM sample stub with double-sided carbon adhesive tape followed by air blowing and the sputtering of gold.

## 8. Technique for large and bulk samples

Samples of rocks, wood, plastics, metals and electronic components, which are large enough to be handled with general laboratory forceps do not need any processing. However, ensure the use of suitable large SEM sample holders and the sample height to conform with the specifications of the SEM (Fig. 15b). The samples must be secured and stuck firmly on the holder to avoid any possible movement or vibrations. For larger samples, securing of the samples should be at the base and also across the samples to hold down the sample firmly on the SEM stub, carefully avoiding the areas to be studied in the SEM (Fig. 15b). Double-sided carbon sticky tapes and tabs (Figs. 7a and 7b) have been found to be the most suitable for the attachment of the sample on the SEM sample stubs, although other materials such as colloidal silver paint, carbon paints and Leit-C (Fig. 15a) can be employed. Leit-C can be very useful in positioning the sample to obtain suitable orientations for imaging in the SEM. Sputtering of the sample is necessary for non-conductive samples [1, 2, 4].

Materials Needed:

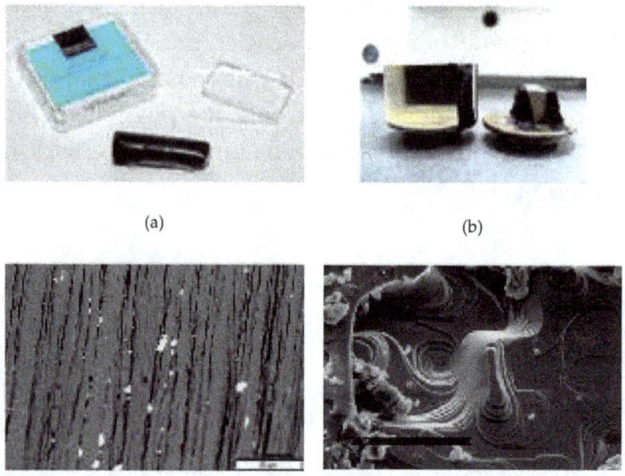

(a) (b)

(c) (d)

(c) Courtesy of EM Unit, Universiti Sains Malaysia, (d) Courtesy of EM Unit, Universiti Sains Malaysia

**Figure 15.** (a) Leit-C, which is flexible like plasticine clay, is useful in positioning the sample to obtain the required orientation large samples for imaging. (b) Large sample on stub. Double-sided carbon adhesive tape has been used to secure the sample on the sample stub. (c) Copper wire with lead particles, secured on the SEM sample stub with double-sided carbon adhesive tape. No processing is required. Being a conductive sample, no metal sputtering is required. (d) Nanocomposite block material, secured on the SEM sample stub with double-sided carbon adhesive tape. Being a non-conductive sample, sputtered with gold, but no other processing is required.

## 9. Reagents

References for section 9 are from [1, 2, 3, 4, 10, 11]

**1.** *Phosphate Buffer: (Sörensen)*

To prepare a 0.2M phosphate buffer of pH 7.2 as follows:

$Na_2HPO_4$..... 20.44 g

$NaH_2PO_4$..... 6.72 g

Add distilled water to a final volume of 1000 ml. Dilute to obtain other concentrations.

**2.** *Cacodylate Buffer :*

Prepare a 0.4M solution of Sodium Cacodylate with

$Na(CH_3)_2AsO_2.3H_20$..... 21.4 g

Add distilled water to make 250 ml.

Prepare 0.2M Cacodylate Buffer, pH 7.2 (approx.)

0.4M Sodium Cacodylate..... 50 ml

0.2M HCl..... 8 ml

Add distilled water to a final volume of 100 ml.

The pH of the buffer is adjusted, if necessary, to the required value with HCl.

**3.** *4% Glutaraldehyde:*

Prepare the fixative as follows:

0.2M buffer..... 50 ml

25% glutaraldehyde..... 16 ml

Add distilled water to a final volume of 100 ml

**4.** *McDowell and Trump's Fixative (1976):*

Prepare the fixative as follows:

0.2M buffer..... 50 ml

37% formaldehyde..... 11 ml

25% glutaraldehyde..... 4 ml

Add distilled water to a final volume of 100 ml

This fixative contains 4% formaldehyde and 1% glutaraldehyde in 0.1M buffer (4F : 1G). Other concentrations are obtained by using different quantities of the constituents. The pH may change on adding the aldehydes.

**5.** *Karnovsky's Fixative (1965) :*

Prepare 20 ml of a 10% solution of paraformaldehyde by dissolving 2.0 g of paraformaldehyde powder in 20 ml of distilled water and heating to about 60°C (in a fume hood). Add a few drops of 1M Sodium Hydroxide until the solution becomes clear. Allow the solution to cool before use.

Prepare the fixative as follows:

0.2M buffer..... 50 ml

10% paraformaldehyde..... 20 ml

25% glutaraldehyde..... 10 ml

Add distilled water to a final volume of 100 ml

This fixative contains 2% paraformaldehyde and 2.5% glutaraldehyde in 0.1M buffer. Other concentrations are obtained by using different quantities of the constituents. The pH may change on adding the aldehydes.

If necessary, adjust the osmolarity of the fixative with sucrose, glucose or sodium chloride.

If required, add sufficient calcium or magnesium chloride to give a final concentration of 1–3 mM, taking care to avoid the formation of precipitates with phosphate buffers.

6. *4% Glutaraldehyde in Seawater :*

Prepare the fixative as follows:

25% glutaraldehyde..... 16 ml OR

50% glutaraldehyde..... 8 ml

Add artificial seawater/membrane filtered seawater to a final volume of 100 ml

7. *Osmium tetroxide solutions*

Notes:

1. Osmium tetroxide is a contact and inhalation hazard. Do not break the ampoule in the open, as the crystals may sprinkle around and the vapour from the crystal is very harmful.
2. The preparation of osmium tetroxide must be performed in a chemical fume hood.
3. Solutions should be prepared in very clean bottles, as the presence of any organic substance tends to react with osmium (and changes colour over time).
4. Store in a fridge (5 deg), by placing the smaller 100 ml bottle containing the stock osmium solution in a larger wide mouth bottle with a tight screw cap, e.g. Kilner Jars.
5. If possible, prepare at least 1 day before use to ensure complete dissolution of the crystals

Items needed to prepare 2% Aqueous Osmium tetroxide:

a. 1g Osmium tetroxide in an ampoule (or any other amount but recalculate for 2%)
b. 50 ml distilled water in a clean glass measuring cylinder
c. A clean 100 ml thicker walled glass bottle to contain the 2% stock solution
d. A glass scorer (any types)
e. A thick glass rod (about 8" long)
f. Get ready all these items in the fume hood

Method

- Take a 1 g osmium vial and score the neck with a glass scorer (making sure not to score too hard such that the glass could break)
- *[If the ampoule has a label sticker on it, remove the sticker, wash the gum remnants on the glass with a little soap solution if necessary, and rinse the ampoule with a few changes of distilled water before proceeding to score and use. Gum material sticking on the glass would react with osmium to deteriorate the stock solution.]*
- From the measuring cylinder containing 50 ml of distilled water, add about 10 ml of the water into the 100ml glass bottle
- Place the ampoule into the bottle

- Use the glass rod to break the ampoule. Once broken, crush the ampoule with the glass rod into a couple of pieces.
- Now add the remaining 40 ml of distilled water into the bottle (while also washing down the glass rod into the bottle)
- Close the bottle tightly and swirl the bottle for a while to enhance the dissolving of the osmium in the water.
- Now place this 100 ml bottle containing the 2% Osmium into a larger wide mouth screw capped glass container and store it in a fridge.
- Before use the next day, swirl the bottle for about a minute. This swirling is needed only for using it for the first time.
- [It is alright to leave the ampoule glass pieces in the 100 ml bottle. Sometimes, when the 50 ml of osmium solution has been used up, I break another ampoule in the same bottle (as described above) without removing the earlier glass pieces. I do this for about 3 ampoules before washing and removing the glass pieces for a complete wash of the bottle, provided the osmium solutions remain clear all the time]
- [Sometimes, if the osmium is needed to be used on the same day as when prepared, the bottle containing the stock solution is left in the ultrasonic bath for a few minutes before use.]
- *The working fixative is prepared just before use, by mixing equal parts of 2% aqueous stock osmium tetroxide and 0.2M phosphate buffer – resulting in 1% osmium tetroxide in 0.1 M phosphate. [However, there are some researchers known to use other concentrations of osmium tetroxide and buffers].*

## Author details

Mogana Das Murtey[1*] and Patchamuthu Ramasamy[2]

*Address all correspondence to: moganbaadshah_071185@yahoo.com

1 Cluster of Integrative Medicine, Advanced Medical and Dental Institute (AMDI), Universiti Sains Malaysia, Penang, Malaysia

2 Quest International University Perak (QIUP), Ipoh, Perak Darul Ridzuan, Malaysia

## References

[1] Bozzola, J. J. and Russel, L. D. (1999) *Electron microscopy: Principles and techniques for biologists.* Jones and Barlett Publishers, Sudbury, MA.:48–71

[2] Dykstra, M. J. (1992) *Biological electron microscopy, theory, techniques and troubleshooting.* Plenum Press, New York and London.: 5-77, 237–242

[3] Hayat, M. A. (Ed.), (2000) *Principles and techniques of electron microscopy: Biological applications*. Cambridge University Press, 4th ed.:1-80, 400–431

[4] Hayat, M. A. (1984). *Fixation for electron microscopy*. Academic Press, Inc. NY. 320–331

[5] Allen, T. D. (1983) *The application of scanning electron microscopy to cells in culture: selected methodologies*. Scanning electron microscopy. 1983;(Pt4):1963–721

[6] Watson, L. P., McKee A. E., Merrell B. R. (1980) *Preparation of microbiological specimens for scanning electron microscopy*. Scanning Electron Microscopy. 1980;(Pt.2):45-56

[7] Wurzinger-mayer, A., Shipway, J. R., Kristof, A., Scwaha, T., Cragg, S.M. and Wanninger, A. (2014). *Developmental dynamics of myogenesis in the shipworm Lyro pedicellatus (Mollusca: Bivalva)*. Frontiers in Zoology, 11(90)

[8] Glauert, A. M. (Ed.), (1980) *Fixation, Dehydration and embedding of biological specimens, in: Practical methods in electron microscopy*, North-Holland, Amsterdam

[9] Saito, N., Sato, F., Oda, Kato M., Takeda, H., Sugiyama, T., Asaka, M. (2002) *Removal of mucus for ultrastructural observation of the surface of human gastric epithelium using pronase*. Helicobacter, 2002 Apr;7(2):112–5, PubMed

[10] Walker, M. H. and Roberts, E. (1982). *The protozooans epizooties found on the gills of Gammarus pulex*. Hydrobiologia 88, 171–176

[11] Karnovsky, M. J. (1965). *A formaldehyde-glutaraldehyde fixative of high osmolality for use in electron microscopy*. Journal of Cell Biology 27:137A

[12] McDowell, E. M., and Trump, B.F. (1976). *Histologic fixatives suitable for diagnostic light and electron microscopy*. Archives Pathology and Laboratory Medicine. 100 : 405

[13] Nation, J. L. (1983). *A new method using Hexamethyldisilazane for preparation of soft insect tissues for scanning electron microscopy*. Stain Technology Vol. 58, No.6

[14] Dey, S., Basu, T. S., Roy, B. and Dey, D. (1989) *A new rapid method of air-drying for scanning electron microscopy using tetramethylsilane. Journal of Microscopy 156(2): 259–261*

[15] Thomasson, S. A. and Thomasson, J. R. (2011) *A Comparison of CPD (Critical Point Drying) and HMDS (Hexamethyldisilazane) in the preparation of Corallorhiza spp Rhizomes and Associated Mycorrhizae for SEM (Scanning Electron Microscopy)* Transactions of the Kansas Academy of Science 114(1&2):129–134

[16] Lindroth, M., Bell, P.B.Jr. and Fredriksson, B.A. (1988) *Comparison of the effects of critical point drying and freeze drying on cytoskeletons and microtubules*. Journal of Microscopy 151(Pt.2):103–114

[17] Jusman, Y., Ng, S. C., and Abu Osman, N. A. (2014) Investigation of CPD and HMDS Sample Preparation *Techniques for Cervical Cells in Developing Computer-Aided Screening System Based on FE-SEM/EDX*. Scientific World Journal, Vol. 2014, Article ID 289817

[18] Botes, L., Price, B., Waldron, M. and Pitcher, G. C. (2002) *A simple and rapid scanning electron Microscope preparative technique for delicate "gymnodinioid" dinoflagellates.* Microsc. Research and Technique15;59(2):128–130

[19] Celine Barre, David O'Neil and Bricelj, V. M. (2006) *Preparation of large bivalve specimens for scanning electron microscopy using Hexamethyldisilazane (HMDS).* Journal of Shellfish Research 25(2):639–641

[20] Neděla, V., Tihlaříková, E., Hampl, A., and Sedláčková, M. (2012) *SEM and ESEM Observation of stem cells. Comparative study of human embryonic stem cell morphology.* Imaging & Microscopy, Darmstadt: GIT Verlag, 2012, roč. 2012, č. 4, s. 32–34

[21] Lee, J. T. Y., and Chow, K. L. (2011) *SEM sample preparation for cells on 3D scaffolds by freeze-drying and HMDS.* SCANNING Vol. 33, 1–14

[22] Al-Salihi, K. A. (2009) *In vitro evaluation of Malaysian Natural coral Porites bone graft substitutes (CORAGRAF) for bone tissue engineering: a preliminary study.* Brazilian Journal of Oral Science 8(4):210–216

[23] Mazia, D., Schatten, G. and Sale, W. (1975) *Adhesion of cells to surfaces coated with polylysine. Applications to electron microscopy.* Journal of Cell Biology 66(1):198–200

3

# Evaluation of the Glomerular Filtration Barrier by Electron Microscopy

Diogo Benchimol de Souza, Bianca Martins Gregório,
Marlene Benchimol and Fernanda Amorim de Morais Nascimento

**Abstract**

The plasma filtration and formation of the urine is a very complex process necessary for the elimination of metabolites, toxins, and excessive water and electrolytes from the body. The initial process of urine formations is done by the glomerular filtration barrier inside the glomeruli. This specialized barrier consists of three layers, fenestrated endothelium, basement membrane, and podocytes, which ensure that water and small molecules pass through while cells and large molecules are retained. The glomerular filtration barrier is found with abnormal morphology in several diseases and is associated with renal malfunction; thus, it is interesting to study these structures in different experimental and clinical conditions. The normal glomerular barrier and its alterations in some conditions (hypertension, diabetes, and fetal programming) are discussed in this chapter. Furthermore, some methods for studying the glomerular filtration barrier by electron microscopy, both by qualitative and quantitative methods, are present.

**Keywords:** Electron microscopy, glomerulus, filtration barrier, morphometry

## 1. Introduction

The human kidney is responsible for many functions, such as filtration of the plasma, blood pressure control, and hormonal production among others. Kidneys are bean-shaped organs located in the retroperitoneal space, irrigated by the renal artery, and covered by its fibrous capsule. Renal parenchyma can be distinguished into cortex and medullar regions, with different anatomic–histological characteristics. Inside a human kidney, there are about one million nephrons, which are composed of tubules and renal corpuscles. The corpuscles contain several tortuous arterioles covered by podocytes. The set of the endothelial layer of these arterioles, its basement membrane, and the podocytes correspond to the glomerular filtration barrier [1, 2].

The ultrafiltration of the plasma by this barrier and formation of the primary urine requires normal glomerular morphology. It is well documented that the podocytes are important to guarantee the selectivity during the filtration process, preventing macromolecules that pass through the slit diaphragm formed between its foot processes. Many studies suggest that alterations on these cells or on the vessels, as those during hypertension, diabetes, or lupus, can reduce the oxygen flux, leading to hypoxia stage, being responsible for podocyte alteration and death [3, 4].

The glomerular filtration barrier morphology has been extensively studied in several experimental and clinical conditions. For this purpose, electron microscopy (both transmission and scanning) suits perfectly for observing the fine structure of this barrier, indicating with accuracy morphological alterations when present [5, 6].

Thus, in this chapter, we aimed (a) to describe the elements of the glomerular filtration barrier, (b) to present scientific examples of how the glomerular filtration barrier elements are related to different clinical and experimental conditions, and (c) to show how the elements of the glomerular filtration barrier can be analyzed by scanning and transmission electron microscopy with qualitative and quantitative methods.

## 2. Main body

### 2.1. The normal glomerular filtration barrier

The glomerulus is a highly irrigated structure that performs selective filtration of the plasma. Inside the Bowman's capsule, several tortuous arterioles receive the blood and filtrate it forming the primary urine, which then passes to the proximal tubule. The glomerular capillaries are lined by a fenestrated endothelium, covered externally by specialized cells, called podocytes. Between the cell layers, there is a basement membrane, which also has an important filtering function. Together, the endothelial cells, the basement membrane, and the podocytes form the glomerular filtration barrier [1, 3].

The glomerular filtration barrier is highly permeable to water and small molecules. Moreover, it is slightly permeable to macromolecules and acts as a physical and electrical barrier for the filtration process. These characteristics are dependent of the cellular structures, and its function is influenced by factors such as molecular weight and electric charge [2]. Moreover, changes in the cell junctions of the glomerular barrier prejudice the glomerular function [7].

Internally, the glomerular filtration barrier is constituted by the glomerular fenestrated endothelium. These endothelial cells have a glycocalyx over its luminal surface, which form a highly negatively charged coating [2]. Thus, glycocalyx covers the endothelium and promotes a first selection of molecules passing through the barrier by electric charge. The endothelium coating the glomerular capillaries is very thin and has several fenestrae with 70–90 nm of diameter. These pores can filtrate only large molecules and blood cells [1].

The endothelial cells are supported by a basement membrane (about 100–150 nm), which is the only continuous layer of the glomerular filtration barrier. As this membrane is thought to be the fusion of the endothelial and epithelial basement membranes, two laminae lucida (interna and externa) and, between them, a lamina dense are found [1]. The basement membrane is composed of a complex network of glycosaminoglycans and fibrous proteins (laminin and collagen type IV), which are continuously produced and deposited by podocytes and mesangial cells [8]. These proteins adhere to the cell membranes by surface receptors and form the glomerular filtration barrier [5].

Podocytes are specialized epithelial cells found in the external layer of the glomerular filtration barrier and exhibit several long cellular processes, which bears various secondary processes, named as foot processes. The foot processes involve capillaries by interdigitations, and small gaps are left in between. These small gaps measure about 20–30 nm and is named as slit diaphragm being responsible for passage of small molecules, whereas larger ones are retained [6]. Further, the slit diaphragm is filled with nephrin and podocin, the transmembrane proteins that are also important for the correct function of glomerular filtration barrier. Experimental studies have shown that mutations in expression of these proteins can alter the filtration barrier and probably is the cause of nephrotic syndrome [9].

In addition to adhesion receptors from integrin family, there are also proteoglycan transmembrane receptors such as those of syndecan family. Also, the junction of podocytes membrane is seen by electron microscopy, showing zipper-like structure where binding proteins form cell junctions and small pores, as previously mentioned on filtration slit [6]. Thus, it is important to consider all glomerular ultrastructure in the filtration process and renal function. Some diseases provoke changes in glomerular ultrastructure, which promote irregular filtration rates, proteinuria, and even kidney failure, further discussed here. Electron microscopy and correlate techniques allow better to understand these pathological changes and become an effective and important method in the study of renal function.

### 2.2. The altered glomerular filtration barrier

Literature describes well several diseases that affect the glomerular filtration barrier. Some conditions such as diabetes mellitus, hypertension, and maternal nutritional changes during critical periods of development, known as fetal programming, are well recognized as important risk factors for developing chronic kidney disease.

Diabetes mellitus is one of the most chronic diseases emerging on twenty-first century, in which hyperglycemia is a major indicator, generating microvascular damage such as retinopathy, neuropathy, and kidney disease [4]. Diabetic nephropathy is a chronic progressive disease that affects 20–40% of patients with diabetes mellitus [10]. Histopathologically, this disease in humans courses with the thickening of glomerular and tubular glomerular basement membrane, podocytopenia, mesangial expansion, glomerular, and arteriolar hyalinosis. The earliest clinical manifestation of diabetic nephropathy is microalbuminuria, a strong predictor of renal and cardiovascular disease in patients with type 1 and type 2 diabetes mellitus [11, 12]. These modifications contribute to the abnormal stimulation of resident kidney cells, which increases the production of TGF-β1 and causes collagen (types I, IV, V, and VI), fibronectin, and laminin depositions in the extracellular matrix of the glomerulus. Thus, this structural disorganization of the glomerular slit diaphragm enhances renal damage and chronic kidney disease progression [13, 14].

Likewise, hypertension is directly related to renal failure [15]. Hypertensive nephropathy, a consequence of chronic increase of blood pressure, is secondary to diabetic nephropathy in terms of diagnosis and is considered the last stage of renal disease. Some complications are associated with hypertensive nephropathy such as glomerular damage resulting in impaired renal function [16]. Hypertension causes an increase of numerous local factors, such as angiotensin II, which may contribute to the development of renal fibrosis. It is reported that angiotensin II stimulates the gene expression of TGF-β1 and the release of this protein. The presence of TGF-β1 activates the conversion of fibroblasts into myofibroblasts, producing large amounts of extracellular matrix components, and induces renal fibrosis [17].

Moreover, angiotensin II regulates the number and integrity of podocyte. In high quantities, it promotes the disintegration and breakup of these highly specialized cells and causes glomerular endothelial cells hypertrophy since it raises intraglomerular pressure, preferably affecting the afferent glomerular arterioles [18]. The increase of blood volume is also capable to reduce the levels of podocin (abundant protein in the podocyte body) and nephrin (a structural component of the slit diaphragm) in the glomerulus, which enhances the glomerular podocyte injury and albuminuria [19].

Recently, experimental and epidemiological studies report that several metabolic disorders manifested in adulthood have their roots dating embryonic periods [20, 21]. Protein or energy restriction during pregnancy induces low birth weight and, consequently, the injury of nephrogenesis, increasing the incidence of chronic kidney disease in adulthood [22]. As previously mentioned, the glomerulus (the most important filtering apparatus in the body) is a highly specialized structure formed by four types of cells: mesangial, endothelial, visceral (podocytes), and parietal epithelial cells. Accordingly, the intrauterine growth restriction induced by maternal protein restriction may cause morphological and functional changes in the glomerulus, thereby decreasing the filtration barrier efficiency with consequent glomerulosclerosis [23].

Just as occurs in diabetic nephropathy, low birth weight promotes glomerulosclerosis [24], expansion of mesangial matrix, hyalinosis, and podocytopenia, which compromises glomerular filtration and facilitates progressive kidney dysfunction [25]. The loss of podocytes may represent the starting point for an irreversible glomerular injury, characterized by proteinuria and glomerular scarring. Transmission electron microscopy analysis revealed that the offspring from maternal low protein diets at 26 weeks of age and at 16 weeks old and presents the effacement of pedicels, the absence of the slit diaphragm, and an increase of glomerular basement membrane thickness, denoting a reduced of barrier efficiency filtration [24, 26].

Another metabolic programming model related to the occurrence of chronic kidney disease in adult life is maternal vitamin D restriction. This vitamin is essential for the development of the nervous system, immune function, skeletal formation, and fetal kidney [27, 28]. Studies in rats showed that vitamin D restriction during critical periods of development has increased the number of glomerulus, reduced the size of renal corpuscles, and delayed glomerular maturation, further lower expression of WT1 and podocin in adult offspring [28, 29]. These structural and functional adaptations in the glomerulus may progress to chronic kidney disease

## 2.3. Glomerular filtration barrier evaluation

### *2.3.1. Sample preparation*

A very good sample preparation for EM must follow careful cares because small mistakes will be big ones when analyzed by EM. Thus, we can use either old or new approaches when we intend to study not only the glomerulus but also any biological sample.

However, the glomerulus is particularly difficult due to different osmolarities found in each segment of the nephron. Cryofixation, in general, needs much smaller samples sizes than conventional chemical fixation. A freeze-depth of about 10–15 μm is achieved; one exception is high-pressure freezing (HPF). Nowadays, there are new techniques and good, although expensive, equipments commercially available such as high-pressure freezing (HPF), slam freezing, cryoultramicrotomy, freeze-fracture, freeze-etching, and so on. New scanning electron microscopies with high resolution and extreme high resolution that appeared recently provided beautiful and illustrative aspects of all different cell types. Concerning transmission EM, new methodologies allow 3-D reconstruction such as tomography of thick or semithick sections, serial sections, or high-voltage TEMs. In addition, spectacular images were recently obtained with new Helium ion (proton) microscopes.

When sample preparation for the evaluation of the glomerular filtration barrier by electron microscopy follows the standards of routine biological sample preparations, the person should manipulate the tissue with the animal alive, under anesthesia, for obtaining a small fragment with a size of 0.5 mm or even less and immediately immerse it in the fixative. Alternatively, perfusion gives very good results.

When studying biopsy samples of human kidney or whenever the perfusion is not possible or recommendable, fixation by immersion is an adequate option. For this, a small fragment of the renal cortex is collected, washed in buffered solution, and immediately immersed in the fixation solution [30]. This fragment should be collected from the outer part of the renal cortex, avoiding juxtamedullary region where glomeruli are sparser [1]. As routine preparation, fixation of the kidney is done with glutaraldehyde alone or in combination with freshly prepared formaldehyde from paraformaldehyde powder in warm water. The formaldehyde molecule is a very small aldehyde and has the advantage to penetrate rapidly into tissue, although it is not a strong linker for proteins. Thus, a good procedure is its combination with glutaraldehyde since it penetrates more slowly than formaldehyde and provokes cross-links with proteins. However, glutaraldehyde is a slow fixative, and it takes some seconds or even some minutes to cause death to cells and tissues. The small fragments are fixed in 2.5% glutaraldehyde or 4% PF with 2.5% glutaraldehyde, diluted in 0.1 M cacodylate buffer, pH 7.2, overnight at room temperature. Most important, buffered solutions should be used for better preservation of the cellular aspects of the glomerular filtration barrier [30, 31]. In the next day, samples are post-fixed in 0.1 M cacodylate buffer containing 1% $OsO_4$ and 0.8% potassium ferricyanide for 1 h. It is important to point out that potassium ferricyanide is an important approach to better visualize the trilaminar structure of cell membranes. Then samples are dehydrated in graded series of acetone or alcohol. Ultrathin sections obtained after epoxy resin embedding are stained with uranyl acetate and lead citrate and observed in transmission electron microscope.

When studying samples collected from large animal models (i.e., pig, sheep, dog, etc.), the whole kidney can be adequately fixed by perfusion through the renal artery. When studying small species of animals (i.e., rats, mouse, etc.), the whole animal can be perfused through the left ventricle or aorta. In both cases, the researcher may take into account that some fixatives could hamper some other analysis, as for example, perfusing the kidney with glutaraldehyde may prejudice immune-labeling analysis, unless used in very low concentrations such as 0.5% glutaraldehyde or less [31]. We recommend fixation by perfusion in a mixture of glutaraldehyde and freshly prepared formaldehyde because the latter is a fast fixative allowing not only good preservation but also a quick cell death. Whenever using perfusion fixation, the osmolarity of the fixative solution should be taken into account. Preferentially, the osmolarity of the solution for kidney perfusion fixation should be of 420 mOs [32]. After organ perfusion, a small fragment of the renal cortex should be collected and immersed in the fixative solution for postfixation and dehydration as describe by immersion above. It is important to remember removing the renal capsule whenever it is present, allowing the fixative penetrate into the tissue of interest. Also, cryofixation can be an option for sample fixation for glomerular filtration barrier evaluation. One advantage of high-pressure freezing technique is that larger fragments with edges as large as 1 mm can be used, being a good choice as a cryofixation method [31].

As the object of interest is present in the glomerulus, we may take into account that our samples should have at least one glomerulus for examination. The glomerulus of rats has an average area (in its maximum cross section) of 0.02 $mm^2$ and occupies around 7.5% of the surface area on renal cortex [33]. Thus, for improving the chance of having a glomerulus on our sample, the examined surface area of the sample should be of at least 0.5 $mm^2$ ($0.02 \times 200 / 7.5$). Based on this, the cortical fragment collected for glomerular filtration barrier analysis with electron microscopy should be about 1 × 0.5 × 0.5 mm in size. Furthermore, as there is no guaranty of having glomeruli in the sample, several fragments must be collected, processed, stored, or fixed.

It is important to point out one step that should not be overtaken: obtaining and observation of semithin sections before the ultrathin sections. Semithin sections are important because it is necessary (a) to confirm that there is a glomerulus in the cutting surface of the sample, (b) to determine whether the pyramid in the block face is correct in size and shape or should centered over a glomerulus, and (c) to observe the hole cutting surface under low magnification and obtain a copy in light microscopy that will be useful for searching glomeruli when this sample is observed with electron microscopy. It is very often that when the semithin sections are observed, the block is taken away as no glomerulus was found on its surface. This reinforces the importance of collecting several fragments of each animal/patient for electron microscopy analysis.Alternatively, electron tomography can be used. For this, thick sections (>200 μm) and intermediate (200–300 kV) or high-voltage (750–1000 kV) EM equipped with a eucentric tilting stage are used. Colloidal gold particles (10 nm) are deposited on surface of the grids, to be used as fiducial markers for alignment of the tilted views. Single-axis tilt series from –60° to 60° images are collected in 1° increments using TEM coupled to a 4k × 4k CCD camera. About 400 images are obtained and three dimension reconstruction and data analyses are performed using proper software package as recommended by the EM company.

Sample preparation for SEM requires the same care as for TEM. Aldehyde fixation is adequate, but in order to improve lipid stability, contrast, and electron beam scanning, osmium tetroxide is also used. Sample is dehydrated with the same agents for TEM; however, we prefer to use alcohol starting from 7.5% to three times in 100% to avoid problems of drying artifacts. Next, tissue will critical point dried and coated with a very thin layer of gold or other conductive metal, producing beautiful images but without high resolution.

### *2.3.2. Helium ion microscopy*

Recently, the group of Rice et al. [34] used helium ion microscopy (HIM) to analyze the rodent glomeruli and obtained wonderful images of uncoated samples by SEM. The authors used transcardial perfusion with aldehyde fixatives, vibratome sectioning, gradual dehydration in methanol series, and a very careful critical point drying. There is no need to apply conductive coatings to the samples prior to imaging, and thus sample surface information is preserved. Even at low magnification, the high quality and depth of field of HIM images is impressive. The images obtained with HIM are important and beautiful. The authors depicted some nanoprojections of the foot processes membrane (Figure 1). Also, podocytes and endothelial fenestrae within the filtration slit diaphragm were observed with much more detail (Figure 2). Nanoprotrusions originating from the processes are clearly seen projecting into urinary space. The future will provide much more resolution for biological samples.

### *2.3.3. Qualitative analysis of the elements of the glomerular filtration barrier*

The glomerular filtration barrier can be easily evaluated by simple observation with electron microscopy. One aspect commonly evaluated is the loss of characteristic fenestration of the endothelial layer, which has been reported in some medical conditions previously described. By transmission electron microscopy, the endothelial cells are seen lying over the basement membrane and gaps of this layer, which correspond to fenestrae are fewer when compared to normal conditions [1].

The glomerular basement membrane is another part of the barrier easily evaluated with transmission electron microscopy. By routine examination, the membrane thickness can be qualitatively assessed. As the thickness of the membrane is of great importance for proper functioning of the filtration barrier, this is of interest in many studies. In addition, alterations found in basement membrane are highly associated with several diseases, as previously mentioned, reinforcing the importance of study this structure.

Finally, the podocytes, the third layer of the glomerular filtration barrier, are probably the most studied structure of this barrier. One important aspect of these cells is the interlacement formed

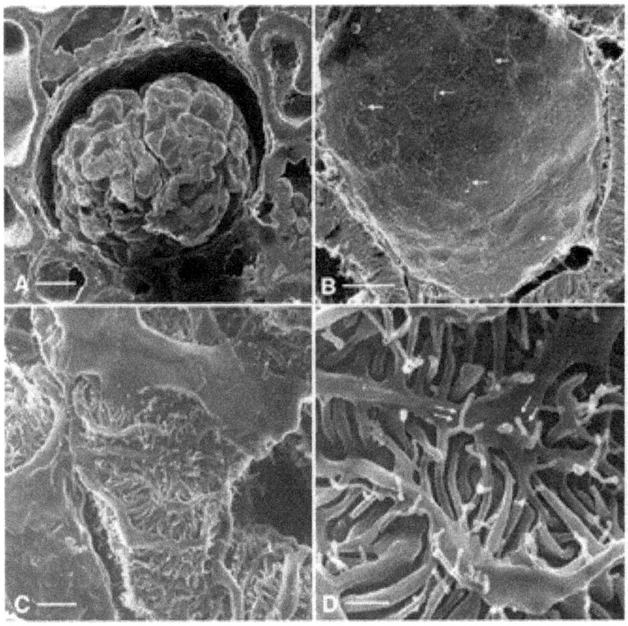


**Figure 1.** Microscopic images of glomerulus of rats as seen by scanning helium ion microscopy. In image A, we see the glomerulus in low magnification and several cut open tubules surrounding the glomerulus. Bar = 20 mm. In image B, the interior of a Bowman's capsule from which the glomerular tuft was removed is observed. Each parietal cell displays a single, long central cilium (arrows) that is very well preserved and visualized without heavy metal coating commonly used in scanning electron microscopy but unnecessary for ion microscopy. Bar = 10 mm. In image C, with intermediate magnification of the surface of a glomerular tuft, it is showed the complex interdigitations of podocytes and their foot processes. Bar = 2 mm. Finally, under higher magnification, it is possible to observe in image D that the podocyte processes are decorated by fine, thread-like protrusions (arrows). Bar = 0.5 mm. Image courtesy of Prof. Dennis Brown [34], available at doi:10.1371/journal.pone.0057051.g001.

by its foot processes, with a small slit diaphragm in between [35]. The alteration of the normal morphology of podocytes and foot processes is named effacement, which has been demonstrated in several conditions [36]. Podocyte effacement is characterized by loss of the normal interdigitations pattern of foot processes, leaving a thinner cell covering a large area outside the basement membrane. As consequence, morphological modifications as fewer and wider foot processes of podocytes are found and can be observed either with transmission or scanning electron microscopy. Among the advantages of observation with scanning electron microscopy are the following: (a) deep field, (b) possibility of analysis of larger areas, allowing the visualization of several podocytes and foot processes in the same sample, and (c) beautiful images of the glomeruli within Bowman's capsule, which are easy to understand and interpret, as shown in figures 3 and 4.

Qualitative differences found in glomerular filtration barrier from individuals submitted to any experimental or clinical condition are only possible when comparison is done with health organisms. However, only qualitative studies do not allow a deeper analysis and statistics and thus lacks scientific excellence. In addition, qualitative observation leads to problems of the

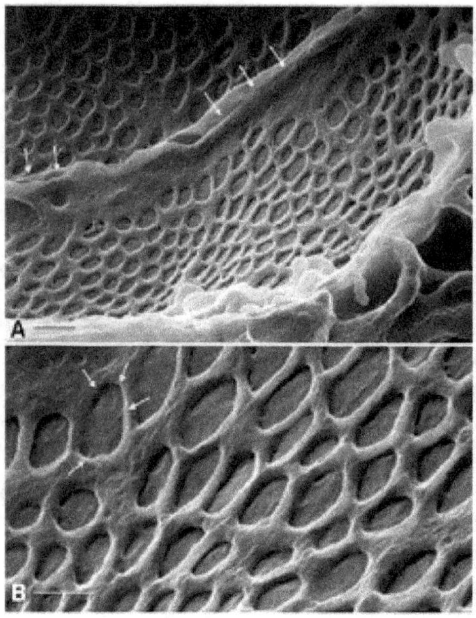


**Figure 2.** Images of glomerular endothelial cells as seen from the capillary luminal side using scanning helium ion microscopy. In image A, note the numerous, round fenestrations present over the entire cell surface. The raised ridges (arrows) represent the location of the tight junction between the two cells. Bar = 175 nm. In image B, under higher magnification, it is possible to observe the details of the fenestrations. In some of them, a substructure consisting of faint spokes like a bicycle wheel can be seen (arrows). Bar = 80 nm. Image courtesy of Prof. Dennis Brown [34], available at doi:10.1371/journal.pone.0057051.g004.

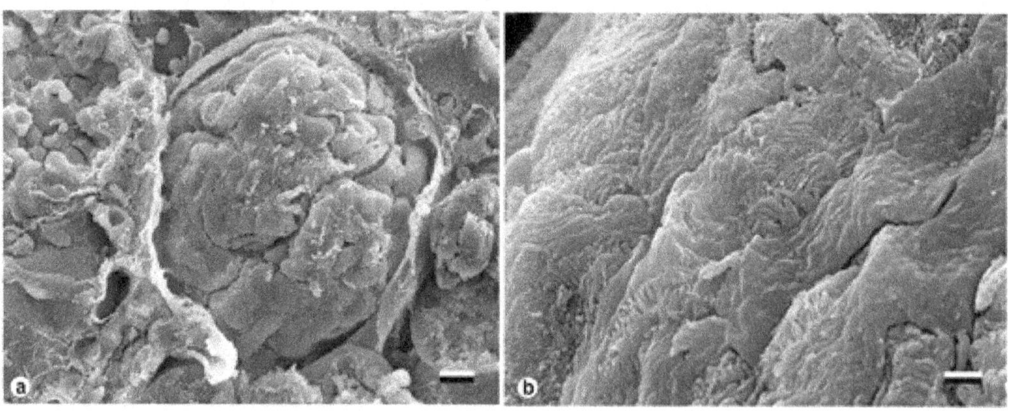


**Figure 3.** Scanning electron microscopic images of glomerulus of rats, showing the podocytes and its foot processes. In image A, under low magnification, we can observe the entire glomeruli with its Bowman's capsule (×950; 15 kV, scale bar represent 10 μm). In image B, we can observe several podocytes and foot processes (×6500; 15 kV, scale bar represent 1.5 μm).

interpretation due observer expertise [37]. Thus, it is clear that quantitative methods have advantages over those qualitative ones.

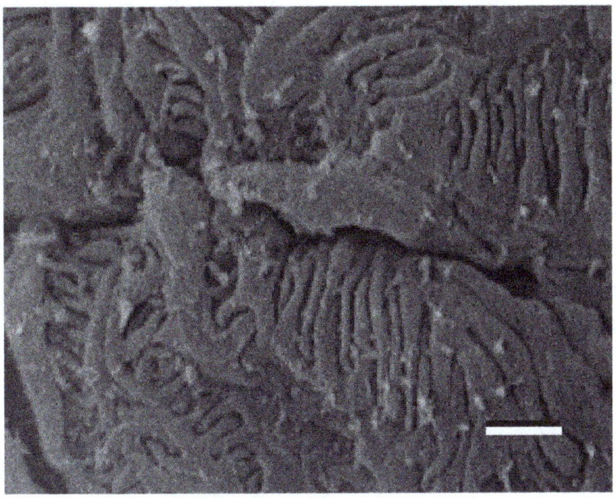

**Figure 4.** Scanning electron microscopic images showing the fine detail of the interdigitating pattern of the foot processes of podocytes at high magnification (×10,000; 15 kV, scale bar represent 1 μm).

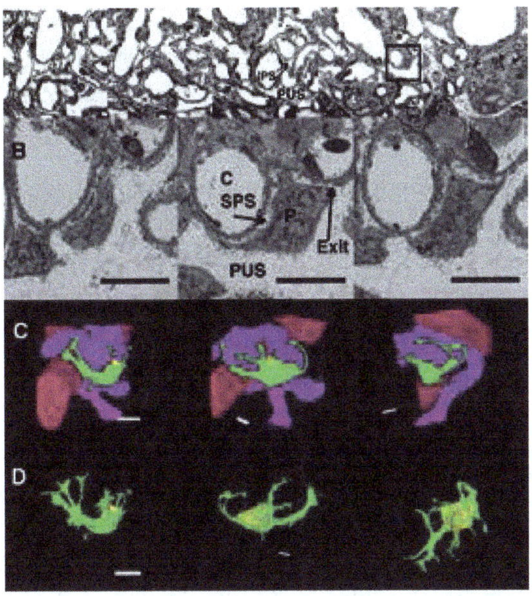


**Figure 5.** Scanning electron microscopy of a human glomerulus after serial block face slicing. In image A, we observe figures that resemble transmission electron micrographs by inverting the signal intensity of the backscattered electron. We see micrographs of complete field of view, with slices 15 μm apart one from another (z). In image B, we observe micrographs of a podocyte of a higher magnification of image A, illustrating the subpodocyte space. In image C, we observe the same podocyte as in image B and adjacent capillaries after highlighting and 3D reconstructing. Finally, in image D, the same podocytes of image C is separated from capillaries after 3D reconstruction, showing the subpodocyte space. P = podocytes, SPS = subpodocyte space, C = capillary lumen, PUS = periphery urinary space, IPS = interpodocyte space, and Exit is an exit region for an SPS. Purple/pink = capillaries, green = podocyte cell body, and yellow = SPS (under the main cell body). Scale bar is 10 μm in all cases. Image courtesy of Prof. Kenton Arkill [38], available at doi:10.1186/1471-2369-15-24.

The glomerular basement membrane was recently observed in three dimensions by different methodologies [38]. In their paper, the authors compared the images and the 3D reconstruction of the glomerular filtration barrier as performed using serial block face scanning electron microscopy, focused ion beam milling scanning electron microscopy, and transmission electron tomography. They reported that the transmission electron tomography technique had the advantage of a higher resolution, with the disadvantages of limited field of view and anisotropic shrinkage. Focused ion bean and serial block face with scanning electron microscopy had greater field of view with lower resolution. The most interesting aspect of these techniques is the reconstruction and digital dissection of cells as demonstrated in Figure 5.

### 2.3.4. *Quantitative analysis of the elements of the glomerular filtration barrier*

The quantification of morphological structures is highly recommended for studying biological alterations in several tissues and different situations. The quantification of morphological aspects of the glomerular filtration barrier is possible and desirable. Translating the morphological aspects of tissues in numbers is very useful from a scientific point of view. It improves the understanding of the modifications of the barrier provoked by different conditions and allows statistical comparisons with other specimens, subjected to different conditions or at different stages of the disease [37]. Based on these premises, methods have been used for different purposes in the study of the glomerular filtration barrier morphology, and important scientific knowledge has been generated from these analyses.

Most quantitative analyses use digital microscopic images that can be digitally acquired or scanned from conventional images. For absolute values (linear measures, for example), it is very important that a scale is maintained on the image. Most electron microscopes automatically print a scale bar on each image, and this is very suitable for morphometric purposes [30].

Several software are eligible for morphometric analyses. In this chapter, we present the steps used in ImageJ 1.37v software. ImageJ is an open source, free software produced and distributed by the National Institutes of Health and can be downloaded at http://imagej.nih.gov/ij/index.html.

The different elements of endothelial layer of the glomerular filtration barrier can be objectively evaluated, and the possible loss of fenestrations can be easily assessed by linear measurements in transmission electron microscopy images. For this, we should first calibrate the software for the correct magnification. The straight-line tool should measure the scale bar of the image, and then we should use the option "analyze—set scale." In the box opened, the distance in pixels should represent the size of the line over the scale bar, and one should insert the real size of the bar at "known distance" space. The "unit of length" space should be fulfilled with the unit of the scale bar (μm or nm). When one intend to measure several images, all of them should be acquired under the same conditions (magnification, resolution, size, etc.), and the option "global" can be marked. By this, the same configuration may be used for all following micrographs. Figure 6 shows the above-mentioned steps.

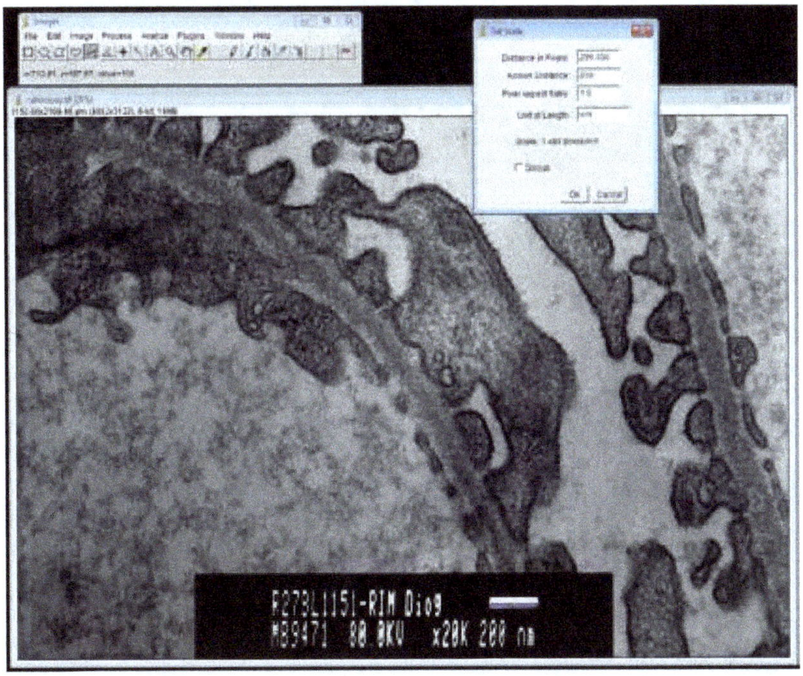

**Figure 6.** Transmission electron microscopy used to demonstrate how to calibrate the software for correct magnification of the micrograph with ImageJ software. The straight-line tool (in blue) is used to measure the scale bar, which here is 200 nm. Next, in the "set scale" box, one can see the distance in pixels of the blue line, and the space of "known distance" is fulfilled with 200 whereas "unit of length" with nm. After clicking "OK," any measurement in this image will be at nanometers calibrated for the real size.

After the software calibration for the magnification, the straight-line tool is used to measure ("Analyze—measure") the linear distance of endothelial cells and their fenestrations. The results, expressed in the unit of the scale bar, appear sequentially in the "results" window, and can be easily copied to spreadsheet or statistic software, as seen in Figure 7. One can further compare (a) the size of fenestrations, (b) the endothelial layer, and (c) their proportion (size of fenestrations/size of the endothelial layer). It is also possible to count the number of fenestrations observed on each image and divide by the linear size of the entire filtration barrier present on the image and thus obtain the number of fenestrations per distance. In addition, the thickness of endothelial cells can be measured by applying the straight-line perpendicular to basement membrane, in some randomly chosen points. One should consider that several measurements should be performed to obtain the whole thickness of endothelial layer, as it can greatly vary in the same glomerulus [39].

Furthermore, the glycocalyx that covers the endothelial cells can be studied with some special techniques. Because conventional methods do not stain the glycocalyx since it is mainly composed of proteoglycans, glycosaminoglycan, and hyaluronan, some techniques were developed for its identification. Perfusion with negatively charged lipid particles allows observation of these intralipids inside the capillary lumen. The measurement of linear distance from these intralipids to the endothelial inner membrane reflects glycocalyx thickness. When

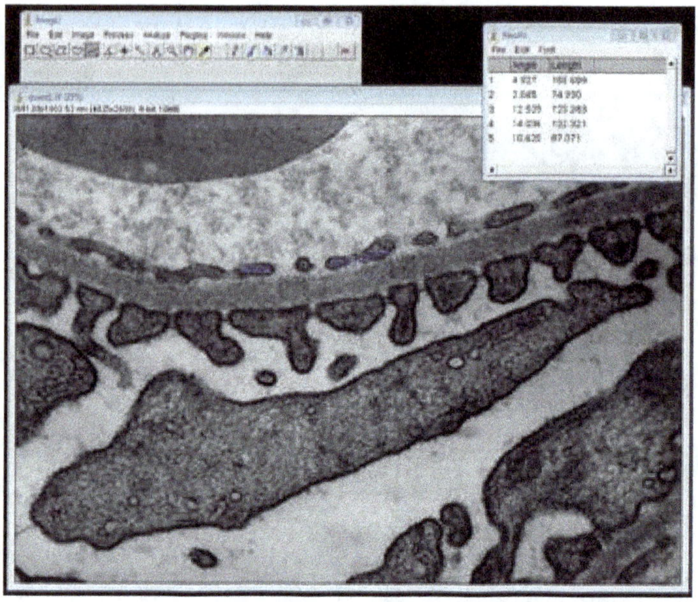

Figure 7. The linear distance of the endothelial cells can be measured with the straight-line tool of ImageJ software. In this electron micrograph, blue lines measure some endothelium segments, and the result is seen in the upper right corner. In the results box, the length of each measurement is expressed in nanometers because the image was previously calibrated for the real size.

reduction of glycocalyx thickness occurs, as measured by this method, correlation with proteinuria is thought, as this structure is one of the responsible for the filtration of plasma proteins [40,41].

Furthermore, the number of the endothelial cells per glomerulus can be measured using stereological methods. This needs determination of the glomerular volume (by the Cavalieri principle, dissector technique, or volume-weighted methods), the cellular density of glomerulus (counting cell nuclei by the Weibel and Gomez point-counting method), and the proportion of cell types of the glomerulus (also by the point-counting method) [40,41]. These measurements can be assessed with light microscopy with the benefits of a faster and cheaper method, or by electron microscopy with the advantage of being a more accurate method.

Regarding the basement membrane, despite its importance for the plasma filtration, the most commonly measurement performed is membrane width. This directly reflects the membrane thickness that occurs in several diseases and is easily and accurately assessed with transmission electron microscopy images. For this, after calibration for the magnification, the straight-line tool of ImageJ should be used perpendicularly to the membrane, as seen in Figure 8.

The effacement of the podocytes, which is commonly qualitatively assessed, can be objectively assessed by two methods [33]. In transmission electron microscopy images, the linear size of the foot processes touching and parallel to the basement membrane can be measured by the straight-line tool, as shown in Figure 9a. As the podocyte effacement is characterized by the loss of the normal interdigitating pattern, with fewer and larger foot processes, in this

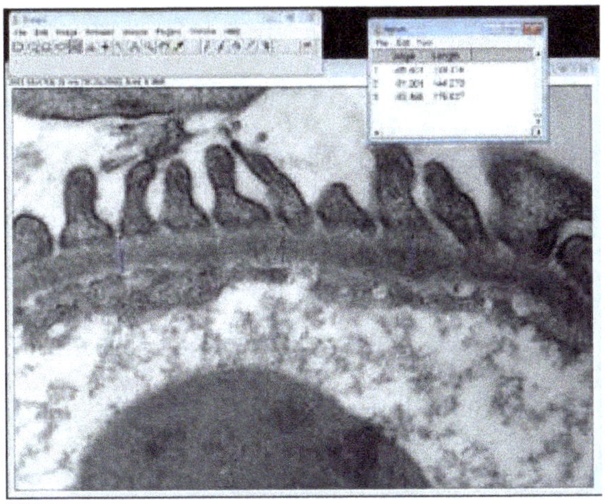

Figure 8. The basement membrane width easily measured in transmission electron micrographs using the "straight line" tool of the ImageJ software. In this image, blue lines measure the basement membrane width, and the result is seen in upper right corner of the image. After calibration of the software, each basement membrane measurement is expressed in nanometers in the results box.

condition, these processes present a higher area touching the basement membrane. Thus, the increased linear size measured by this method numerically represents the podocyte effacement. In addition, one can determine the number of slit diaphragm divided by the linear size of the entire filtration barrier present on the image, and the number of slit diaphragm per distance is obtained (Figure 9b), which is also reduced in podocyte effacement. Also, the number of foot processes per distance can be measured by the same method, as used by Jonsson et al. [42], who quantitatively demonstrated the podocytes effacement, as observed in Figure 10. The size of slit diaphragm can be also easily measured using the straight-line tool applied parallel to the basement membrane, after software calibration.

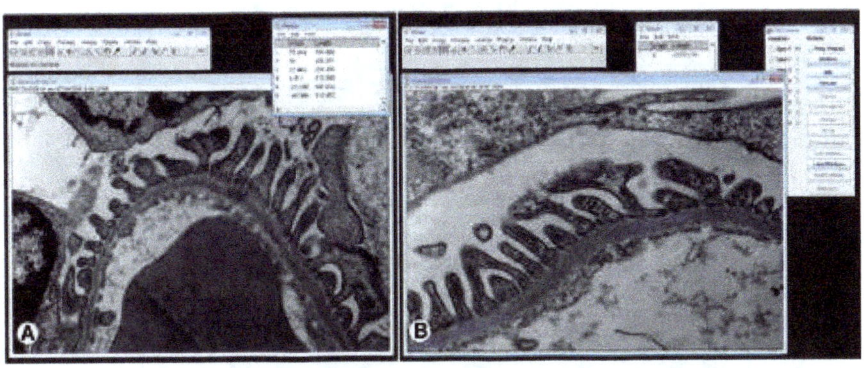


Figure 9. Measurement of foot process parameters that relate with podocyte effacement using transmission electron micrographs. In image A, the linear size of the foot processes that touch the basement membrane is measured by the straight-line tool. Several foot processes can be measured for this parameter. In image B, slit diaphragms can be counted using the "cell counter" tool of ImageJ and the linear size of basement membrane, measured with the "Segmented line selections" tool.

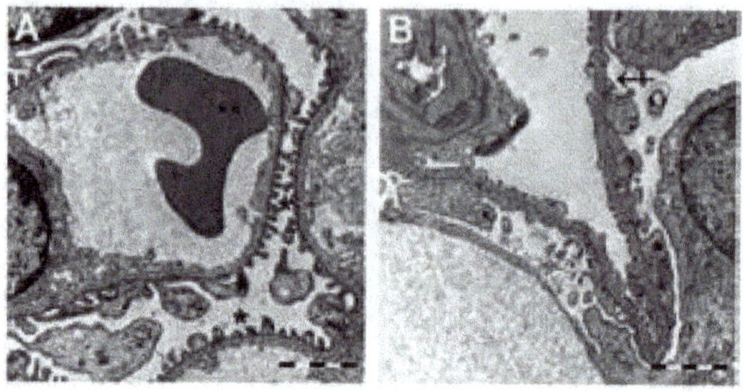


**Figure 10.** Transmission electron microscopic images of control mice and animals with passive Heymann nephritis. In image A, we see the normal glomerular filtration barrier (*) in control animals. In image B, from mice that received an injection of Anti-Fx1A IgG antibody to induce passive Heymann nephritis, it is possible to observe a high degree of foot process effacement (arrow). Scale bar = 2 mm. Image courtesy of Prof. Annika Lindskog Jonsson [42], available at doi:10.1371/journal.pone.0087816.g004.

In scanning electron microscopy images, the number of foot processes per podocyte can also be measured. This parameter only could be determined in podocytes that can be entirely visualized in the scanning image, and special attention should be taken for not count the foot process of the adjacent cell. This makes this method time consuming but can be an option when only scanning electron microscopy is available. In addition, the size of the foot process can be measured in these images, however, with fewer accuracy than when measured in transmission electron microscopy images (Figure 11). Finally, the number of podocytes per glomerulus can be determined using the same method described for determination of endothelial cells per glomerulus.

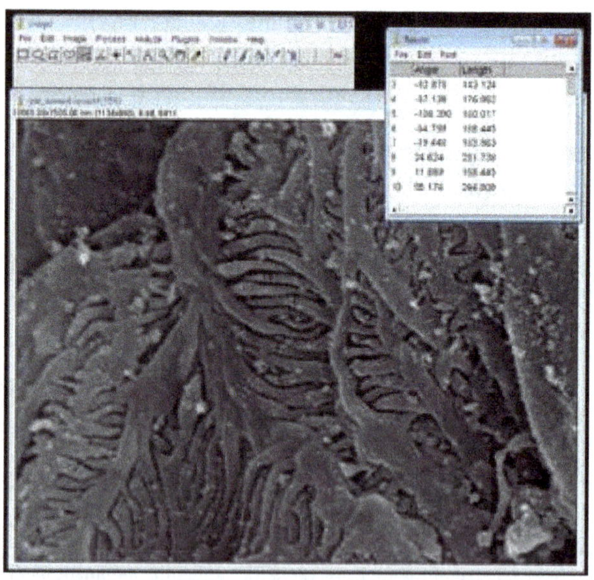

**Figure 11.** The foot processes can also be measured in scanning electron micrographs with the straight-line tool of ImageJ software after calibration for the correct magnification.

## 3. Conclusions

The glomerular filtration barrier is a main component for the filtration of the plasma and formation of primary urine. It is composed of specialized cells and noncellular structures that, together, can avoid the loss of important plasma components but permit the passage of water and undesirable molecules. For this functionality, this barrier has a specific morphology with a fenestrated endothelium covered with glycocalyx, a basement membrane, and a set of slit diaphragms formed by the foot processes of podocytes.

The glomerular filtration barrier morphology has been studied in several diseases and is directly associated with kidney malfunction. Furthermore, it is very important to study this barrier under different clinical and experimental situations. Morphological alterations of components explain some physiopathological findings in clinical setting and correlate with kidney function. For this, scanning and transmission electron microscopy suits perfectly for obtaining high-quality images of this barrier.

Several studies qualitatively described the alterations of glomerular filtration membrane. This gives valuable information, especially when reporting clinical cases or individual lesions that could not be compared. However, when the purpose of the study is to compare the results of a group of individuals with another, quantitative analysis is more appropriate. In this chapter, we presented some objective methods for easy evaluation glomerular filtration membrane. Results obtained with these methods generate numerical data that can be statistically compared with other groups, in different phases of the disease, after some treatment, etc. Thus, whenever possible, quantitative analysis of the glomerular filtration barrier should be favored.

## Author details

Diogo Benchimol de Souza[1*], Bianca Martins Gregório[1], Marlene Benchimol[2] and Fernanda Amorim de Morais Nascimento[1,3]

*Address all correspondence to: diogobenchimol@gmail.com

1 Urogenital Research Unit, Department of Anatomy, State University of Rio de Janeiro, Rio de Janeiro, Brazil

2 Unigranrio University, Rio de Janeiro, Brazil

3 Basic Nutrition and Dietetics Nucleus, Federal University of Rio de Janeiro, Macaé, Brazil

## References

[1] Bloom W, Fawcett DW. A Textbook of Histology. 10th ed. Philadelphia: Saunders; 1975.

[2] Jarad G, Miner JH. Update on the glomerular filtration barrier. Curr Opin Nephrol Hypertens. 2009;18:226–232. DOI: 10.1097/MNH.0b013e3283296044

[3] Mundel P, Kriz W. Structure and function of podocytes: an update. Anat Embryol (Berl). 1995;192:385–397. DOI: 10.1007/BF00240371

[4] Wild S, Roglic G, Green A, Sicree R, King H. Global prevalence of diabetes: estimates for the year 2000 and projections for 2030. Diabetes Care. 2004;27:1047–1053. DOI: 10.2337/diacare.27.5.1047

[5] Mould AP, Craig SE, Byron SK, Humphries MJ, Jowitt TA. Disruption of integrin-fibronectin complexes by allosteric but not ligand-mimetic inhibitors. Biochem J. 2014;464:301–313. DOI: 10.1042/BJ20141047

[6] Relle M, Cash H, Brochhausen C, Strand D, Menke J, Galle PR, et al. New perspectives on the renal slit diaphragm protein podocin. Mod Pathol. 2011;24:1101–1110. DOI: 10.1038/modpathol.2011.58

[7] Inoue K, Ishibe S. Podocyte endocytosis in the regulation of the glomerular filtration barrier. Am J Physiol Renal Physiol. 2015;309:F398-F405. DOI: 10.1152/ajprenal.00136.2015

[8] Lennon R, Randles MJ, Humphries MJ. The importance of podocyte adhesion for a healthy glomerulus. Front Endocrinol (Lausanne). 2014;5:160. DOI: 10.3389/fendo.2014.00160

[9] Craici IM, Wagner SJ, Weissgerber TL, Grande JP, Garovic VD. Advances in the pathophysiology of pre-eclampsia and related podocyte injury. Kidney Int. 2014;86:275–285. DOI: 10.1038/ki.2014.17

[10] Gaede P, Lund-Andersen H, Parving HH, Pedersen O. Effect of a multifactorial intervention on mortality in type 2 diabetes. N Engl J Med. 2008;358:580–591. DOI: 10.1056/NEJMoa0706245

[11] Kannel WB, Stampfer MJ, Castelli WP, Verter J. The prognostic significance of proteinuria: the Framingham study. Am Heart J. 1984;108:1347–1352. DOI: 10.1016/0002-8703(84)90763-4

[12] Yuyun MF, Dinneen SF, Edwards OM, Wood E, Wareham NJ. Absolute level and rate of change of albuminuria over 1 year independently predict mortality and cardiovascular events in patients with diabetic nephropathy. Diabet Med. 2003;20:277–282. DOI: 10.1046/j.1464-5491.2003.00940.x

[13] Schena FP, Gesualdo L. Pathogenetic mechanisms of diabetic nephropathy. J Am Soc Nephrol. 2005;16 Suppl 1:S30–33. DOI: 10.1681/ASN.2004110970

[14] Tryggvason K, Patrakka J, Wartiovaara J. Hereditary proteinuria syndromes and mechanisms of proteinuria. N Engl J Med. 2006;354:1387–1401. DOI: 10.1056/NEJMra052131

[15] Pistoia F, Sacco S, Degan D, Tiseo C, Ornello R, Carolei A. Hypertension and stroke: epidemiological aspects and clinical evaluation. High Blood Press Cardiovasc Prev. DOI: 10.1007/s40292-015-0115-2

[16] Bidani AK, Griffin KA. Long-term renal consequences of hypertension for normal and diseased kidneys. Curr Opin Nephrol Hypertens. 2002;11:73–80. DOI: 10.1097/00041552-200201000-00011

[17] Lan HY. Diverse roles of TGF-beta/Smads in renal fibrosis and inflammation. Int J Biol Sci. 2011;7:1056–1067. DOI: 10.7150/ijbs.7.1056

[18] Kato T, Mizuguchi N, Ito A. Blood pressure, renal biochemical parameters and histopathology in an original rat model of essential hypertension (SHRSP/Kpo strain). Biomed Res. 2015;36:169–177. DOI: 10.2220/biomedres.36.169

[19] Le Jemtel TH, Rajapreyar I, Selby MG, Payne B, Barnidge DR, Milic N, et al. Direct evidence of podocyte damage in cardiorenal syndrome type 2: preliminary evidence. Cardiorenal Med. 2015;5:125–134. DOI: 10.1159/000375130

[20] 20. Barker DJ, Forsen T, Uutela A, Osmond C, Eriksson JG. Size at birth and resilience to effects of poor living conditions in adult life: longitudinal study. BMJ. 2001;323:1273–1276. DOI: 10.1136/bmj.323.7324.1273

[21] Gregorio BM, Souza-Mello V, Carvalho JJ, Mandarim-de-Lacerda CA, Aguila MB. Maternal high-fat intake predisposes nonalcoholic fatty liver disease in C57BL/6 offspring. Am J Obstet Gynecol. 2010;203:495 e1–8. DOI: 10.1016/j.ajog.2010.06.042

[22] Hoy WE, Douglas-Denton RN, Hughson MD, Cass A, Johnson K, Bertram JF. A stereological study of glomerular number and volume: preliminary findings in a multiracial study of kidneys at autopsy. Kidney Int Suppl. 2003;63:S31-S37. DOI: 10.1046/j.1523-1755.63.s83.8.x

[23] Ikezumi Y, Suzuki T, Karasawa T, Yamada T, Hasegawa H, Nishimura H, et al. Low birthweight and premature birth are risk factors for podocytopenia and focal segmental glomerulosclerosis. Am J Nephrol. 2013;38:149–157. DOI: 10.1159/000353898

[24] Sene Lde B, Mesquita FF, de Moraes LN, Santos DC, Carvalho R, Gontijo JA, et al. Involvement of renal corpuscle microRNA expression on epithelial-to-mesenchymal transition in maternal low protein diet in adult programmed rats. PLoS One. 2013;8:e71310. DOI: 10.1371/journal.pone.0071310

[25] Hodgin JB, Rasoulpour M, Markowitz GS, D'Agati VD. Very low birth weight is a risk factor for secondary focal segmental glomerulosclerosis. Clin J Am Soc Nephrol. 2009;4:71–76. DOI: 10.2215/CJN.01700408

[26] Villar-Martini VC, Carvalho JJ, Neves MF, Aguila MB, Mandarim-de-Lacerda CA. Hypertension and kidney alterations in rat offspring from low protein pregnancies. J Hypertens Suppl. 2009;27:S47-S51. DOI: 10.1097/01.hjh.0000358838.71675.5e

[27] Lapillonne A. Vitamin D deficiency during pregnancy may impair maternal and fetal outcomes. Med Hypotheses. 2010;74:71–75. DOI: 10.1016/j.mehy.2009.07.054

[28] Nascimento FA, Ceciliano TC, Aguila MB, Mandarim-de-Lacerda CA. Maternal vitamin D deficiency delays glomerular maturity in F1 and F2 offspring. PLoS One. 2012;7:e41740. DOI: 10.1371/journal.pone.0041740

[29] Maka N, Makrakis J, Parkington HC, Tare M, Morley R, Black MJ. Vitamin D deficiency during pregnancy and lactation stimulates nephrogenesis in rat offspring. Pediatr Nephrol. 2008;23:55–61. DOI: 10.1007/s00467-007-0641-9.

[30] Maunsbach AB, Afzelius BA. Biomedical Electron Microscopy. Illustrated Method and Interpretations. London: Academic Press; 1999.

[31] Souza W. Técnicas de microscopia eletrônica aplicadas às ciências biológicas. 3rd ed. Rio de Janeiro: Soc. Bras. Microsc; 2007.

[32] Bone Q, Denton EJ. The osmotic effects of electron microscope fixatives. J Cell Biol. 1971;49:571–581. DOI: 10.1083/jcb.49.3.571

[33] Souza DB, Costa WS, Cardoso LE, Benchimol M, Pereira-Sampaio MA, Sampaio FJ. Does prolonged pneumoperitoneum affect the kidney? Oxidative stress, stereological and electron microscopy study in a rat model. Int Braz J Urol. 2013;39:30–36. DOI: 10.1083/jcb.49.3.571

[34] 34. Rice WL, Van Hoek AN, Paunescu TG, Huynh C, Goetze B, Singh B, et al. High resolution helium ion scanning microscopy of the rat kidney. PLoS One. 2013;8: e57051. DOI:10.1371/journal.pone.0057051

[35] Shirato I. Podocyte process effacement in vivo. Microsc Res Tech. 2002;57:241–246. DOI: 10.1002/jemt.10082

[36] Wagner MC, Rhodes G, Wang E, Pruthi V, Arif E, Saleem MA, et al. Ischemic injury to kidney induces glomerular podocyte effacement and dissociation of slit diaphragm proteins Neph1 and ZO-1. J Biol Chem. 2008;283:35579–35589. DOI: 10.1074/jbc.M805507200

[37] Felix-Patricio B, De Souza DB, Gregório BM, Costa WS, Sampaio FJ. How to quantify penile corpus cavernosum structures with histomorphometry: comparison of two methods. Biomed Res Int. DOI: 10.1155/2015/832156

[38] Arkill KP, Qvortrup K, Starborg T, Mantell JM, Knupp C, Michel CC, et al. Resolution of the three dimensional structure of components of the glomerular filtration barrier. BMC Nephrol. 2014;15:24. DOI: 10.1186/1471-2369-15-24

[39] Hjalmarsson C, Johansson BR, Haraldsson B. Electron microscopic evaluation of the endothelial surface layer of glomerular capillaries. Microvasc Res. 2004;67:9–17. DOI: 10.1016/j.mvr.2003.10.001

[40] Andersson M, Nilsson U, Hjalmarsson C, Haraldsson B, Nystrom JS. Mild renal ischemia-reperfusion reduces charge and size selectivity of the glomerular barrier. Am J Physiol Renal Physiol. 2007;292:F1802-F1809. DOI: 10.1152/ajprenal.00152.2006

[41] Steffes MW, Schmidt D, McCrery R, Basgen JM. Glomerular cell number in normal subjects and in type 1 diabetic patients. Kidney Int. 2001;59:2104–2113. DOI: 10.1046/j.1523-1755.2001.00725.x

[42] Jonsson AL, Granqvist A, Elvin J, Johansson ME, Haraldsson B, Nystrom J. Effects of melanocortin 1 receptor agonists in experimental nephropathies. PLoS One. 2014;9:e87816. DOI: 10.1371/journal.pone.0087816

# 4

# Electron Diffraction

Mohsen Asadi Asadabad and Mohammad Jafari Eskandari

**Abstract**

Electron microscopes are usually supplied with equipment for obtaining diffraction patterns and micrographs from the same area of a specimen and the best results are attained if the complete use is to be made of these combined facilities. Electron diffraction patterns are used to obtain quantitative data including phase identification, orientation relationship and crystal defects in materials, etc. At first, a general introduction including a geometrical and quantitative approach to electron diffraction from a crystalline specimen, the reciprocal lattice and electron diffraction in the electron microscope are presented. The scattering process by an individual atom as well as a crystal, the Bragg law, Laue conditions and structure factor are also discussed. Types of diffraction patterns such as ring pattern, spot pattern and Kikuchi pattern, and general and unique indexing diffraction patterns are explained. The procedure for indexing simple, complicated and imperfect patterns as well as Kikuchi lines and a combination of Kikuchi lines and spots is outlined. The known and unknown materials are identified by indexing patterns. Practical comparisons between various methods of analysing diffraction patterns are also described. The basic diffraction patterns and the fine structure in the patterns including specimen tilting experiments, orientation relationship determination, phase identification, twinning, second phases, crystallographic information, dislocation, preferred orientation and texture, extra spots and streaks are described in detail. Finally, electron diffraction patterns of new materials are investigated.

**Keywords:** Electron diffraction pattern, Spot and ring pattern, Kikuchi line, Phase identification

## 1. General introduction

In quantum mechanics, electrons may be considered as particles or waves. Electrons are used in transmission electron microscopy (TEM) because the wavelength of electrons is shorter than the visible light. For this reason, high magnifications can be achieved in TEM. In TEM, tungsten filament is usually used to produce a monochromatic beam of electrons by thermionic or field emission processes. Electrons are accelerated by applied voltage and focused by the objective lens. These particles with negative charge travel the spiral path when passing through the electromagnetic lenses. Then, this beam of electrons is transmitted through very thin specimen (thickness about 100–300 nm) and magnified by the electromagnetic lens, forming the electron diffraction pattern. Electrons are accelerated to close to the speed of light at high voltages. So, relativistic effects should be considered in equations of electron beam wavelength in electron microscopy at high accelerated voltage. The modified relativistic wavelength is

$$\lambda = \frac{h}{\left(2m_e Ve\left(1 + eV / 2m_e c^2\right)\right)^{1/2}} \tag{1}$$

where $h$ is Planck's constant, $m_e$ is the rest mass of electrons, $e$ is charge of electrons, $V$ is accelerated voltage of electrons and $c$ is the velocity of light. In fact, electron diffraction directly demonstrates the reciprocal lattice of the crystalline lattice of the selected area from a sample [1].

### 1.1. Elastic scattering of electrons by individual atom

When a beam of electrons transmits through a thin specimen, different interactions can occur such as ionization, secondary emission and excitation with loss of energy and can be scattered by nuclei and electron cloud without loss of energy. The elastic scattering process by an isolated atom is illustrated in Figure 1.Some of the electrons are backscattered and the rest are scattered by nucleus and electron cloud (Rutherford scattering). The atomic scattering amplitude for electrons $f_\theta$ (atomic diffraction factor: a measure of the diffracting capability of an isolated atom) is given by

$$f_\theta = \frac{m_e e^2}{2h^2}\left(\frac{\lambda}{sin\theta}\right)^2\left(Z - f_x\right) \tag{2}$$

where $\theta$ is the scattering angle, $Z$ is the atomic number of isolated atom (Rutherford scattering) and $f_x$ is the atomic scattering factor for X-rays. The elastically scattered electrons' main contribution is in the form of diffraction patterns. Most of the particles are scattered within $\pm 5°$ of the direct incident beam [2].

### 1.2. Scattering by an ideal crystal structure

For understanding the essence of electron diffraction by a three-dimensional crystal structure of a material, the principles of diffraction of a monochromatic light by Young's double slit experiment should be investigated. First of all, diffraction by two slits is investigated, then, diffraction is generalized to an infinite number of slits and, finally, diffraction from a regular arrangement of atoms is investigated. In Figure 2, the plane waves collide with a barrier with

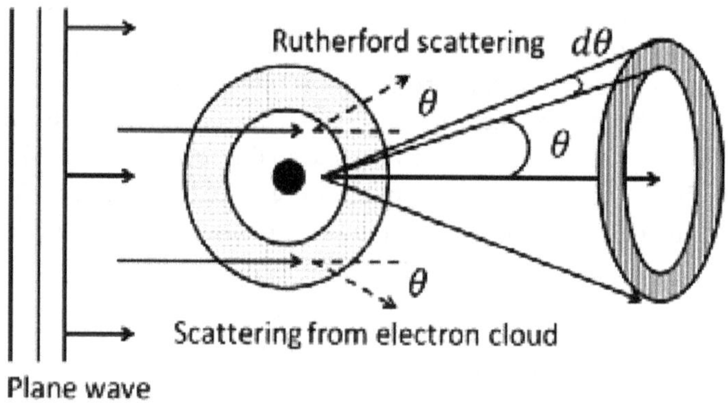


**Figure 1.** Electrons are as plane wave and black circle is an isolated atom. Electrons are scattered from electron cloud and nucleus of a single atom at the angle $\theta$.

two slits in which the length and width of slits are $l$ and $a$, respectively, with respect to $l \gg a$. Also, bright and dark fringes are formed on the screen which is placed at a distance $S$, with respect to $S \gg a$. If the waves are in-phase when passing through the slits, there is relative phase difference between two secondary sources. Two beams of electrons have constructive interference if their relative phase difference is an integer multiple of $\lambda$ $(\delta = n\lambda)$, then the bright fringes are formed on the screen. Similarly, two beams of electrons have unconstructive interference if their relative phase difference is not an integer multiple of $\lambda$ $(\delta = n\lambda/2)$. Therefore, dark fringes are formed on the screen, in which intensity is very low or zero. Alternating bright and dark lines are formed on the screen. The intensity of the bright fringes at the center is very high and away from the center as $\theta$ increases, their intensity and width are decreased. In a real three-dimensional crystal lattice, diffraction of electron beams occurs by regular spacing between atoms, which creates an interference pattern.

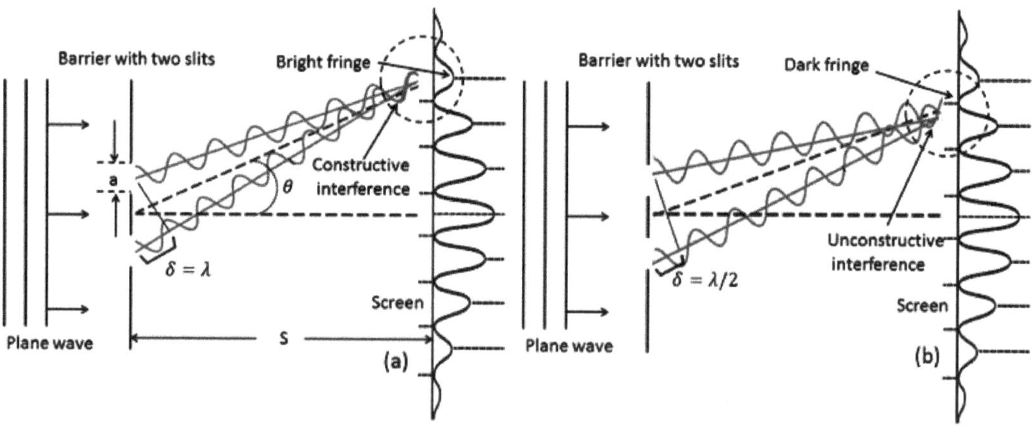


**Figure 2.** Diffracted beam of electrons by barrier with two slits (Young's slits). (a) Lines bright (constructive interference), (b) Lines dark (unconstructive interference).

In general, electron diffraction is according to kinematical theory and some assumptions must also be considered [2, 3].

### 1.3. The Bragg law

For studying the scattered and transmitted beams, a cross-section of a thin specimen with a perfect crystal lattice is considered. Incident coherent electron and in-phase beams are radiated from the top surface of the thin specimen. This electron beam is collided with two atoms, each of which belongs to a plane (two adjacent planes) with different *hkl* (Miller indices) with interplanar spacing of crystal lattice being equal to $d$. Then, the electron beam is diffracted by elastic scattering. These waves are coherent and in-phase after passing through the sample if the path difference of the electron beam is an integer number of wavelength (constructive interference), that is, geometric relationships $AC + AD = n\lambda$ and $AC = AD = d sin\theta$ are satisfied, as can be seen in Figure 3. So, the following relationship is established:

$$2d_{(hkl)} sin\theta = n\lambda \tag{3}$$

where $d$ is the interplanar spacing, $\theta$ is the angle between incident and diffracted beams, $\lambda$ is the electron wavelength and $n$ is the integer number of order diffraction. This relationship is known as the Bragg law. Basically, first-order diffraction is $n = 1$ and Miller indices are used for higher orders $n \geq 2$.

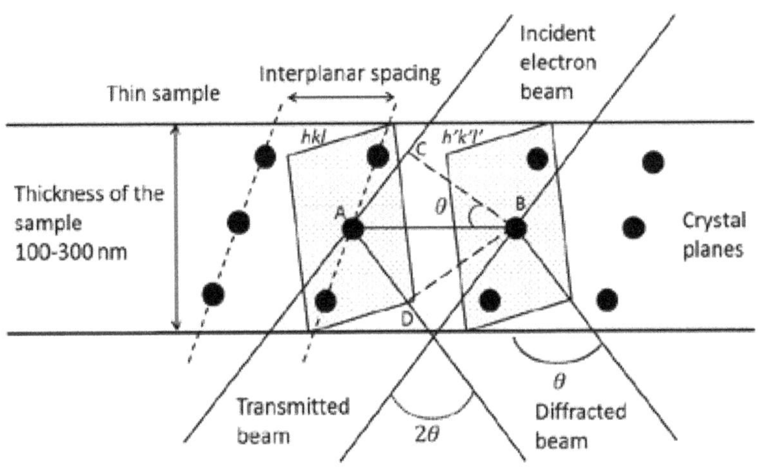


**Figure 3.** Incident, transmitted and diffracted electron beams in a thin specimen for the Bragg law.

In the Bragg law, electrons are collided with the crystal planes. Some of them are diffracted and the rest are transmitted through the specimen, which does not participate in the formation of the diffraction pattern [1–3].

### 1.4. The Laue conditions

Diffraction can be considered as a total scattering of individual atoms. So, diffraction is mathematically expressed in terms of total scattering of atoms. The distance from atom A to atom B is described by vector $r$ in three-dimensional space. Also, incident and diffracted beams are described by unit vectors $\boldsymbol{p_0}$ and $p$, respectively. With the use of these vector notations, we can write paths difference in vector notation as $AC = r.p_0$ and $AD = r.p$. According to the vector relationship in Figure 4, the path difference of wave should be an integer multiple of wavelength and we may write as

$$CAD = \boldsymbol{r}.\boldsymbol{P} = n\lambda \tag{4}$$

The vector $\boldsymbol{r}$, defined in spherical coordinates, is a converted form of vector components $x$, $y$ and $z$ in Cartesian coordinates, and $\boldsymbol{a}$, $\boldsymbol{b}$ and $\boldsymbol{c}$ are unit vectors defined in the direction of coordinates axes which can be considered as distances between atoms. So, we may write Equation (4) as

$$\begin{aligned} \boldsymbol{P}.\boldsymbol{a} &= h\lambda \\ \boldsymbol{P}.\boldsymbol{b} &= k\lambda \\ \boldsymbol{P}.\boldsymbol{c} &= l\lambda \end{aligned} \tag{5}$$

These algebraic relations are known as Laue conditions. These relations must be satisfied when strong diffraction occurs [2, 3].

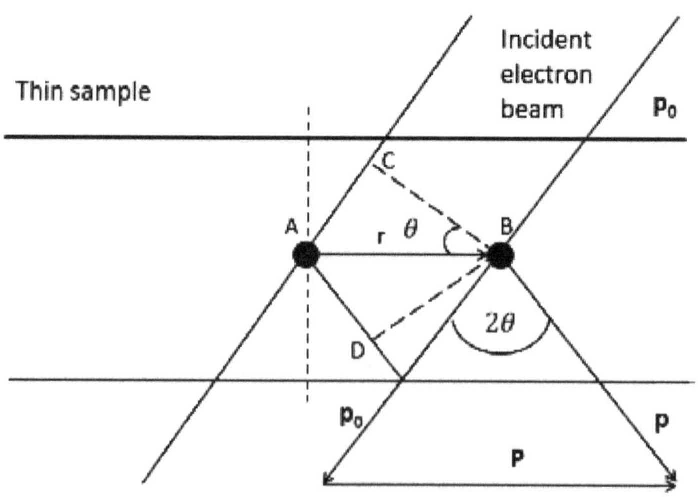


**Figure 4.** Incident and diffracted waves of atoms in the thin specimen are $p_0$ and $p$, respectively, the angle between them is $2\theta$ and the distance from atom A to atom B is described by vector $r$.

### 1.5. The structure factor

Diffraction intensity is different for each of the crystalline planes because the distribution of atoms per unit area is not the same for individual planes. Using the kinematical theory of electron diffraction, a set of crystal planes can be determined for which the diffraction intensity is zero. The structure factor can be defined as a mathematical function stating the amplitude and phase of electron beam diffracted from crystallographic planes. In the structure factor, the location of atoms in the reflection plane and atomic specifications is considered to describe the diffraction process. Also, the structure factor is the sum of the scattered amplitudes of single atoms $f_n$ and the sum of the phase differences, that is,

$$F_{hkl} = \sum_{n} f_n exp\{2\pi i(hx_n + ky_n + lz_n)\} \tag{6}$$

where $x_n,\ y_n,\ z_n$ are positions of the atom in the Cartesian coordinates. The intensity of the diffracted wave is

$$I \propto |F|^2 \propto f^2\left[1 + cos(\pi(h+k+l))\right]^2 + f^2\left[sin(2\pi(h+k+l))\right]^2 \tag{7}$$

In the above relationship, the intensity is sometimes zero, which belongs to any diffraction not existing in these planes and is called a forbidden reflection. By use of the Bragg law and structure factor, diffracted planes in the crystal can be determined. For intermetallic compounds, the diffraction intensity is different because atomic scattering factors of individual metals forming the intermetallic compound are not the same. For example, intermetallic compounds with an AB structure have diffracted intensity according to the following relations:

$$\begin{aligned} I &\propto (f_A + f_B)^2 \text{ when } h+k+l \text{ is even} \\ I &\propto (f_A - f_B)^2 \text{ when } h+k+l \text{ is odd} \end{aligned} \tag{8}$$

Diffraction rules for some of the conventional crystalline structures are presented in Table 1 [2].

| Crystal structure | Reflection absent if |
|---|---|
| simple cubic | all present |
| f.c.c | *h, k, l,* mixed odd and even |
| b.c.c | *h, k, l* odd |
| c.p.h | *h*+2*k*=3*n* and *l* is odd |
| b.c.t | *h*+ *k*+ *l* odd |
| Zinc blende | *h, k, l,* mixed odd and even |
| Sodium chloride | *h, k, l,* mixed odd and even |
| diamond | *h, k, l,* all even and *h*+ *k*+ *l* not divisible by four, or *h, k, l* mixed odd and even |

**Table 1.** Diffraction rules for conventional crystalline structures

### 1.6. The reciprocal lattice

The reciprocal lattice is an array of points in which each point corresponds to a special plane in the crystal lattice. In fact, each of the planes in real crystal lattice is represented by a point in the reciprocal lattice located at distance $1/d_{hkl}$ from the center O. The distance of a point in the reciprocal lattice to the center is illustrated by the vector $g_{(hkl)}$, which is called the diffraction vector. Diffraction pattern and reciprocal lattice are related to each other and this relation is used for the interpretation of different diffraction patterns. The reciprocal lattice has two special properties:

**a.** The diffraction vector $g_{(hkl)}$ of reciprocal lattice is perpendicular to the plane of the crystal lattice

**b.** $g_{(hkl)}=1/d_{hkl}$

The Ewald sphere displays the relation between the reciprocal lattice and the diffraction pattern with a radius of $1/\lambda$. The formation of the Ewald sphere in the reciprocal lattice and the diffraction pattern are depicted in Figure 5. Also, the algebraic relations between incident, transmitted and diffracted beams are shown in this figure. The incident beam of electrons is collided with the thin specimen and then, a certain percentage of the incident beam is transmitted and the rest is diffracted. Using Figure 5, the geometrical relations for distances and angle may be determined from the relation

$$tg2\theta = R / L \tag{9}$$

where $\theta$ is the angle between transmitted and diffracted beams, $R$ is the distance between collision points of transmitted and diffracted beams with the screen and $L$ is the distance between the specimen and the screen (the effective camera length). Using the Bragg law and with the assumption of a small $\theta$, relation (9) can be written as follows

$$Rd_{hkl} = L\lambda \tag{10}$$

where $L\lambda$ is a camera constant. The effective camera length and wavelength of the electron are constant and depend on the characterization of transmission electron microscopy [2, 3].

## 2. Types of electron diffraction patterns

Electron diffraction patterns give crystallographic information about a material and determine different types of materials which can be amorphous, single crystalline or polycrystalline. There are three types of electron diffraction patterns and the formation of each pattern depends on the different conditions of the specimen such as thickness, crystal structure and so on.

**1.** The polycrystalline materials exhibit ring pattern

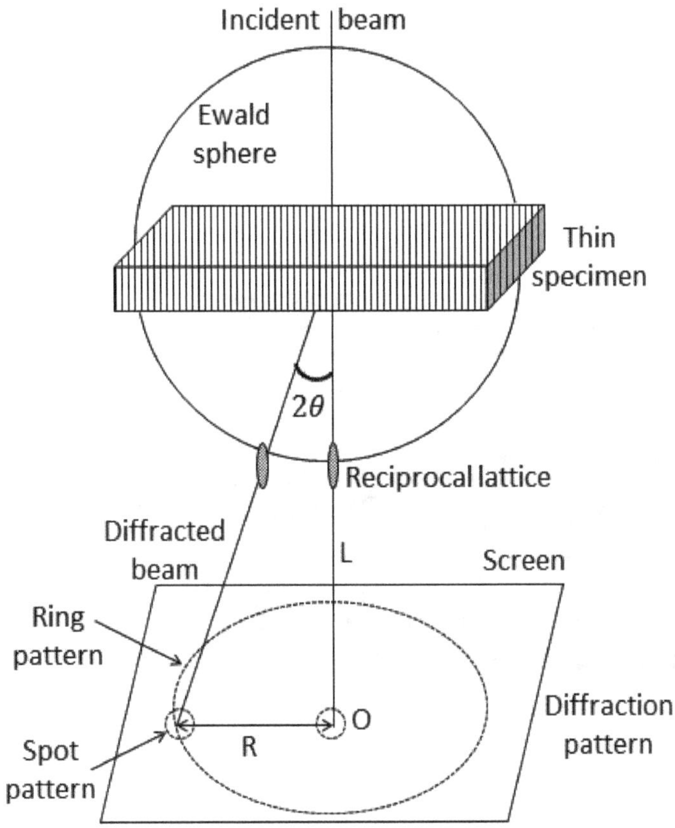


**Figure 5.** The Ewald sphere is drawn in reciprocal lattice. The formation of a diffraction pattern is shown geometrically. The relations between incident, transmitted and diffracted beams, the Ewald sphere and different diffraction patterns are illustrated.

**2.** The single crystalline materials show (a) spot pattern or (b) Kikuchi line pattern or (c) a combination of spot and Kikuchi line patterns

The spot and Kikuchi line patterns are obtained from a special area of specimen which is called the 'selected area' [5–8]. Selected area electron diffraction (SAED) is a technique in TEM to obtain diffraction patterns that result from the electron beam scattered by the sample lattice.

## 2.1. Ring pattern

These patterns are created by ultrafine grains of polycrystalline materials. Basically, phases in various polycrystalline materials are determined by interpretation of their ring patterns. For this purpose, we must use a reference specimen for identification of phases as well as specifying interplanar spacing and Miller indices of crystalline planes. Polycrystalline specimens such as pure gold (Au, f.c.c crystal structure with lattice parameter $a=4.07$ Å) or pure aluminum (Al, f.c.c crystal structure with lattice parameter $a=4.04$ Å) as reference specimens are used to index diffraction patterns of various materials and specify the camera length. To obtain a reference specimen with a ring diffraction pattern, at first, a copper grid with amorphous carbon coating is provided. Then, by use of a sputter coating device, a thin layer of pure gold with a thickness of about 20 nm is coated on the grid. Finally, the diffraction pattern of the specimen is taken which is in a ring shape and continuous, as can be seen in Figure 6. The planes of the gold specimen are specified by Miller indices. The pure gold sample is known as standard sample and is used for identification of crystalline planes and measurement of interplanar spacing of unknown materials with ring patterns and determination of phases in alloys.

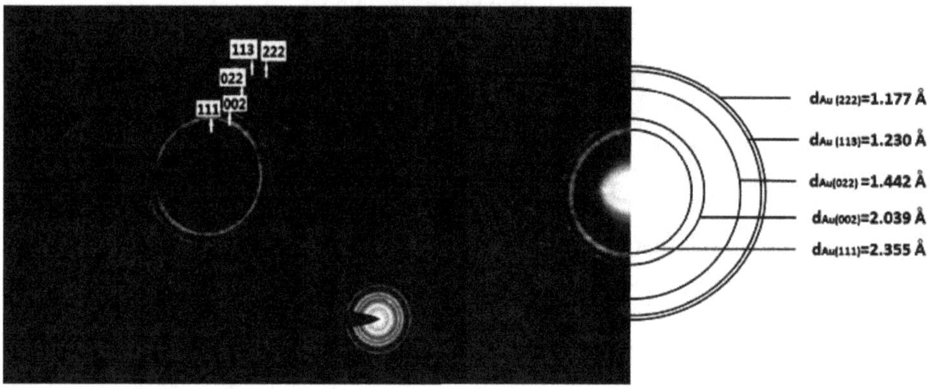


**Figure 6.** The ring diffraction pattern from a polycrystalline pure gold film with an f.c.c crystal structure. Crystal planes and interplanar spacing are shown by Miller indices. Camera lengths are 180 and 88 mm, respectively [4].

Analysis of ring patterns in polycrystalline materials (ultrafine grain) leads to identification of phases in materials. Diffraction patterns of nanoparticles produced by different methods form a ring pattern. In fact, the ring patterns are created when the nanoparticle is formed. Using the radius of each ring, we can specify the distance between the planes or interplanar spacing. Also, XRD analysis is used to determine the Miller indices for a set of planes. So, XRD analysis confirms the results of diffraction pattern from TEM for certain materials. A ring diffraction pattern from a polycrystalline gold specimen is shown in Figure 6. The interplanar spacing and lattice parameter can be calculated by measuring the radius of each diffraction ring (using Equation (10) and Table 2). Also, indexing ring patterns can be performed by XRD analysis [4–10].

### *2.1.1. Indexing ring patterns*

In this chapter, an EM208S (Philips) transmission electron microscope operating at an accelerating voltage of 100 kV with a wavelength $\lambda = 3.7 \times 10^{-3}$ nm and camera length $L = 180, \quad 88$ mm are used. One thing to note is that, accuracy and focus of TEM are very important to obtain an accurate diffraction pattern. Indexing methods used for ring diffraction patterns are as follows:

**a.** For known materials

**1.** Using the gold standard diffraction pattern, we define a scale on the picture of patterns to measure the radius diffraction pattern of specimens.

**2.** The first solution, with known lattice parameters, interplanar spacing is obtained from Equation (10) and Miller indices can be obtained using Table 2. The second solution, the ratio of outer ring to the first ring is equal to the reverse ratio of their interplanar spacing with possible Miller indices.

$$\frac{R_{outer}}{R_{first}} = \frac{d_{first}}{d_{outer}}$$

These possible Miller indices for planes are correct if the result of proportional relation above is almost the same.

**b.** For unknown material

**1.** Measure the radius of diffraction pattern like in the previous section.

**2.** Knowing the camera constant, interplanar spacing is obtained from Equation (10).

**3.** Compare interplanar spacing of unknown material with the ASTM index* to identify phases in diffraction patterns, Miller indices are determined for crystalline planes of phases in alloys [2, 6, 8].

*ASTM index to the powder diffraction file.

| Crystal structure | Proportional relation for interplanar spacing | Possible values of algebraic relations Miller indices | Standard |
|---|---|---|---|
| Simple cubic | $\frac{1}{d^2} = \frac{h^2+k^2+l^2}{a^2} = \frac{N}{a^2}$ | $N$ an integer except 7 or 15 | Proportional relation of squares of radius $\propto N$ |
| f.c.c | $\frac{1}{d^2} = \frac{h^2+k^2+l^2}{a^2} = \frac{N}{a^2}$ | N=3, 4, 8, 11, 12, 16, 19, 20 | Proportional relation $\propto N$ |
| b.c.c | $\frac{1}{d^2} = \frac{h^2+k^2+l^2}{a^2} = \frac{N}{a^2}$ | N=2, 4, 6, 8, 10, 12, 14, 16, 18, 20 | Proportional relation $\propto N$ |
| Diamond structure | $\frac{1}{d^2} = \frac{h^2+k^2+l^2}{a^2} = \frac{N}{a^2}$ | N=2, 8, 11, 16, 19 | Proportional relation $\propto N$ |
| Tetragonal | $\frac{1}{d^2} = \frac{h^2+k^2}{a^2} + \frac{l^2}{c^2}$ | $h^2+k^2$ =1, 2, 4, 5, 8, 9, 10, 13, 16, 17, 18, 20 | Proportional relation often is 2 |
| Hexagonal | $\frac{1}{d^2} = \frac{4}{3}\frac{h^2+hk+k^2}{a^2} + \frac{l^2}{c^2}$ | $h^2+hk+l^2$ =1, 3, 4, 7, 9, 12, 13, 16, 19 | Proportional relation often is 3 |

**Table 2.** Proportional relations for interplanar spacing, Miller indices and lattice constant for different crystal structures [2]

### 2.2. Spot patterns

There are two basic parameters in spot diffraction patterns which are used to interpret and index such types of patterns. These parameters include

1. $R$ is the distance between the diffracted and transmit (center spot) beams in the diffraction pattern screen. Also, this distance can be considered as a normal vector to the plane reflection.

2. The angles such as $\theta$ between two vectors drawn from the center to two adjacent points. In fact, each of these spots represents a set of planes, as can be seen in Figure 7.

The zone axis vector is parallel to the incident beam and is almost parallel with a set of reflected planes that is shown by $z=[uvw]$ with components $u$, $v$ and $w$ along the axis. The spots are in symmetry about the center of the pattern and, using the rules of vectors and the basic parallelogram, we can index spot patterns.

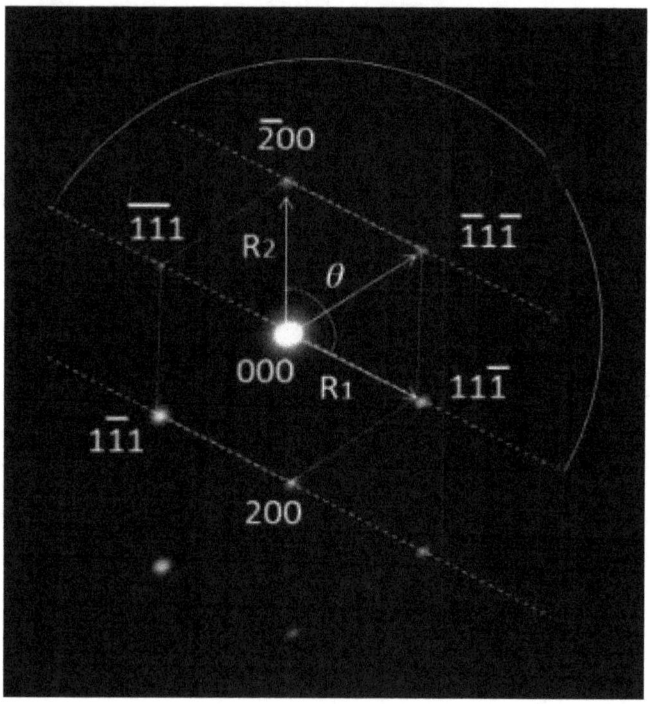


**Figure 7.** The spot diffraction pattern from a single crystal along zone axis $z=[011]$ of Al 1050 alloy with 10 passes of straight rolling.

### 2.2.1. *Indexing spot patterns*

For indexing spot patterns, indices of the spots and zone axis of single crystal materials should be determined. Here, we use the same indexing methods utilized for ring patterns as described in the previous section. In the experimental method, we measure distances of different spots from the center spot as well as angles on the micrograph of patterns and compare with patterns in the International Standard [17]. So, indices of spot and zone axis in pattern can be determined. The zone axis $z=[uvw]$ may be specified by the relations

$$\begin{aligned} u &= k_1 l_2 - k_2 l_1 \\ v &= l_1 h_2 - l_2 h_1 \\ w &= h_1 k_2 - h_2 k_1 \end{aligned} \tag{11}$$

where $h_1k_1l_1$ and $h_2k_2l_2$ are coordinates of each spot in the diffraction pattern [2, 5, 6, 8–12].

### 2.3. Kikuchi patterns

Kikuchi line pattern may happen when the thickness of the specimen is more than normal and almost perfect. These patterns occur by electrons scattered inelastically in small angles with a small loss of energy. Then, this beam of electrons is scattered elastically and creates Kikuchi lines in the patterns. Kikuchi lines in the pattern are pairs of parallel dark and bright lines. The distance between pairs of dark and bright lines is obtained by the following relation:

$$Dd_{hkl} = L\lambda \tag{12}$$

where $D$ is the distance between pairs of Kikuchi lines. Also, the angle between Kikuchi lines in the pattern is in accordance with the angle between the diffraction planes because these lines are parallel with reflecting planes. The pairs of dark and bright lines, sets of reflecting planes and distance of paired lines are shown in Figure 8. The dashed lines are traces of the intersection of reflecting planes. By tilting the specimen, the Kikuchi line pattern changes by the displacement of paired lines. By increasing the sample thickness, the intensity of the spot pattern decreases and the intensity of Kikuchi line pattern increases. Most of the time, spot and Kikuchi line patterns exist simultaneously in micrographs, such as Figure 8. Basically, Kikuchi line patterns present more detailed information than the spot pattern. The appearance of explicit Kikuchi line patterns is a sign of crystal perfection.

#### *2.3.1. Indexing Kikuchi line patterns*

To study the crystal defects and to find out the orientation relationships, a tilting experiment should be used. The position of the Kikuchi line and spot patterns can be seen in Figure 9 in Al 7075 alloy in which a thermomechanical processing has been performed. To index the paired Kikuchi lines in the pattern, the distance between the pair of the Kikuchi lines may be measured. So, interplanar spacing can be determined using Equation (12). Consequently, a set of planes can be specified using the interplanar spacing and the type of material crystal structure.

## 3. Structural characterization of electron diffraction patterns

### 3.1. Orientation relationship

Relations between phases are determined by orientation relationships. Orientation relationships are indicated by a pair of parallel directions and a pair of parallel planes in two-phase

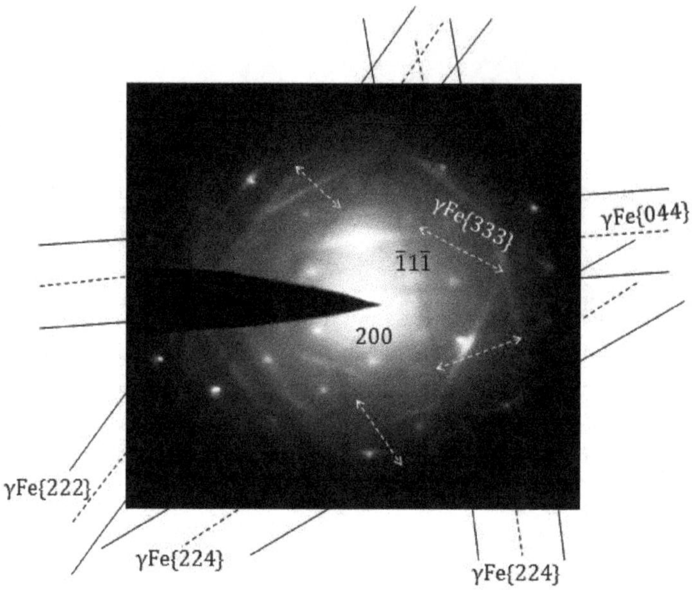


**Figure 8.** The spots, Kikuchi lines and the distance between paired Kikuchi lines in the pattern of $\gamma Fe$ are determined.

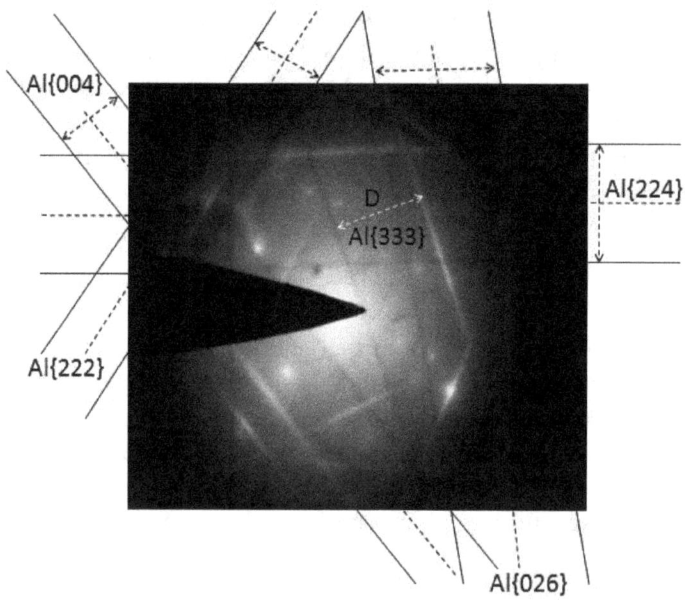


**Figure 9.** Spot and Kikuchi lines in the pattern of thermomechanically processed Al 7075 alloy. The distance between paired Kikuchi lines is presented as dark and bright lines.

materials. With the use of the spot pattern and Kikuchi line pattern methods, orientation relationships can be determined. Orientation relationships in two-phase alloys depend on different factors such as types of crystal structural matrix, crystal structural precipitates, shape and size of precipitates. The spot diffraction pattern of quenched and tempered steel 0.1C-2.5V-2W-4.5Cr with chemical composition is illustrated in Figure 10a, which includes matrix spots ($\alpha Fe$) and precipitate spots (carbide$M_{23}C_6$). In the spot pattern of two-phase alloys, there exists two kinds of spots, as shown in Figure 10b. The precipitates and matrix are shown by small spots and big spots, respectively. Matrix has a b.c.c crystal structure and precipitates have a f.c.c crystal structure. So, the orientation relationship may be represented as

$$[\bar{1}11]_{\text{matrix}} // [\bar{1}11]_{\text{precipitate}}$$
$$\{011\}_{\text{matrix}} // \{022\}_{\text{precipitate}}$$

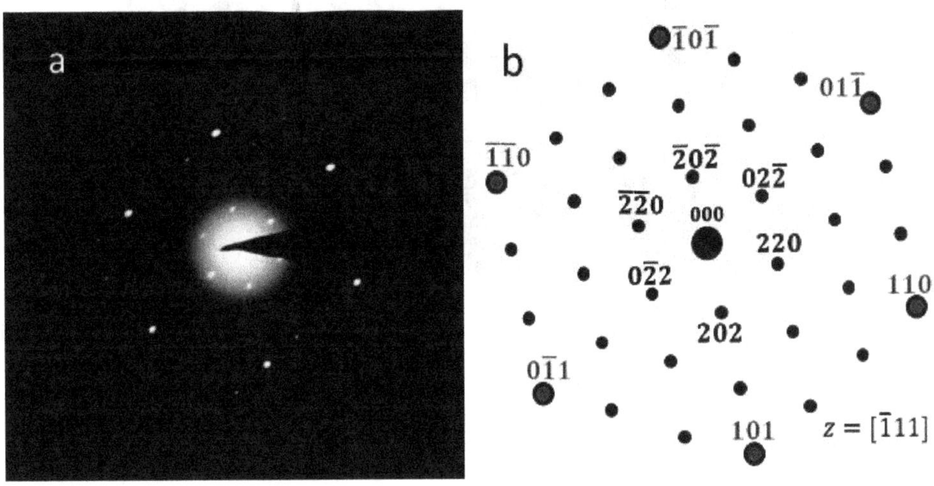


**Figure 10.** Spot diffraction pattern of quenched and tempered steel with 0.1C-2.5V-2W-4.5Cr composition that includes matrix spots ($\alpha Fe$) and precipitate spots ($carbide\ M_{23}C_6$). Small spots and big spots illustrate planes of matrix and planes of precipitate, respectively.

### 3.2. Phase identification

#### *3.2.1. Phase identification in spot pattern*

Interpretation and analysis of diffraction pattern is a method to identify the phases. These patterns can be composed of rings or spots. Generally, identification of phases using diffraction pattern is difficult, and other methods such as XRD should be used to verify the results. To identify the phases in a spot pattern, a spot diffraction pattern of Ti-6Al-4V alloy processed by forging is considered. This alloy contains $\alpha$ and $\beta$ phases with c.p.h and b.c.c crystal structures, respectively. This pattern contains three different phases as shown in Figure 11. In this pattern, the bright and big spots are related to the $\alpha$ phase and the dim and small spots are related to the $\beta$ phase.

**Figure 11.** Spot diffraction pattern of forged $Ti-6Al-4V$ alloy. Two phases are observed in the pattern, namely, $\alpha$ and $\beta$ with $Z=[\bar{1}2\bar{1}6]$ and $Z=[012]$, respectively.

### *3.2.2. Phase identification in the ring pattern*

Carbon nanotubes (CNTs) added to a metal matrix can exhibit significant properties. The main challenge in CNTs-reinforced composites is the uniform dispersion of CNTs in the matrix. Several methods such as ball milling have been developed for distribution of the CNTs in metal matrices. For example, a milled sample of 20 wt% multi-wall carbon nanotubes (MWCNT) + Al powder was investigated using the EDP method. In addition, their corresponding EDP is shown in Figure 12a. To separate ring patterns of CNTs and aluminum, a ring pattern of Al was attached to the EDP of Al-CNTs composite, as illustrated in Figure 12b. This technique helps to better identify the two phases [7].

## 3.3. Twinning

Twinning is one of the crystalline defects that appear mainly as two parallel planes. Additional spots are created around the main spots of diffracted planes in the pattern by twinning because orientation of twinning is different from the crystal lattice. The direction of the reflected planes inside the twinning is not the same as the whole crystal structure and extra spots become visible in the diffraction pattern.

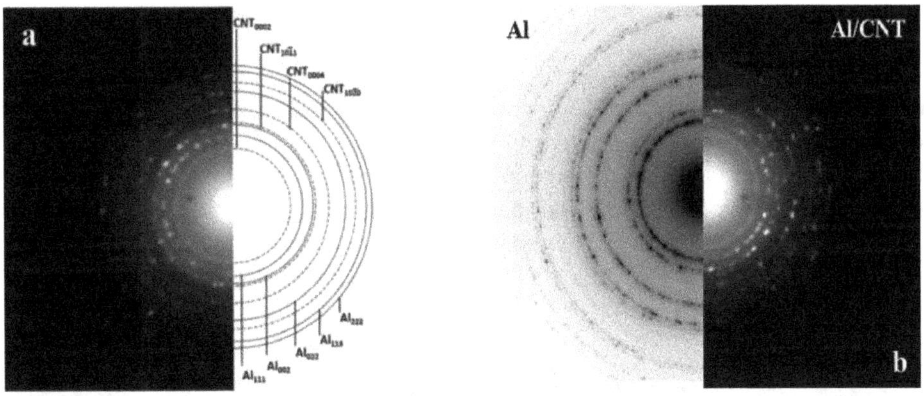


**Figure 12.** The SAED patterns of CNT-reinforced Al powders, (a) planes of Al and CNT are determined and indexed (b) Al/CNT and Al patterns are in a single pattern to identify phases.

For indexing a twining spot pattern in a cubic crystal structure

1. The main spots of the material are identified and indexed in accordance with the previous section.
2. Twinning spots are determined and indexed by 180° rotation around the {111} and {112} planes for f.c.c and b.c.c crystal lattices, respectively.

Basically, f.c.c crystal structure twins on {111} planes and b.c.c crystal structure on {112}. Twin spots in γFe pattern with a f.c.c crystal lattice are indexed according to Figure 13. The spot diffraction pattern of matrix and twin are mirror reflections across the $(1\bar{1}1)$ plane.

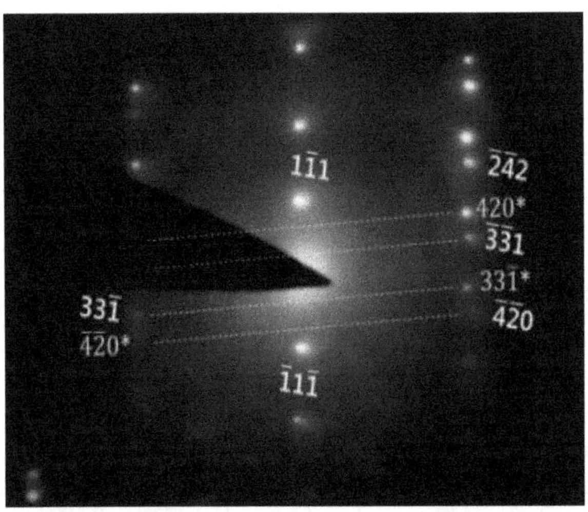


**Figure 13.** Twinning spots pattern from γFe with f.c.c crystal lattice which index matrix and twin spots with zone axis $z=[\bar{1}23]$, twinned on $(1\bar{1}1)$, hkl* planes related to twin planes and hkl planes related to main reflected planes.

### 3.4. Dislocation

The equal channel angular pressing (ECAP) and cryo-cross-rolling process have been performed on Al 5083 alloy and Al 1050 alloy with two passes and ten passes, respectively. These processes create a high dislocation density in Al matrix, which affects the shapes of spots in the diffraction pattern. Also, the diffraction patterns of these alloys illustrate streaks on spots due to high accumulation of dislocations and many partial rings due to preferred orientation. Spots in the pattern deform from a usual shape (circular shape) to stretched and irregular spots (disk-type halo). Dislocations change the crystal orientation locally and diffraction spots are extended along the diffraction ring as seen in Figure 14a,b.

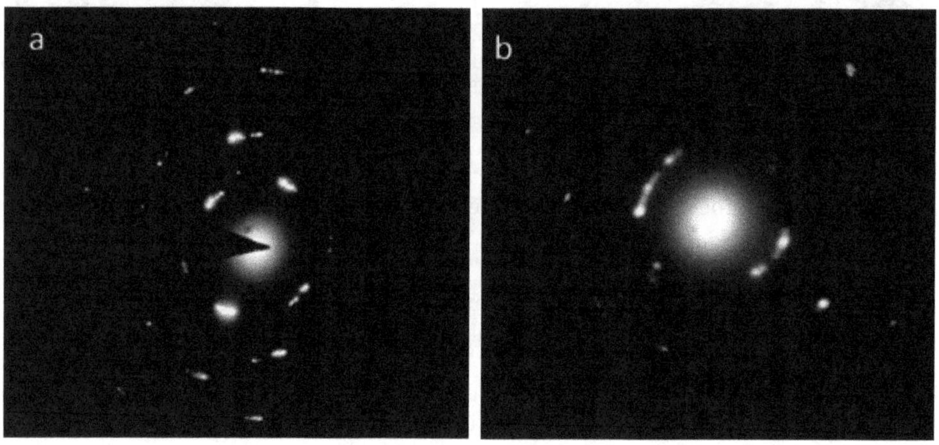


**Figure 14.** (a) Diffraction pattern of Al 5083 alloy with two passes of equal channel angular pressing and (b) diffraction pattern of Al 1050 alloy with ten passes of cryo-cross-rolling process illustrate streaks on spots due to high density of dislocations.

### 3.5. Preferred orientation and texture

In certain specimens, preferred orientation of planes occur by some mechanical processes such as various types of rolling and ECAP. If the crystal structure in the specimen is oriented in a favored and preferred direction, the SAED pattern will be formed from many partial rings, as shown in Figure 15. Diffraction pattern obtained from the texture can be considered as an intermediate case between the diffraction from a single crystal and a polycrystalline material. The texture created in alloys may be investigated by interpretation of their diffraction patterns. The preferred orientation {110} [001] is created in Al 2024 alloy by ECAP process.

### 3.6. Streaks

Presence of fine structure such as streaks and extra spots in many patterns is an indication of the presence of crystal defects which include dislocations, different types of precipitates, twins and staking faults. So, the type of streaks arising in patterns depends on several factors such as various structural defects in the lattice, specimen conditions and diffraction conditions. The

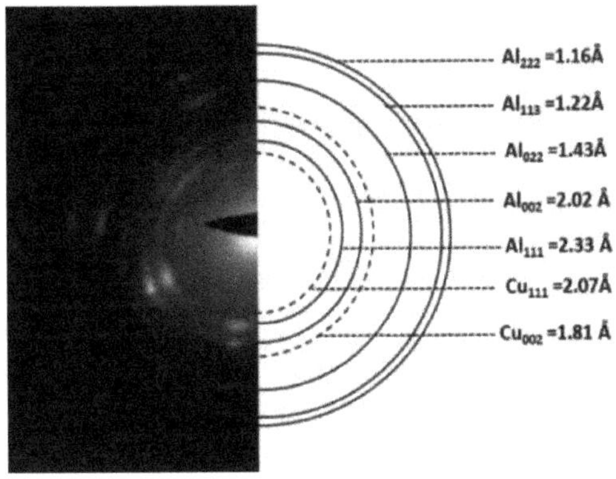


**Figure 15.** The diffraction pattern for Al 2024 alloy with 4 passes of ECAP, showing the preferred orientation of specimen.

main factors in streaks are precipitates, stacking faults, twins and dislocations. Different shapes of precipitate determine the shapes of reciprocal lattice. So, the final shape of spot patterns depends on the types of precipitates. Streaks on spot pattern are created by stacking faults of quasi-sphere and rod particles of carbide $M_7C_3$, as shown in Figure 16a. The corresponding spot diffraction pattern illustrates long streaks (diffuse scattering). In fact, streaks in the diffraction pattern can be created by stacking faults in any crystal structure. Depending on the diffraction conditions, streaks or enlarged spots and extra spots or pseudo-satellites will be created, as shown in Figure 16b.

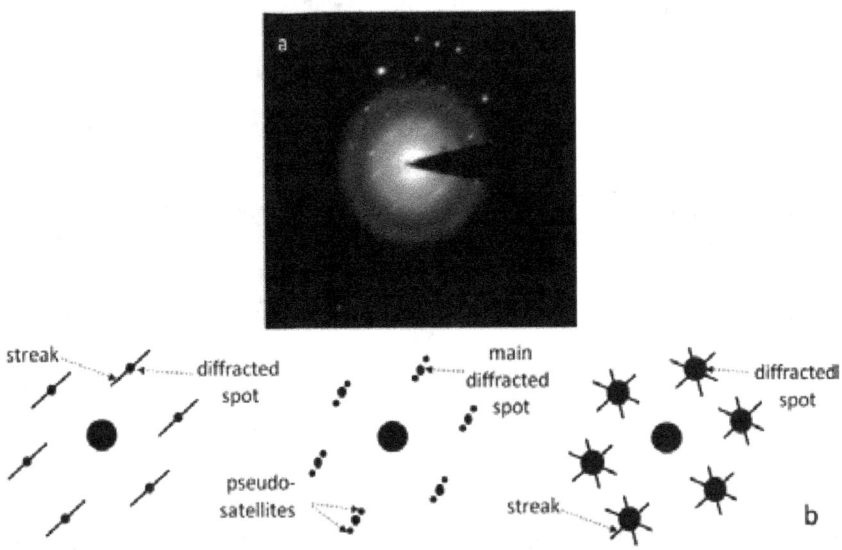


**Figure 16.** (a) Streaks on spots due to stacking faults of carbide $M_7C_3$ quasi-sphere and rod particles are created, (b) different types of streaks in the spot diffraction pattern in various conditions of diffraction.

# 4. Electron diffraction pattern of new materials

## 4.1. Highly crystalline Multi-Wall carbon Nanotubes (MWNTs)

It should be noted that the structure of CNTs depends substantially on the synthesis methods. A ring pattern of the sample demonstrates a crystal structure corresponding to a graphite ring pattern. As a result, the interplanar spacing of MWCNTs is almost similar to the interplanar spacing of graphite. Thus, the interplanar spacing of nanotubes is indexed using the graphite crystal structure as shown in Figure 17a. With this interpretation, crystal structure of MWCNTs is hexagonal and its lattice parameters are $a$=2.41 Å and $c$=6.61Å. In MWCNTs, many amorphous carbons exist as impurities which are created during the production process. Also, the CNTs are complicated and in the shape of a coil. For this reason, the shape of their EDP is a coaxial ring as well as the halo shown in Figure 17b. The EDP of MWCNTs have many rings, each one corresponds to a set of atomic planes. Highly crystalline MWCNTs are a type of MWCNTs with identical chiralities of zigzag type that do not have any impurity and are constructed from monochirality graphite shells. These kinds of CNTs are synthesized by a low-temperature chemical vapor deposition process in plasma. The EDP of highly crystalline MWCNT made of two simple hexagonal patterns overlapping each other is seen in Figure 17b. It is noteworthy that the EDP is taken from an area on the wall of the nanotube. Their spot pattern shows that all of the layers have almost the same chirality.

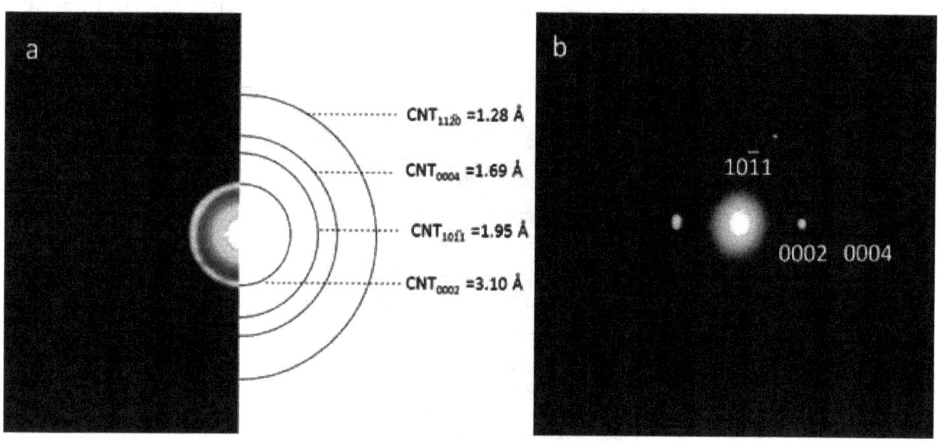


**Figure 17.** (a) The ring diffraction pattern of coiled MWCNT, (b) the spot diffraction pattern of highly crystalline MWCNT.

## 4.2. Palladium doping into MWCNTs

Various methods are available for doping nanoparticles into CNTs which give specific features to them and changes the optical, transport, magnetic, electronic and chemical properties of CNTs. In this investigation, a couple of Pd nanoparticles doped into MWCNTs by chemical methods are selected. The EDP of the selected area on the sample is shown in Figure 18. The ring and spot patterns belong to CNTs and palladium nanoparticles, respectively. Atomic planes can be determined by measuring distances and angles between spots. According to the results of measurements, palladium has a f.c.c crystal structure and the lattice parameter and zone axis of the sample are 3.90 $\AA$ and $z=[\bar{1}22]$, respectively.

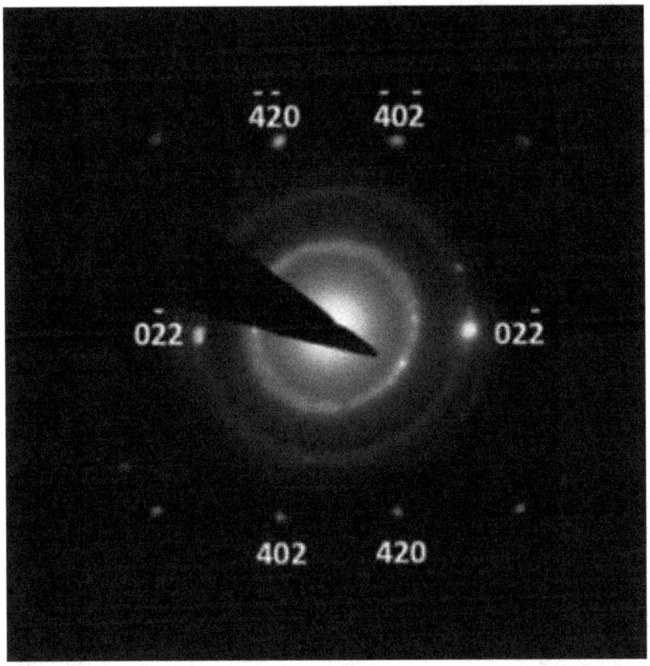


**Figure 18.** The ring pattern of MWCNTs superimposed on the spot patterns of palladium.

### 4.3. Vanadium Oxide Nanotubes (VONTs)

In order to study the organic nanotubes, VONTs were selected. It is noteworthy that using EDP method and its interpretation results, structure of various nanotubes will be predictable. Figure 19 shows a ring diffraction pattern of VONTs. Interplanar spacing and lattice parameter can be determined by analysing the ring diffraction pattern. The ring pattern with the Miller indices according to the interplanar spacing of VONTs crystal structure is given. Based on the results of the measurements, VONT has a f.c.c crystal structure and its lattice parameter was 3.92 $\AA$. In addition, the results of EDP technique have been approved by XRD analysis.

### 4.4. Amorphous coating

When the grain size of the specimen is extremely fine or completely amorphous, the feature of concentric rings in the pattern disappears and a halo is left around the bright center spot, which shows that the electrons are scattered randomly by the amorphous structure of specimen. The amorphous and glassy materials are identified by this method. Diffraction

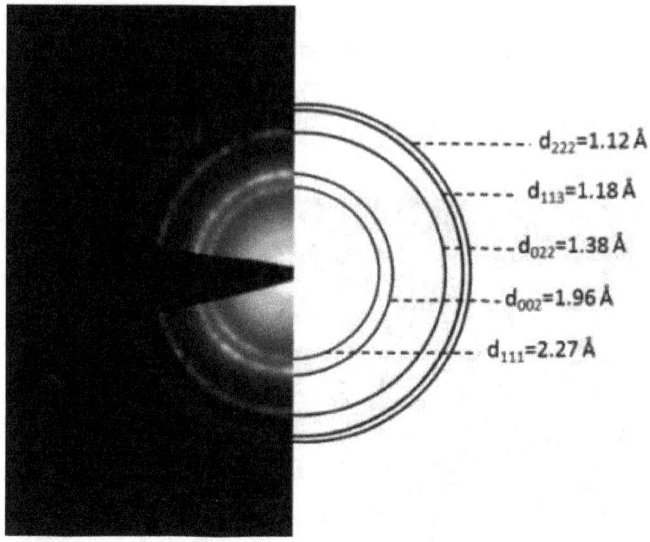


**Figure 19.** The ring diffraction pattern of VONT, interplanar spacing and planes are determined.

pattern for amorphous coating of silica-zinc oxide multilayered nano-porous membrane on $\alpha$-alumina substrate prepared by sol-gel method is shown in Figure 20.

**Figure 20.** The diffraction pattern of amorphous coating $SiO_2-10\%$ ZnO prepared with sol-gel method on $\alpha$-alumina substrate.

## 5. Appendix 1

| Accelerating voltages (kV) | λ (Å) |
|---|---|
| 50 | 0.0536 |
| 100 | 0.0370 |
| 200 | 0.0251 |
| 500 | 0.0142 |
| 1000 | 0.0087 |

**Table 3.** Electron wavelength $\lambda$ for applied accelerating voltages in electron microscopy

## Author details

Mohsen Asadi Asadabad* and Mohammad Jafari Eskandari

*Address all correspondence to: asadimohsen@gmail.com

Materials Research School, NSTRI, Isfahan, Iran

## References

[1] Cowley J M. Diffraction Physics, third revised edition, Elsevier; 1995. DOI:10.1016/B978-0-444-82218-5.50024-9.

[2] Edindton J W. Electron Diffraction in the Electron Microscope. London, Macmillan; 1975.

[3] K. W. Andrews. Interpretation of Electron Diffraction Patterns. London, ADAM HILGER LTD, 1971.

[4] Asadi Asadabad M, Jafari Eskandari M. Transmission electron microscopy as best technique for characterization in nanotechnology. Synthesis and Reactivity in Inorganic, Journal of Metal-Organic, and Nano-Metal Chemistry. 2015; 45, 323–326. DOI: 10.1080/15533174.2013.831901.

[5] Weirich T E, Lábár J L, Zou X. Electron crystallography. Novel Approaches for Structure Determination of Nanosized Materials. Springer; 2004.

[6] Beeston B E P, Horne R W, and Markham R. Electron Diffraction and Optical Methods in Electron Diffraction Techniques. Elsevier, North Holland. 1972.

[7] Asadi Asadabad M, Jafari Eskandari M, Tafrishi R, Emamalizadeh M. Transmission electron microscopy characterization of diffraction nanotubes, Submitted.

[8] Hammond C. The Basic of Crystallography and Diffraction. Oxford, Oxford Science Publications; 1997.

[9] M. Jafari Eskandari, R. Tafrishi, M. Asadi Asadabad. Study of the nanostructure evolution in aluminum 1050 sheet by cryo-cross-rolling process, Submitted.

[10] Hirsch P B, Howie A, Nicholson R B, Pashley D W, and Whelan M. Electron Microscopy of Thin Crystals, London, Butterworth; 1965.

[11] Hren J J, Goldstein J I, and Joy D C. Introduction to Analytical Electron Microscopy. Plenum; 1979.

[12] Ball C J. An Introduction to the Theory of Diffraction. Pergamon, Oxford; 1971.

[13] Rymer T B, Electron Diffraction, Methuen, London; 1970.

[14] Vainshtien B K, Structure Analysis by Electron Diffraction. Pergamon Press, Oxford; 1964.

[15] Asadi Asadabad M, Jafari Eskandari M, Tafrishi R, Bagherzadeh M. The effect of cross-rolling process on nanostructure of Al 1050 alloy. Journal of Nanoanalysis. 2014; 1: 93–98.

[16] Microbeam analysis, analytical electron microscopy, selected-area electron diffraction analysis using a transmission electron microscope, International Standard, ISO 25498, First edition, 2010.

[17] David B, Williams C. Barry Carter. Transmission electron microscopy. A Textbook for Materials Science. Springer Science Business Media; LLC 1996, 2009.

[18] Mogilevsky P, et al., Evolution of texture in rhabdophane-derived monazite coatings. Journal of the American Ceramic Society. 2003; 86: 1767–1772. DOI: 10.1111/j.1151-2916.2003.tb03552.

[19] Fillingham P J, Leamy H J, and Tanner L E. Electron Microscopy and Structure of Materials, University of California Press; 1972.

[20] Hollox G E, Rowcliffe D J, and Edington J W, Electron Microscopy and Structure of Materials. University of California Press; 1972.

[21] Thomas G, Modern Diffraction and Imaging Techniques in Materials Science. North-Holland; 1970.

# 5

# Immunogold Techniques in Electron Microscopy

Mogana Das Murtey

**Abstract**

Ever since electron microscopy became an important tool in the scientific research, the focus had been mainly on ultrastructural analysis with little success in the development and application of suitable techniques for the localization of macromolecules in cells. The emergence of immunogold techniques in the 1960s managed to fill this gap in serving this function. The aim of this chapter is to equip researchers, postgraduate students, and technicians with essential knowledge to utilize immunogold techniques for ultrastructural investigations in the life sciences. The principles and factors involved have been highlighted to give researchers a quick review of the techniques before embarking on their ultrastructural localization procedures. The advantages and limitations of the four types of immunogold labeling techniques have been discussed.

**Keywords:** Immunogold labeling, scanning electron microscopy, transmission electron microscopy, double labeling, pre-embedding immunogold labeling, post embedding immunogold labeling

## 1. Introduction

Immunogold labeling is a very powerful technique for identifying active sites and the presence of biomarkers in the cells. It is probably the only technique that can probe the cells ultrastructurally because of the processes devised to attach gold probes to secondary antibodies and consecutively to primary antibody to reveal the presence of an antigen. A primary antibody is designed to bind onto a specific antigen in the cells. Thus, the gold probe with its excellent electron scattering property is an important element for immunohistochemistry in the electron microscope.

Although gold probe solutions have been in use since the mid-1800s, it was in 1961 that the gold probes became popular to be used as an ultrastructural tracer. A decade later,

Faulk and Taylor were the first to use immunogold labeling technique to identify Salmonella antigens.

Immunogold labeling techniques were initially more applicable for transmission electron microscopy (TEM) and were crucial for low-density markers. However, the emergence of high resolution SEMs has made it possible for the application of these techniques in scanning electron microscopy.

The gold probes became a popular choice over peroxidase or other enzyme marker techniques because of their localized spread around the reaction sites with minimal obscuring of ultrastructural details. Over the years, there has been a rapid development of numerous protocols for immunolabeling, including techniques for detecting multiple antigens within a cell and the selective use of different sizes of gold particles. Unlike the enzyme reaction products, gold probes can be quantified, which could be crucial for comparative studies.

The technique also allows routine EM stains such as uranyl acetate and lead citrate to be used to improve visualization of cellular details without obscuring the site of immunogold reactions.

The two most widely used immunogold labeling techniques are the "pre-embedding" and "post embedding" techniques [1-5].

## 2. Principles of immunogold labelling

### 2.1. Uses of the immunogold techniques

Immunogold labeling is being very useful in the localization of target markers in cells and tissues. The high resolution of the technique could provide excellent insight with regards to structure–function relationships in the microenvironment of cells and tissues. It can also be used in the study of protein distribution in cellular and extracellular components. However, success of immunogold labeling techniques is very much dependent on the quality of protein antigen preservation, antigen-primary antibody specificity, and the antibody's ability to infiltrate cells and tissues. Good handling of samples which involves fixation and dehydration as well as an appropriate selection of embedding media are crucial for immunogold labelling analysis [1, 3].

### 2.2. Antigen-antibody reactions

Antigen-antibody reaction is an important biochemical interaction between antibodies produced by white blood cells and antigens. It is a basic reaction where the body is protected from complex foreign entities. The antibody binds with the antigen through a specific site called epitope. Immunogold labeling relies solely on this antibody-antigen reaction in order to get intended results. There are two types of antibodies: monoclonal and polyclonal antibody. Monoclonal antibodies are derived from cellular reaction of plasma cells while polyclonal antibodies are derived from the repetition of antigen stimulation. Generally, immunogold labeling is focused more on indirect patterns (Figure 1) whereby gold conjugated secondary antibodies bind with specific primary antibodies in a microenvironment. This indirect pattern is more favorable than the direct pattern for two reasons: (a) higher density of secondary antibody albeit for longer incubation time and (b) increased sensitivity, since the secondary antibody is able to bind with multiple sites on primary antibody [3, 6-8].

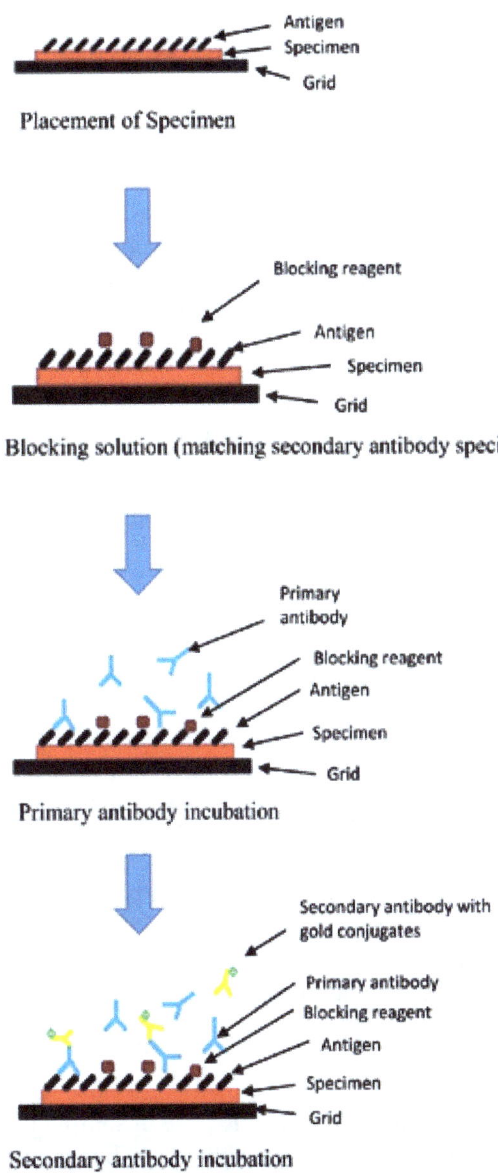


**Figure 1.** This figure shows the whole process of immunogold labeling starting from blocking the sample to secondary antibody incubation.

### 2.3. Gold particles as a probe

Gold became the most reliable choice for immunogold labeling due to its large specific surface area, good biocompatibility and high electron density, which offers easy detection, excellent electrical and thermal conductivity in the electron microscope. The distribution and presence of the gold particles do not hinder the viewing of the images at high resolution modes. The size of gold particles used for immunogold labeling varies from 1 to 40μm and can be chosen according to the type of labeling techniques employed [1, 4-6, 9].

Detecting multiple antigens within a cell may require the selective use of different sizes of gold particles.

### 2.4. Advantages of immunogold techniques

Immunogold labeling is a relatively easy technique to do and suitable for real-time monitoring of ultrastructure and antigens. It can improve the assessment of target markers on cellular microenvironment and provide good intratechnique comparison and reassessment, providing crucial information for further histochemical analysis [1].

## 3. Factors influencing immunogold labelling

### 3.1. Fixatives

Fixation of tissues in immunogold labeling should preserve both the structural integrity and antigenic components of the tissue, without being detrimental to the labelling outcome. The commonly used fixative formulations for immunogold labeling are based on glutaraldehyde and paraformaldehyde combination. Glutaraldehyde is a good ultrastructural fixative but has a lower tissue penetration rate. However, the slowness of glutaraldehyde is compensated with the addition of formaldehyde in the fixative formulations. Formaldehyde penetrates tissues at a much more rapid rate. The author uses a mixture of formaldehyde (2-4%) with glutaraldehyde (1%) for rapid stabilizing with lower cross linking and greater structural preservation [1, 7, 10-11].

It is preferable to avoid the use of osmium tetroxide as a post fixative in immunogold processing technique due to the probability of some of the gold particles being superimposed on a heavily osmicated area of the images. However, osmium tetroxide can still be used if the level of contrast in the images obtained is very poor.

### 3.2. Types of embedding media

Generally there are two types of resin available for use in sample processing. Epoxy resins such as Araldite, Spurs (Figure 2a), and Epon, which are hydrophobic, and acrylic resins such as Lowicryl and LR White (Figure 2b), which can be hydrophobic or hydrophilic. Although epoxy resin gives better ultrastructural preservation and stability, acrylic resins are more suited for immunolabeling because it does not need much heat for polymerization and shows better antigen preservation. Acrylic resins can also retain water for better demonstration of protein components, while giving greater accessibility to the antibodies during incubations [1]. LR White resin is the most popular of the acrylic-based resins. It is easier to section and polymerize at moderate temperature (50ºC), which is good for antigen preservation, tolerates partial dehydration, retains higher labeling density, and is much less toxic than Lowicryl [3, 12].

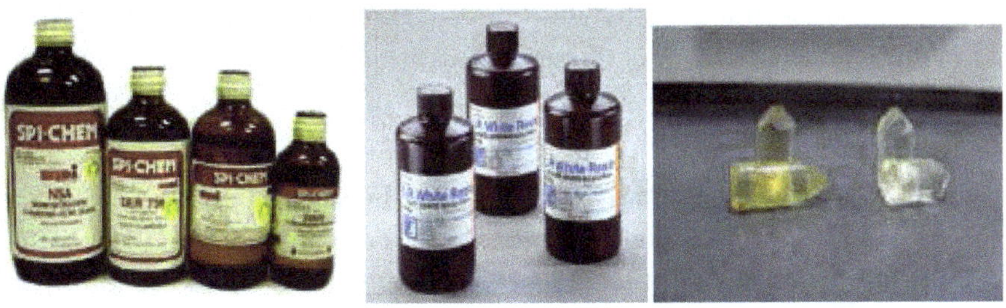


(a) (b) (c)

**Figure 2.** This figure shows two types of embedding media. (a) Bottles of Spur's resin, (b) LR White, and (c) embedding blocks of Spur's resin (yellowish brown blocks) and LR White (white yellowish blocks).

### 3.3. Blocking and reducing agents

Blocking solution, an important step in immunogold labeling, is applied before primary antibody incubation. Blocking solution will close the non-specific reactive sites on tissues and cell surfaces without altering the epitope for antibody binding. It will reduce the non-specific attraction of the primary or secondary antibodies to the tissue components or the resin material. Non-specific reactions or background occur because of the general physical chemical properties of both the specimen and the antibodies [3, 8].

### 3.4. Controls

Controls are always used to confirm the integrity of specific labeling in samples. The control should be carefully chosen as it plays a crucial role in troubleshooting the contamination of samples or poor performance of an individual reagent in any part of immunogold labeling procedure [3, 7-8]. The types of controls normally used are as follows.

Positive control

A known positive sample tested with same immunogold labeling technique to avoid false negative and test the effectiveness of the labeling procedure.

Negative control

Omitting of the primary antibody step and incubating only in secondary antibody. This is to determine the secondary antibody labeling capacity.

### 3.5. Gold probes

Over the years, being the choice metal for immunogold labeling, gold particles have been produced to be smaller and smaller, more penetrating, and highly sensitive. The choice of gold particle size depends on the magnification at which the cells are to be studied in the electron microscopy. In principle, smaller gold particles (2 nm or 5 nm) produce a higher labeling intensity and lower steric hindrance due to their greater concentration in solution. Larger particle sizes (10 nm or more) basically reduce the potential labeling intensity due to their sheer size but are more easily seen at lower magnifications [7, 13]. Detecting multiple antigens within a cell may require the selective use of different sizes of gold particles.

### 3.6. Grids

The three types of grids commonly used for electron microscopy are copper grids (Figure 3a), gold grids (Figure 3b), and nickel grids (Figure 3c). For immunogold labeling, the buffers used can influence the choice of grids. Nickel grids can be a good choice for immunogold labeling for its stability and not being affected by chemical oxidation. However, handling nickel grids can be difficult due to its magnetic properties. Gold grids are stable, non-magnetic, and also a good choice for immunogold labeling but expensive. Copper grids are not suitable for immunogold labeling because Tris Buffer Saline (TBS) Tween buffer react with copper grids. A support film such as formvar, pioloform, or collodion coated on the grid may be necessary to hold the sections firmly during the immunogold labeling process [14].

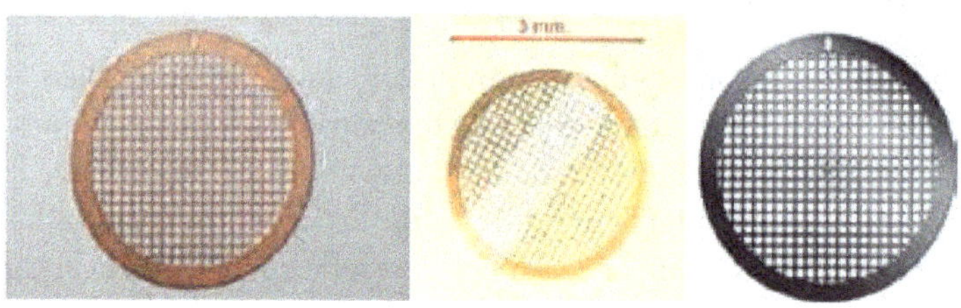

(a) (b) (c)

**Figure 3.** This figure shows three types of grids: (a) copper grids, (b) gold grids, and (c) nickel grids.

### 3.7. Temperature and pH

The temperature and pH of reagents could influence the outcome of immunogold labeling. In most of the protocols, buffers and reagents are kept at low alkaline conditions to ensure that the gold probes could combine firmly with the target protein. This is due to the electrostatic reaction between negative charges of the colloidal gold with the positive charges of the target protein. Generally, the temperature of section incubations is kept at ambient room temperature around 16–20°C to avoid aggregation of the gold particles. The temperature of the sections can affect the stabilization of gold conjugates. Aggregation of gold conjugates can occur due to the repulsive energy forces that separate gold conjugates being lesser than the primary attraction forces (van der Waals forces). This phenomenon will eventually cause the aggregation of gold conjugates [10, 15].

# 4. Immunogold labelling methods

## 4.1. Pre-embedding immunogold technique

### *4.1.1. General concepts*

The pre-embedding approach is to perform the immunolabeling on the sample before embedding and the sectioning process.

### *4.1.2. Materials*

1. McDowell-Trump fixative
2. Phosphate Buffer, Sorensen 0.2M pH7.2
3. Phosphate Buffer, Sorensen 0.2M pH7.6
4. TBS Tween (0.05M TBS, 0.05% Tween 20, pH7.6)
5. Sucrose solution (0.2 M)
6. London White Resin Medium Grade
7. Ultra V Block Serum free
8. Primary antibodies
9. Secondary antibody
10. 1% toluidine blue with sodium borax
11. Ultra-antibody diluent

### *4.1.3. Standard protocol*

Initial sample preparation steps are different for animal/human tissues and cultured cells.

**a. Animal and human tissue**

Cut the tissue into small pieces about 1 $mm^3$ and place in vial containing McDowell-Trump fixative for at least 24 hours.

**b. Cultured cells**

*[Note: All centrifugation in this technique should be at 1,000–2,000g for 15 minutes.]*

**i.** Use 6 well plates to grow adherent cells. (For suspension cells, you can proceed directly to fixation step).

**ii.** Incubate the cells (inside well) with 3 ml of fixative (for each well) for 10 minutes. Discard the fixative and use Trypsin to detach the cells. [Leave the plate in incubator with temperature 37°C for 10 minutes – depends on type of cell line used.]

iii. Neutralize the Trypsin with culture medium and transfer the solution into 15 ml Falcon Tube.

iv. Centrifuge the sample and discard the supernatant. **[From now on, the steps are similar for adherent and suspension cells]**

v. Resuspend the pellet with McDowell Trump fixative prepared in 0.1M phosphate buffer (pH 7.2) for at least 15 minutes.

vi. Centrifuge the resuspended sample. Discard the supernatant [transfer the pellet to an Eppendorf tube at this stage, if needed]. Resuspend the pellet in 0.1M phosphate buffer.

vii. Centrifuge the resuspended sample. Discard the supernatant and resuspend the pellet in 0.1M phosphate buffer.

viii. Centrifuge the resuspended sample, discard the supernatant and place this tube containing a pellet of fixed cells in a water bath at 45°C for about 15–30 minutes depending on the amount of sample.

ix. Prepare a 2% solution of agarose by dissolving the agarose in boiling distilled water. Pour the solution into a test tube while it is still molten, and place the tube in the water bath at 45°C. At this temperature the agarose remains liquid.

x. After the temperature of both the agarose and the pellet have equilibrated to 45°C, transfer a small drop of the agarose to the tube containing the pellet of cells (with a warm pipette), and stir up the pellet, just enough to break the pellet into small blocks, and to suspend these small blocks in the agarose.

xi. Immediately pour the agarose with the suspended pellet blocks on to a glass microscope slide.

xii. After the agarose has set (1–2 minutes), cut the solidified agarose containing the cells into small cubes, about 1 $mm^3$, with a sharp razor blade and place them in a vial containing 0.1M phosphate buffer.

**Procedure for immunolabeling**

The animal/human tissue will be in small pieces while cultured cells will be held together in 2% agarose in the form of small cubes. The following steps will be similar for both tissues and cultured cells in agarose cubes.

1. Rinse the sample in 0.1M phosphate buffer for 10 minutes (3X).
2. Incubate the sample in 0.2 M sucrose solution for 15 minutes (2X).
3. Rinse the sample briefly in 0.1M phosphate buffer.
4. Incubate the sample in blocking reagents for 30 minutes at room temperature.
5. After 30 minutes, incubate the sample in primary antibody for two hours at room temperature. [Dilutions of 1:10 were used for primary antibody incubation.]

6. Wash the sample in TBS-Tween 20 buffer 5 times, for 5 minutes for each wash. After the washing step, incubate the sample in a secondary antibody with gold conjugates for one hour at room temperature with a dilution of 1:50.
7. Wash the sample again in TBS Tween 20 solution 5 times, for 5 minutes for each wash.
8. Prepare to incubate the sample for following dehydration steps. *[Follow the steps carefully.]*
    a. 70% ethanol for 30 minutes (2X)
    b. London White Resin and 70% ethanol mixture (2:1) for 1 hour
    c. Two changes of pure London White Resin (1 hour each)
    d. Fresh London White Resin change overnight at room temperature
9. Take out the sample from vial and drop it at the bottom of BEEM capsules and fill it up with London White Resin to the brim.
10. Let the capsule to polymerize at 50°C for 72 hours.
11. Semi-thin and ultrathin sectioning of the resin-embedded samples is performed in the ultramicrotome using glass or diamond knives.
12. Collect ultrathin sections of about 60–70nm in thickness by using 200–300 mesh nickel grids. Gray, silver, or gold sections were collected.
13. Let the grids with the sections to dry.
14. Rinse the grids for 2 minutes each in two large droplets of TBS Tween 20.
15. The grids can then be stained in uranyl acetate and lead citrate for 10 and 5 minutes, respectively. If the sections stain heavily, the times can be varied. In some instances, one of the stains can be omitted to provide acceptable "gold particles to tissue" contrast levels.

#### *4.1.4. Precautions*

It is preferable to use samples of smaller size (about 1 mm$^3$) for pre-embedding immunogold techniques, to achieve greater sensitivity and better ultrastructural preservation.

#### *4.1.5. Troubleshooting*

1. Good permeabilization of gold probes is crucial for this technique. The penetration rate can be increased by using the smallest gold particles possible, while being mindful that smaller particles would require higher magnifications in the TEM.

### 4.2. Post embedding immunogold technique

#### *4.2.1. General concepts*

The post embedding labeling technique is the more widely used technique in comparison to the pre-embedding technique. For post embedding, samples are first fixed, embedded, and sectioned before being subjected to immunolabeling. The indirect labelling method is generally preferred for this technique whereby the primary antibody is unlabeled and the gold is conjugated to a secondary antibody.

### *4.2.2. Materials*

1. McDowell-Trump fixative
2. Phosphate Buffer, Sorensen 0.2M pH7.2
3. Phosphate Buffer, Sorensen 0.2M pH7.6
4. TBS Tween (0.05M TBS, 0.05% Tween 20, pH7.6)
5. Sucrose solution (0.2 M)
6. London White Resin Medium Grade
7. Ultra V Block Serum free
8. Primary antibodies
9. Secondary antibody
10. 1% toluidine blue with sodium borax
11. Ultra-antibody diluent

### *4.2.3. Standard protocol*

**Please refer to 4.1.3 Section A and B [for initial sample preparation for animal/human tissue and cultured cells].**

**Procedure**

1. Rinse the sample in 0.1M phosphate buffer for 10 minutes (3X).
2. Incubate the sample in 0.2 M sucrose solution for 15 minutes (2X).
3. Rinse the sample briefly in 0.1M phosphate buffer.
4. Prepare to incubate the sample for following dehydration steps. [Follow the steps carefully.]
    a. 70% ethanol for 30 minutes (2X)
    b. London White Resin and 70% ethanol mixture (2:1) for 1 hour
    c. Two changes of pure London White Resin (1 hour each)
    d. Fresh London White Resin change overnight at room temperature
5. Take out the sample from vial and drop it at the bottom of BEEM capsules and fill up with London White Resin to the brim.

6. Let the capsule to polymerize at 50°C for 72 hours.
7. Semi-thin and ultrathin sectioning of the resin-embedded samples is performed in the ultramicrotome using glass or diamond knives.
8. Collect ultrathin sections of about 60–70nm in thickness by using 200–300 mesh nickel grids.
9. Let the grids with the sections to dry.
10. Rinse the grids for 2 minutes each in two large droplets of TBS Tween 20.
11. Incubate the grids in blocking reagents for 30 minutes at room temperature.
12. After 30 minutes, incubate the sample in primary antibody for 2 hours at room temperature. [Dilutions of 1:10 were used for primary antibody incubation.]
13. Wash the sample in TBS-Tween 20 buffer 5 times, for 5 minutes each time. After the washing step, incubate the sample in a secondary antibody with gold conjugates for 1 hour at room temperature with a dilution of 1:50.
14. Wash the sample again in TBS Tween 20 solution 5 times, for 5 minutes for each wash.
15. The grids can then be stained in uranyl acetate and lead citrate for 10 and 5 minutes, respectively. If the sections stain heavily, the times can be varied. In some instances, one of the stains can be omitted to provide acceptable "gold particles to tissue" contrast levels.

#### *4.2.4. Precautions*

1. Resin preparation should be performed carefully with the 70% ethanol added slowly, drop by drop to the London White Resin and gently shaken to avoid the resin mixture becoming milky (cloudy).
2. Sections tend to overlap during the embedding process so it is important to maintain the tissue as flat as possible during the embedding procedure, in order to obtain uniformly labelled sections on grid.
3. It may be preferable to omit osmium tetroxide to preserve better antigenicity.

#### *4.2.5. Troubleshooting*

Gold conjugates tend to aggregate, making it difficult for the localization of markers. Generally, the temperature of section incubations is kept at ambient room temperature around 16–20°C to avoid aggregation of the gold particles. The temperature of the sections can affect the stabilization of gold conjugates. Aggregation of gold conjugates can occur due to the repulsive energy forces that separate gold conjugates being lesser than the primary attraction force (van de Waals force). This phenomenon will eventually cause the aggregation of gold conjugates [10, 15].

## 4.3. Double immunogold labeling technique

### *4.3.1. General concepts*

Double immunogold labeling is a technique that reveals the distribution of multiple antigens simultaneously on the same sample section. This is made possible with the availability of different sizes of gold probes. Post embedding, labeling is the preferred method for this technique.

### *4.3.2. Materials*

1. McDowell-Trump fixative
2. Phosphate Buffer, Sorensen 0.2M pH7.2
3. Phosphate Buffer, Sorensen 0.2M pH7.6
4. TBS Tween (0.05M TBS, 0.05% Tween 20, pH7.6)
5. Sucrose solution (0.2 M)
6. London White Resin Medium Grade
7. Ultra V Block Serum free
8. Primary antibodies
9. Secondary antibody
10. 1% toluidine blue with sodium borax
11. Ultra-antibody diluent

### *4.3.3. Standard protocol*

**Please refer to 4.1.3 for initial sample preparation and 4.2.3 (Step 1 until 12) for sample preparation and continue with following steps:**

1. After incubating the sample in blocking reagent for 30 minutes, incubate the grids in a mixture of two primary antibodies for 2 hours at room temperature.
2. Wash the sample in TBS-Tween 20 buffer 5 times for 5 minutes each. After the washing step, incubate the sample in a mixture of secondary antibody with gold conjugates (different sizes) for 1 hour at room temperature with a dilution of 1:50.
3. Wash the sample again in TBS Tween 20 solution 5 times for 5 minutes of wash each.
4. The grids can then be stained in uranyl acetate and lead citrate for 10 and 5 minutes, respectively. If the sections stain heavily, the times can be varied. In some instances, one of the stains can be omitted to provide acceptable "gold particles to tissue" contrast levels.

*4.3.4. Precautions*

1. Ensure that the primary and secondary antibodies are compatible with each other.
2. Ensure that the size of gold conjugates for each antigen can be differentiated during observation in the TEM. Do not choose gold probes that are similar in size.

## 4.4. Immunogold technique for scanning electron microscopy

*4.4.1. General concepts*

Immunogold labeling in the SEM allows the localization of antigens on the surface areas of samples. The technique can also be performed on internal surfaces of samples after the removal of the external layers to expose the targeted regions. Some prior knowledge of the general areas where the antigen is likely to be located could be useful in obtaining specific results.

*4.4.2. Materials*

1. McDowell-Trump fixative
2. Phosphate buffer, Sorensen
3. 1% osmium tetroxide
4. Graded alcohols
5. Hexamethyldisilazane (HMDS)
6. Ultra V Block Serum free
7. Primary antibodies
8. Secondary antibody
9. Ultra-antibody diluent

*4.4.3. Standard protocol*

Initial sample preparation

a. Animal and human tissue

Fix the animal/human tissue in McDowell-Trump fixative prepared in 0.1M phosphate buffer (PBS) (pH 7.2) at 4°C for 2–24 hours.

b. Cultured cells

   i. iPrepare a solution of 0.1% poly-L-lysine in phosphate-buffered saline (pH 7.2).
   ii. Place a drop of the poly-L-lysine solution on a coverslip and place it in a 6-well plate for about 1 hour.

iii. Rinse the poly-L-lysine coated coverslip with distilled water and fill the well with suitable medium for adherent cells to grow. Let the cells attain 80–90% confluency.

iv. Carefully discard the medium from the well and immediately fill the well with fixative to replace the medium. (Do not let it dry in between the step). Leave it alone for about 10 minutes.

*The following steps are similar for tissues and cells grown on coverslip in the 6-well plate.*

1. Wash tissue/coverslip with cells (**SAMPLE**) in PBS (3 x 10 minutes)
2. Incubate the **SAMPLE** in blocking reagents for 30 minutes at room temperature.
3. Incubate the **SAMPLE** in primary antibody for 2 hours at room temperature.
4. Wash the **SAMPLE** in TBS-Tween 20 buffer 5 times, for 5 minutes each time.
5. Incubate **SAMPLE** in a secondary antibody with gold conjugates for 1 hour at room temperature with a dilution of 1:50.
6. Wash the **SAMPLE** in TBS Tween 20 solution 5 times, for 5 minutes each time.
7. Wash in distilled water (2 x 10 minutes).
8. Dehydrate as follows:

| | |
|---|---|
| 35% ethanol | 15 minutes |
| 50% ethanol | 15 minutes |
| 75% ethanol | 15 minutes |
| 95% ethanol | 15 minutes (x2) |
| 100% ethanol | 20 minutes (x3) |

9. The **SAMPLE** can be subjected to either the "air drying protocol" or the critical point drying (CPD) protocol for drying.

**HMDS method**

Immerse the **SAMPLE** with cells in 1–2 ml of hexamethyldisilazane (HMDS) for 10 minutes.

Decant the HMDS and leave the **SAMPLE** in a desiccator to air-dry at room temperature.

**CPD method**

Transfer the **SAMPLE** into the CPD specimen holder. The specimen holder should contain (or be immersed in) enough acetone to cover the **SAMPLE** (15–30 minutes).

Critical point drying (refer to SEM preparation chapter).

10. The dried **SAMPLE** is then mounted on to a SEM specimen stub with a double-sided sticky tape (or silver paint).
11. View in the SEM

*4.4.4. Precautions*

1. Times recommended are for small pieces of tissues (smaller than 1–2 $mm^3$). For larger pieces times may have to be varied accordingly.
2. Buffer: normally 0.1M of pH 7.2, but requirements may vary for different tissues. Refer to published papers.
3. Never let the tissues to dry at any stage (till the samples are immersed in the HMDS).

4. ADVANTAGES AND LIMITATIONS OF THE IMMUNOGOLD TECHNIQUES

**Pre-embedding immunogold labeling**

**Advantages**

- All antigenic site throughout the specimen are potentially accessible for staining.
- Has superior sensitivity for the localization of sparse antigens.
- Produces more discrete and nondiffusible markers indicating the precise subcellular localization of antigens which are also readily quantified.

**Limitations**

- The labelling procedures require extended incubations in buffers to increase the penetration of probes. As a result, preservation of adequate ultrastructure is often a problem.
- Potential sources of false-positive and false-negative results resulting from
  - Cross-reaction of the primary antibody with unknown proteins
  - Cross-reaction of secondary antibodies with inappropriate species
  - Nonspecific immunolabeling
  - Erroneous transport of tract-tracing agents

**Post-embedding immunogold labeling**

**Advantages**

- Since immunostaining is done after the sample has been fixed, embedded, and sectioned, these preparation steps will preserve antigenicity and reduce problems regarding ultra-structural preservation.
- Better reactivity due to the contact between the antibodies and internal antigens exposed at the surface of thin sections obtained from resin embedded tissues.
- Better visualization of antigenic sites with high resolution allows good quantitative evaluations of labeling intensity and morphometrical determinations.

**Limitations**

- Restriction of antibody staining to epitopes present on the section surface.

- Antibodies cannot penetrate into the resin, and consequently only the antigens that are exposed at thin section surfaces can be labelled.
- Antigens could be affected by fixatives, solvents, resins, and heat during polymerization, thus compromising antigenicity.

**Double immunogold labeling**

**Advantages**

- Ability to be bound to several molecules makes marker being extremely versatile.
- Provide better understanding on relationship and localization of different types of molecules.

**Limitations**

- Chances for section contamination are higher than single labeling procedure.
- Increase in background noise.

**Immunogold with scanning electron microscopy (SEM)**

**Advantages**

- Reveals the surface immunoreactions on a sample that cannot be otherwise resolved with light microscopy techniques.
- A good depth of field in the SEM provides additional information on undulated surfaces.
- Ability to specifically mark and examine one structure and be able to relate with other structures for better understanding.

**Limitations**

- Certain information other than the lateral position of gold particles in the backscatter image is lost.
- Most biological samples need to be dehydrated before being placed in the SEM. This can distort cellular features or create artifacts.

## Author details

Mogana Das Murtey*

Address all correspondence to: moganbaadshah_071185@yahoo.com

Advanced Medical and Dental Institute (AMDI), Universiti Sains Malaysia, Bertam Kepala Batas, Penang, Malaysia

## References


[1] Sirerol-Piquer, M. S., Cebrian-Silla, A., Alfaro-Cervello, C., Gomez-Pinedo, U., Soriano-Navarro, M. and Garcia Verdugo, J. GFP immunogold staining, from light to electron microscopy, in mammalian cells. Micron. 2011. 43(5): 589-599.

[2] Hermann, R., Walther, P. and Muller, M. Immunogold labeling in scanning electron microscopy. Histochem Cell Biol. 1996. 106(1): 31-39.

[3] Ana, L. D. P., Jorge H. M., Juan P. P., Silvina G., Amado A. Q., Cristina A. M. and Alicia, I. T. Immunoelectron Microscopy: A Reliable Tool for the Analysis of Cellular Processes, Applications of Immunocytochemistry, Dr. Hesam Dehghani (Ed.). InTech. 2012. ISBN: 978-953-51-0229-8. DOI: 10.5772/33108.

[4] Glasbey, C. A. and Roberts, I. M. Statistical analysis of the distribution of gold particles over antigen sites after immunogold labelling. J Microsc. 1997. 186(3): 258-262.

[5] Goldberg, M. W. and Fiserova, J. Immunogold labeling for scanning electron microscopy. Methods Mol Biol. Methods Mol Biol. 2010. 657: 297-313.

[6] Lai, C., Zeng, G., Huang, D., Feng, C., Hu, S., Su, F., Zhao, M., Huang, C., Wei, Z. Detection based on immunogold labeling technique and its expected application in composting. Chin J Anal Chem. 2010. 38(6): 909–914.

[7] Bunea, M. and Zarnescu, O. New current aspects on the immunohistochemical techniques. Roum Biotechnol Lett. 2001. 6(3): 177-206.

[8] Su-Yao, M., Lorette, C. J. and Ute, M. K. Immunocytochemical Methods and Protocols, Constance, O. and Maria, C. J. (Eds.), 3rd Ed. Humana Press. 1999. p3-10.

[9] Hayat, M. A. Colloidal Gold: Principles, Methods, and Applications, 1st Ed. Academic Press. Elsevier Science. 1989. p1-536.

[10] Hayat, M. A. Principles and Techniques of Electron Microscopy: Biological Applications, 4th Ed. Cambridge. Cambridge University Press. 2000. 6 p262-263.

[11] Berryman, M. A., Porter, W. R., Rodewald, R. D. and Hubbard, A. L. Effects of tannic acid on antigenicity and membrane contrast in ultrastructural immunocytochemistry. J Histochem Cytochem. 1992 40(6): 845-857.

[12] Philimonenko, V. V. Janácek, J. and Hozák, P. LR White is preferable to Unicryl for immunogold detection of fixation sensitive nuclear antigens. Eur J Histochem. 2002. 46(4): 359-364.

[13] Hainfeld, J. F. and Robinson, J. M. New frontiers in gold labeling. J Histochem Cytochem. 2000. 48(4): 459-460.

[14] Hyatt, A. D. and Eaton, B. Immuno-Gold Electron Microscopy in Virus Diagnosis and Research, 1st Ed. CRC Press. Taylor & Francis. 1992. p1-464.

[15] Della Torre, G., Canevari, S., Orlandi, R. and Colnaghi, M. I. Internalization of a monoclonal antibody against human breast cancer by immunoelectron microscopy. Br J Cancer. 1987. 55(4): 357-359.

6

# Combined Transmission Electron Microscopy – *In situ* Observation of the Formation Process and Measurement of Physical Properties for Single Atomic-Sized Metallic Wires

Hideki Masuda

**Abstract**

This chapter introduces the research of *in situ* high-resolution transmission electron microscope (HRTEM) methods combined with the functions of atomic force microscope and scanning tunneling microscope. Using this method, we demonstrate fabrication, dynamic observation on an atomistic scale, and mechanical and electrical measurements of nanometer-sized contact simultaneously. We evaluate the silver atomic-sized wire (ASW) which appeared at the final stage of the rupture process of the silver nano-contact.

**Keywords:** Silver, atomic-sized wires, atomic force microscopy, high-resolution transmission electron microscopy

## 1. Introduction

Currently, miniaturization of electronics is underway. Now device development is heading toward atomistic and molecular scales [1]. Devices included in these circuits are nanometer-sized contacts (NCs), atomic-sized wires (ASWs), single molecular junctions (SMJs), etc. (Figure 1) [2]. SMJ is a system of single molecule sandwiched by a pair of nanometer-sized metallic electrodes. SMJs enable single electronic operation, high-density integration, and electric power saving [3-7]. To engineer SMJs, we need to reveal that the structure of device configuration includes interfaces between molecules and electrodes and mechanical and electrical properties. Metallic NCs and ASWs are fundamental materials that have the potential for device applications as well as being the key factors required for the application of SMJs [8].

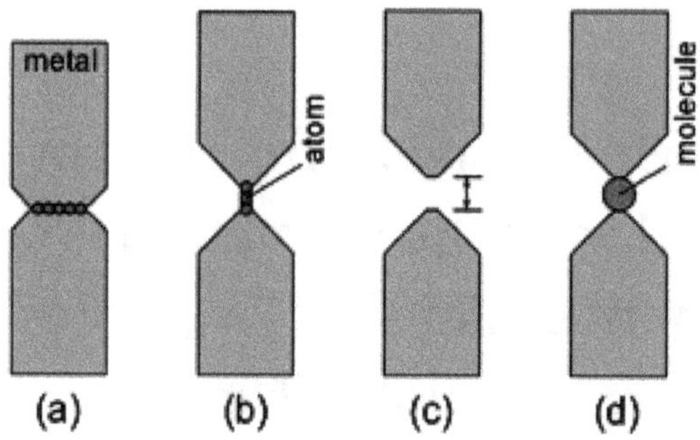


**Figure 1.** Schematics of atomistic-scale devices. (a) NC; (b) ASW; (c) nano-gap structure; and (d) SMJ.

Actually, to observe the formation and deformation of NCs, two *in situ* high-resolution transmission electron microscope (HRTEM) methods were developed; one is the electron beam double holes drilling method. An electron beam drills two holes on a material film with a focused beam. The bridge that was formed between the holes is gradually deformed using a defocused electron beam [9]. The other method is the tip-sample contact method [10], and this fabricates and deforms NCs using a piezo-driven tip. Kizuka and Tanaka observed Zn NCs using this method in HRTEM in 1994 [11]. In 1998, Kizuka et al. directly observed the slip deformation process of Au NCs (Figure 2) [12]. This is the first report that showed atomistic-level observation of a slip deformation process in crystals. After that, Ohnishi et al. fabricated Au NCs and observed the deformation process from a 5-atom width to an 1-atom width in 1998 [13].

In 2001, Kizuka et al. observed the deformation process of Au NCs with atomistic resolution using HRTEM based on the *in situ* method combined with atomic force microscope (AFM) and scanning tunneling microscope (STM) [10]. They measured stress and strain quantitatively and started material mechanics research of metallic NCs. Using this method, the mechanical properties of Cu [14], Ir [15], Pd [16], and Pt [17] NCs were researched.

Further, as Kizuka et al. improved the *in situ* HRTEM method [18]; they were able to observe the structural dynamics of ASWs formation during the tensile deformation process and research electrical and mechanical properties. Using this method, it was revealed that there was a corresponding relationship between structural dynamics and physical properties [19]. Until that point in time they had been researched individually. They observed Au ASW at the final stage of the tensile deformation process of Au NCs. The observed Au ASW is up to 10 atoms in length with an average interatomic distance of 0.27 nm (Figure 3). In the tensile deformation process of ASWs, since tensile stress is concentrated on the contact region, interatomic distances of Au ASWs become longer, up to 0.30 nm. At the same time, conductance of Au ASWs was measured. Resultant conductance greatly decreases when the number of atoms used to construct the ASWs exceeds 4. Moreover, the force acting on the contact was measured. The tensile strength of this ASW was estimated to be from 8 to 17 GPa. This value is several times larger than that of Au NCs and much larger than bulk Au.

At elastic deformation regions undergoing stress–strain, the Young's modulus of Au ASWs was estimated to be between 47 and 116 GPa. This value is remarkably comparable with that of a single crystal Au.

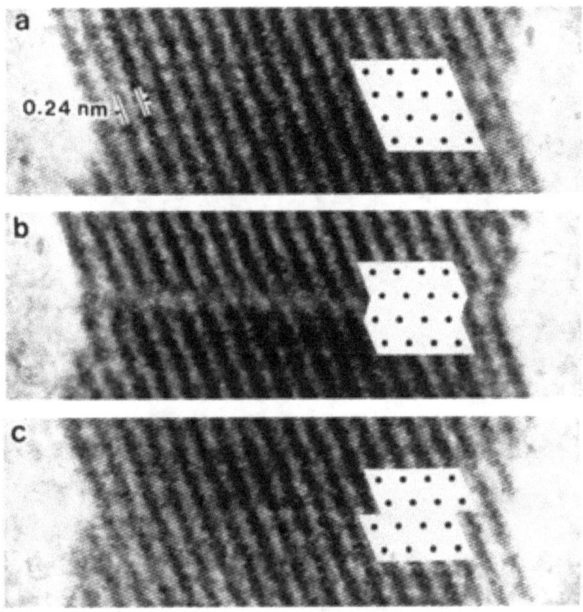


**Figure 2.** Time variation in the elementary steps of slip in the shear deformation of Au NCs [12].

The NCs and ASWs of other materials other than Au have been also observed. Au, Ag [20, 21], Cu [22], Pt [17, 23], Pd [23, 24], Ir [15], and Co [23] ASWs have been observed.

As described above, common problems in research in metallic ASWs and NCs existed, revealing corresponding relationships between structure and electrical properties. As research in ASWs has concentrated on Au, the structural dynamics of ASW formation is uncertain. For some of the metallic ASWs already researched, only the structures that appeared in the tensile deformation process of NCs were observed. Therefore, the stable structure and electrical conductivity of the NCs have not yet been revealed. In order to produce a general rule for the basic phenomenon that appears in metallic NCs and ASWs, it is necessary to examine structural dynamics, electrical conductivity, and mechanical properties, in order to clarify the corresponding relationship between the structure and properties directly. The method is limited to *in situ* HRTEM.

## 2. Combined HRTEM

For the *in situ* tip-sample contact experiment inside the HRTEM, we proposed inserting multiple specimen holders in the specimen chamber. The secondary holder is usually intro-

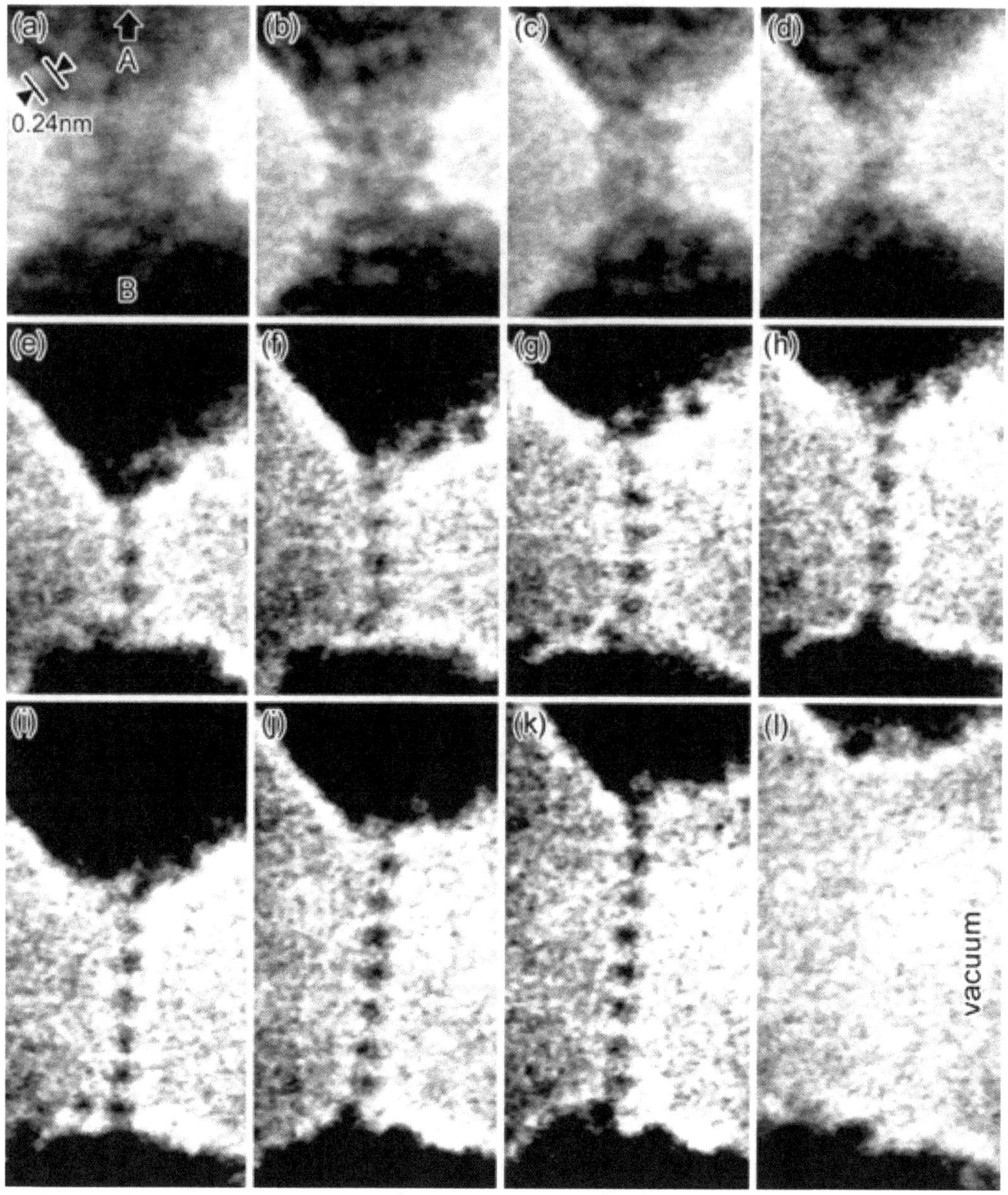


**Figure 3.** Time variation of the Au ASW formation process observed using the *in situ* HRTEM method [19].

duced from the direction perpendicular to the first holder (Figure 4). Since this space is often used for an EDX detector, it is necessary to replace it. In addition, we should consider that this is a space occupying in z direction; the relation between the gap distance of pole pieces and the height of sample holders (Figure 5). When you want to do mechanical tests, for example, you would introduce an AFM cantilever. In this configuration, it is rarely applied to a pole piece for high resolution use. This is because the samples must be precisely adjusted to contact at the standard focal position of the microscope, including the movement range in the $z$ direction of the goniometer, which is attached to each holder.

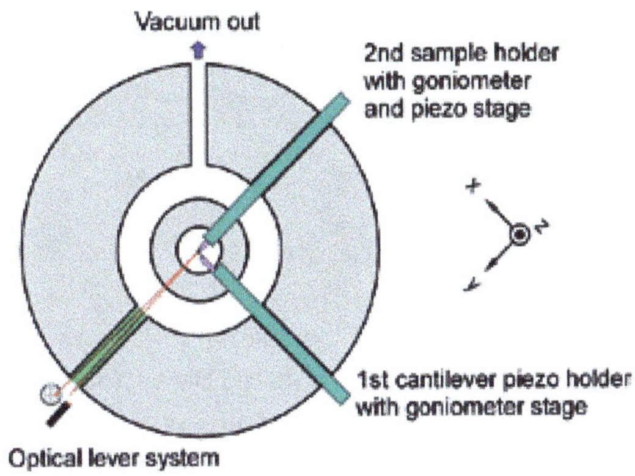


**Figure 4.** Illustration depicting the specimen chamber of the HRTEM combined with atomic force microscope (AFM).

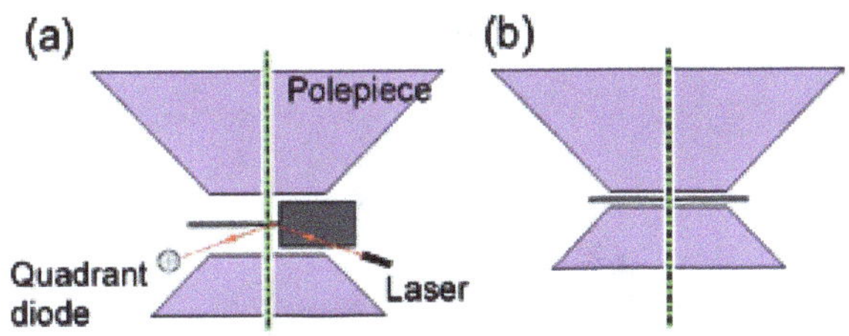


**Figure 5.** Illustrations depicting the configuration of samples and pole pieces for a combined HRTEM.

## 3. Sample preparation

For samples, we use a cut-out metal foil and the metal sputtered tip of the AFM cantilever. In the case of metallic foil, first we cut out the foil with a thickness of below 0.05 mm to a size of 1 mm by 10 mm and sharpen one side of the plate (Figure 6(a)). After that, we polish this side mechanically using emery papers and aluminum or diamond wrapping film. In rare cases, although a burr part of cut is sometimes thin enough for HRTEM observation, the yield is poor and the workability of nano-tip creation processes in a HRTEM is not good because the sample is affected by a stress concentration where there is a sudden change in thickness. This sample is then thinned further by an Ar ion polish of 3–10 kV (Figure 6(b)). It may be that an ionic cleaner is used just before introducing it into a vacuum.

When performing the mechanical tests simultaneously, it is suggested that you use a metallic sputtered Si tip on the cantilever for AFM for one of the samples (Figure 7). For high-resolution (HR) observations, in order to detect the small force needed to deform the fine structure, the

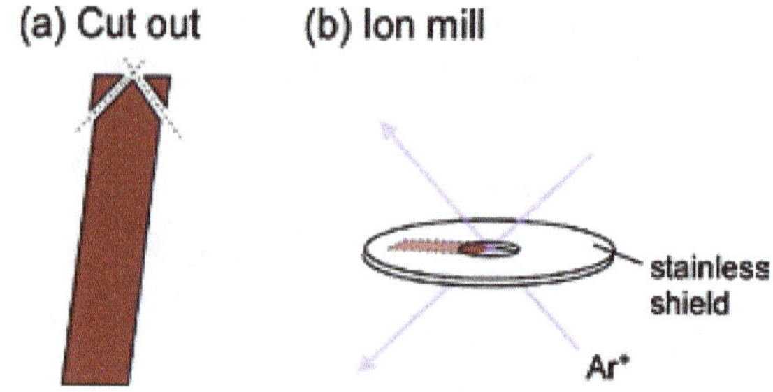


**Figure 6.** Illustrations showing the downsize process of metal foil.

spring constant of the cantilever is required to be small (5 N/m or less is preferred). Metal is sputtered on the cantilever and the tip under a reduced pressure Ar atmosphere. The thickness of metallic film is approximately 20 nm. The requirements in this process are (i) depositing metal atoms on the tip for use in contact with the counter sample and (ii) ensuring the cantilever surface is covered with a uniform continuous metallic film to offer sufficient conductivity. In order to achieve the latter requirement, it would be preferable to sputter in the same way on both sides of cantilever. Moreover, even if the required amount of sputtering is achieved, the spring constant of the cantilever is not changed.

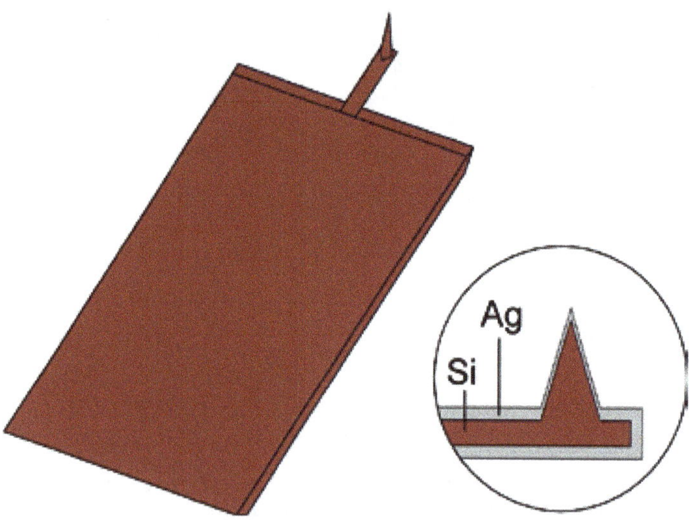


**Figure 7.** Illustration of cantilever preparation for the Ag contact experiment.

Both samples are attached to a specimen holder respectively, then inserted into the HRTEM specimen chambers. Because the sample is exposed to air for several minutes during this preparation process, a natural oxide or sulfide film is deposited on the sample surface. After the first contact with the sample each other, once we should make the contact larger than a width of several tens nanometers. This is because we pull a clean metal part to the surface from inside the samples.

# 4. Observation of Ag ASWs

Figure 8 shows HR images during the process of miniaturization of Ag NCs. The black areas at the top and bottom of each image are Ag, the center is the NC, and the surrounding area is a vacuum. On all of the NCs, a lattice interval of Ag (0.24 nm) has appeared – as indicated by (111) in the figure. That is, the contact region is a single crystal. Therefore, the direction of incidence of the electron beam is indicated on the figure as [0-11] and the upward direction on image is identified as [100] in the figure. The minimum cross-sectional width of the NC is (a) 4 atoms, (b) 3 atoms, and (c–e) 1 atom. Figure 8(f) shows the contact braking.

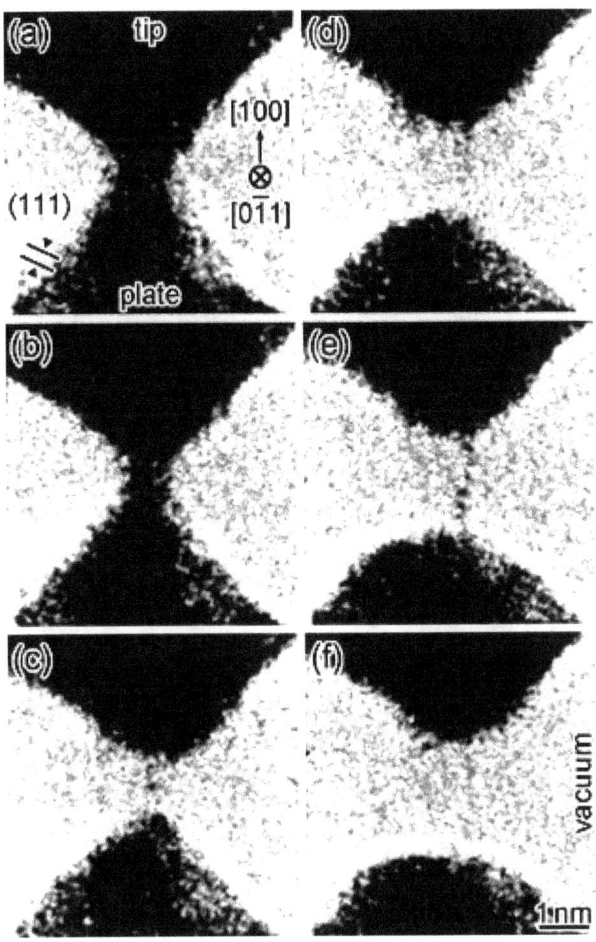


**Figure 8.** Time variation of Ag ASW formation process observed by *in situ* HRTEM.

Figure 9, it shows the calculated image and HR image for the Ag ASW. For the image calculation, we used a model in which both ends of the wire (composed of seven atoms) were connected to two pyramidal Ag crystals (Figure 9(a)). Seven of the Ag ASW atoms were along the [100] crystal pyramid and arranged at intervals of 0.289 nm, which is the nearest neighboring distance of the bulk Ag. Figure 9(b) is a calculated image for the model. In Figure 9(c), the image intensity along the center line of the atomic wires is shown. The centers of the models of the atoms are displaced from the centers of the black point in computational image by only 0–0.02 nm. Since the experimental spatial resolution of the current HRTEM observation is 0.1 nm, the calculated result agree with the experimentally observed image within that resolution. This correspondence is similar to that of Au atoms in a wire [19]. Figure 9(d) is an enlarged view of the Ag ASW shown in Figure 8(e). Figure 9(c and e) shows the image intensity along the center line of the ASWs. The darkest positions are indicated by the arrows. From the results of the image calculation, the atomic positions of the ASWs can be considered to correspond to the darkest position in the image intensity.

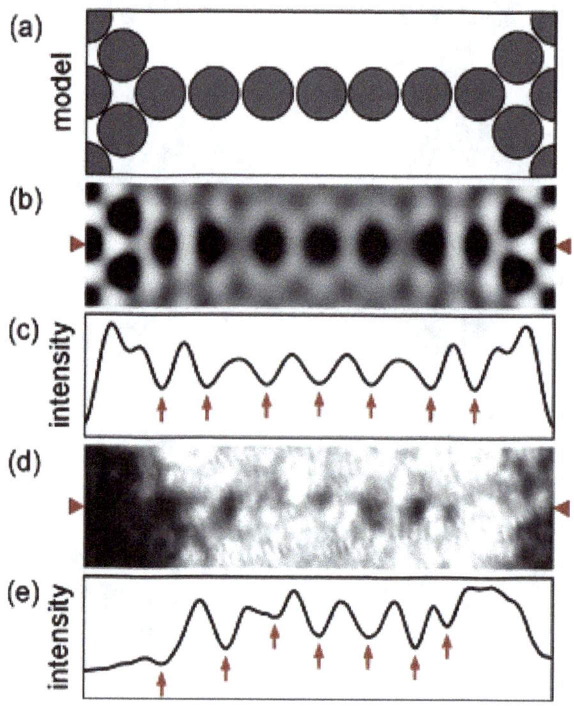


**Figure 9.** HR images of Ag ASWs. (a) Projected model along [110]; (b) calculated image for the structural model of (a); (c) line profile of (b); (d) enlarged experimental image of Figure 8(e); and (e) line profile of (d).

Figure 10 shows the time variation of force and conductance in the Ag ASW formation process shown in Figure 8. Times (a–f) correspond to the times that each image in Figure 8 was observed, respectively. As the NC narrowed, the conductance stepwise reduced (a–b). Similarly, the force acting on the NC also decreased stepwise. Thereafter, when the ASW forms (c–e), both conductance and force were reduced below 0.1 $G_0$ and 0.5 nN.

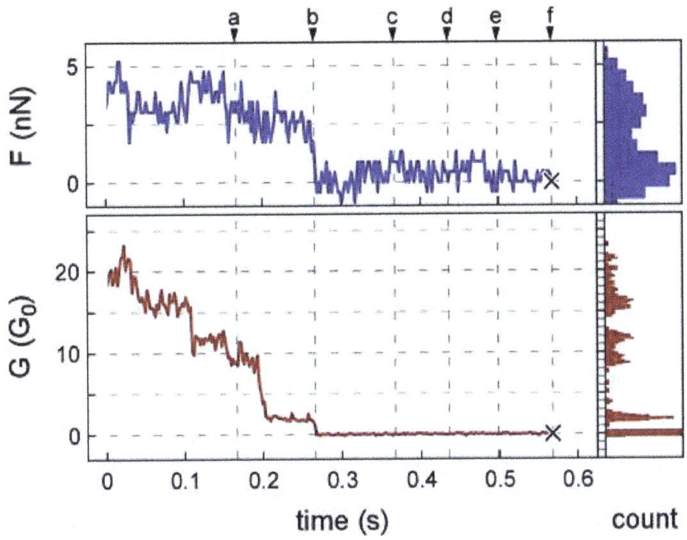


**Figure 10.** Time variation in the force and conductance measured in the deformation process of Figure 8.

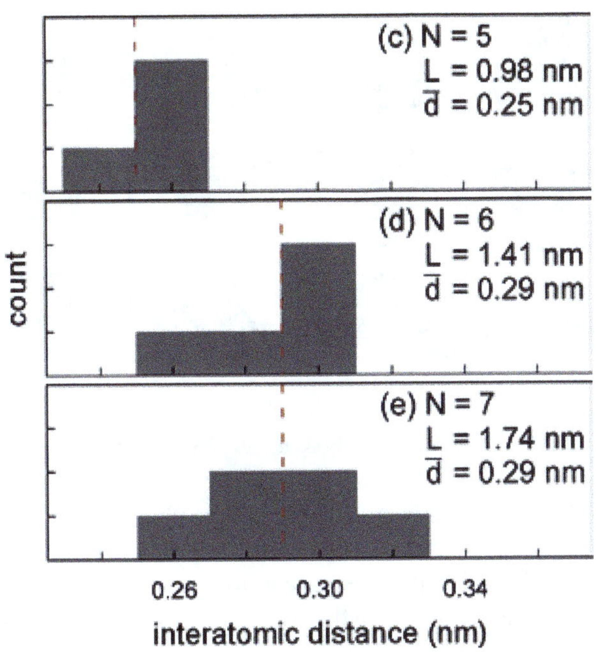


**Figure 11.** Distribution of interatomic distances in Ag ASWs.

Figure 11 shows the distribution of the interatomic distances between atoms in the wire, observed in Figure 8. When the number of atoms in the ASW was five, the interatomic distance was 0.25±0.01 nm. However, this became 0.29±0.05 nm for an ASW comprised of six or seven atoms. In other words, as the ASW becomes longer, the interatomic distance increases.

ASWs are generally considered to be formed by pulling out atoms one by one from the electrode. Bahn and Jacobsen calculated the process of pulling metal NCs (Figure 12) using the density functional method via molecular dynamics calculations and effective medium approximation [25]. In the state where tip atoms, from two pyramid-shaped electrodes, bond to one another, the single atom located at the minimum cross-section part bonds to not only three or four atoms of the electrode side but also atoms at the opposite electrode simultaneously. They estimated the bonding strength between these atoms, then went on to discuss whether this monatomic junction can draw atoms from the electrode without leading to a fracture. As a result, for Au and Pt, the strength of the binding of atoms in the wire part is energetically three times larger than that in the electrode part; it tends to form ASWs. On the other hand, for Cu, Ag, Ni, and Pd, it is difficult to form ASWs in this way.

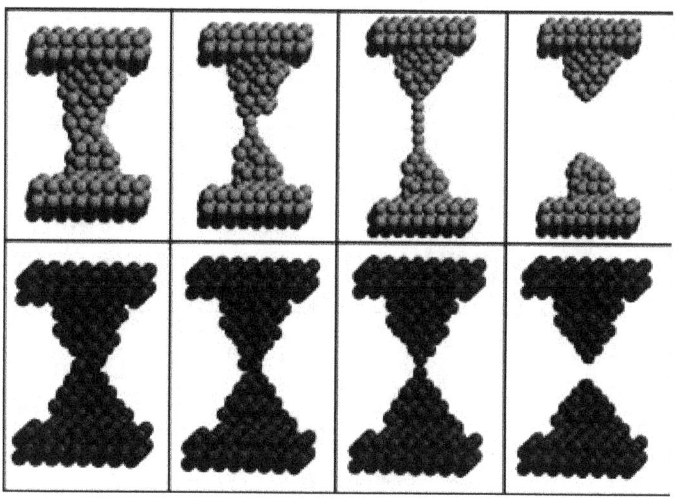

**Figure 12.** MD simulation of the tensile deformation process for Au (upper) and Cu (lower) NCs [25].

In the experiment shown in Figure 8, Ag ASWs were formed by tensile deformation of NCs. Since as soon as the NC is miniaturized to one atom width (Figure 8(b)), it forms four or more atomic length wires, the stretching process from a single atom contact could not be observed. In other words, the structural change from the Ag NC to the Ag ASW was faster than the frame rate available using TEM observations (~17 ms). Moreover, we have applied a voltage of 13 mV for the conductance measurement. The withdrawal of atoms from the electrode for atomic wires involved not only tensile force, but also atom transfer by electron wind force. As an example of this effect, when we apply a voltage of 100 mV to the NC, ASW with a length of ~0.88 nm formed just before breaking without applying tensile force [8]. Therefore, even for some of the elements for which ASWs were hardly expected to form, there remains the possibility of wire formation due to the atomic diffusion effect.

## 5. Conclusion

In this chapter, we introduced combined microscopy techniques which are based on HRTEM, AFM, and STM. This approach was developed to meet the demand for characterization of nanostructures which needs correspondence between the structure and the physical properties directly. In fact, this research has been promoted with respect to the junction between nano-scales or atomistic scales for a single metal. In the future, this method needs to be combined with chemical reaction engineering for single molecular device fabrication, such as catalysts, corrosion, and storage devices.

## Acknowledgements

This work was partly supported by Grant-in-Aid for JSPS Fellows (10J01479).

## Author details

Hideki Masuda*

Address all correspondence to: masuda.hideki@nims.go.jp

National Institute for Materials Science, Tsukuba, Japan

## References

[1] Cuevas, JC, and Scheer, E, *Molecular Electronics: An Introduction to Theory and Experiment*. World Scientific, Singapore, 2010.

[2] Joachim, C, Gimzewski, JK, and Aviram, A: Electronics using hybrid-molecular and mono-molecular devices. 2000; 408: 541. Nature.

[3] Di Ventra, M, Pantelides, ST, and Lang, ND: First-principles calculation of transport properties of a molecular device. 2000; 84: 979. Physical Review Letters.

[4] Kubatkin, S *et al.*: Single-electron transistor of a single organic molecule with access to several redox states. 2003; 425: 698. Nature.

[5] Tao, NJ: Electron transport in molecular junctions. 2006; 1: 173. Nature Nanotechnology.

[6] Danilov, A *et al.*: Electronic transport in single molecule junctions: Control of the molecule-electrode coupling through intramolecular tunneling barriers. 2007; 8: 1. Nano Letters.

[7] Kim, WY, *et al.*: Application of quantum chemistry to nanotechnology: Electron and spin transport in molecular devices. 2009; 38: 2319. Chemical Society Reviews.

[8] Masuda, H, and Kizuka, T: Distance control of electromigration-induced silver nanogaps. 2014; 14: 2436. Journal of Nanoscience and Nanotechnology.

[9] Kondo, Y, and Takayanagi, K: Gold nanobridge stabilized by surface structure. 1997; 79: 3455. Physical Review Letters.

[10] Kizuka, T, *et al.*: Simultaneous observation of millisecond dynamics in atomistic structure, force and conductance on the basis of transmission electron microscopy. 2001; 40: L170. Japanese Journal of Applied Physics.

[11] Kizuka, T, and Tanaka, N: Dynamic high-resolution electron microscopy of diffusion bonding between zinc oxide nanocrystallites at ambient temperature. 1994; 69: 135. Philosophical Magazine Letters.

[12] Kizuka, T: Atomistic visualization of deformation in gold. 1998; 57: 11158. Physical Review B.

[13] Ohnishi, H, Kondo, Y, and Takayanagi, K: Quantized conductance through individual rows of suspended gold atoms. 1998; 395: 780. Nature.

[14] Fujisawa, S, Kikkawa, T, and Kizuka, T: Direct observation of electromigration and induced stress in Cu nanowire. 2003; 42: L1433. Japanese Journal of Applied Physics.

[15] Ryu, M, and Kizuka, T: Structure, conductance and strength of iridium wires of single atom width. 2006; 45: 8952. Japanese Journal of Applied Physics.

[16] Matsuda, T, and Kizuka, T: Structure of nanometer-sized palladium contacts and their mechanical and electrical properties. 2007; 46: 4370. Japanese Journal of Applied Physics.

[17] Kizuka, T, and Monna, K: Atomic configuration, conductance, and tensile force of platinum wires of single-atom width. 2009; 80: 205406. Physical Review B.

[18] Kizuka, T, Umehara, S, and Fujisawa, S: Metal-insulator transition in stable one-dimensional arrangements of single gold atoms. 2001; 40: L71. Japanese Journal of Applied Physics.

[19] Kizuka, T: Atomic configuration and mechanical and electrical properties of stable gold wires of single-atom width. 2008; 77: 155401. Physical Review B.

[20] Rodrigues, V *et al.*: Quantum conductance in silver nanowires: Correlation between atomic structure and transport properties. 2002; 65: 153402. Physical Review B.

[21] Lagos, MJ, *et al.*: Observation of the smallest metal nanotube with a square cross-section. 2009; 4: 149. Nature Nanotechnology.

[22] González, JC, *et al.*: Indication of unusual pentagonal structures in atomic-size Cu nanowires. 2004; 93: 126103. Physical Review Letters.

[23] Rodrigues, V, *et al.*: Evidence for spontaneous spin-polarized transport in magnetic nanowires. 2003; 91: 096801. Physical Review Letters.

[24] Matsuda, T, and Kizuka, T: Palladium wires of single atom width as mechanically controlled switching devices. 2006; 45: L1337. Japanese Journal of Applied Physics.

[25] Bahn, SR, and Jacobsen, KW: Chain formation of metal atoms. 2001; 87: 266101. Physical Review Letters.

7

# Spatiotemporal Calibration of Electron Microscopes

Abed Malti

**Abstract**

This chapter presents a method of calibration for scanning electron microscopes (SEMs). The described calibration method is a twofold step. The first step evaluates the dynamical drift parameters. The second step estimates the parameters of the geometric projection and the static distortion. Both steps process the calibration parameters across the range of magnification scales, thanks to a representation with partial differential equations. The chapter is provided with an example of calibration of the JEOL JSM 820 and an example of application in metrology. The presented method is not the unique way of calibrating an SEM and can be a good start to inspire other methods of calibration.

**Keywords**: Calibration of microscopes, Drift calibration, Distortion calibration, Projection calibration

## 1. Introduction

SEM calibration is required in many applications: 3D reconstruction of microscale and nanoscale specimens [1–3], deformation measurement [4], nanomaterial tracking [3, 5, 6], positioning and handling [7], mobile robot positioning [8], etc. The image acquisition process of a standard SEM has usually three stages: *(i)* The conditioning of the electron beam from the electron gun to the area of the observed specimen. In this stage, the electron beam goes through condensation process by magnetic lenses and a final deflection coils that move the beam in a raster fashion over a rectangular area of the specimen. *(ii)* The interaction between the deflected beam and the observed object, usually called specimen. When the deflected beam hits the specimen, it results in an emission of energy that can be amplified and detected by specific electronic detectors. *(iii)* The conversion of the emitted energy to pixel intensity. In this process, each position of the deflected beam that hits the specimen corresponds to a pixel in the image. An appropriate transform between the detected energy and the gray-scale values is then applied to obtain the corresponding pixel intensity. In the case of a distortion-free imaging process, the obtained image would be the result of the projection of the 3D rastered area onto the image plane. According to an increasing magnification scale, this projection goes from the perspective for small magnification to the weak perspective for average magnification and ends up with orthographic projection with high magnification. However in reality, stages *(i)* and *(ii)* bring some distortion artifacts to these projection transforms. Indeed, the hysteresis effect of the magnetic coils distorts the deflected beam during the observation process. These distortions cause some drifts of the pixels from their original projected positions. For instance, these drifts can exceed tenth of pixels (hundreds of microns) in an hour at 10 k× magnification with a JEOL JSM 820. These drifts can be modeled as a combination of static (spatial) and dynamic (spatiotemporal) mappings which will be composed with the projection mapping to obtain the acquired image.

State-of-the-art studies have established that dynamical drift is time dependent and magnification dependent but space independent [9]. Among estimation methods, the drift was estimated in full field based on digital image correlation (DIC) in [9–12]. In [13], it was estimated in the frequency domain using FFT. The trajectory flow of the drift over time has been modeled using of B-spline curve fitting [9]. A second-order dynamical system embedded in a Kalman filter was tested and validated to model a thermal drift calibration in scanning probe microscopy [14]. The spatial calibration which involves both static distortion calibration and projection calibration was modeled similarly as with classic optical imaging systems [15, 16]. Some state-of-the-art works assumed a non-radial behavior of the static distortion and address this problem using B-splines to fit the spatial evolution of distortion and then warp it to a 3D-to-2D projection. Other works considered a perspective projection in the case of low magnifications (up to 5 k×) and an orthographic projection in the case of high magnifications (more than 5 k×) [1, 9, 17, 18].

In this chapter, we go through an empirical calibration method to fit a system of partial differential equations (PDEs). The partial derivatives of the calibration parameters are estimated with respect to time and magnification. These parameters include the amount of drift in pixel, the amount of static distortion, the focal length, and the principal point. This modeling provides a systematic and flexible solution to this calibration problem. It allows us to smoothly update calibration parameters across the variation of magnification scales and to continuously compensate pixel drifts during experiments.

## 1.1. Notations

Two-dimensional points in homogeneous coordinates are denoted by symbols in typewriter font (e.g., $\mathtt{u} = (u_x, u_y, 1)^T$). Three-dimensional points are indicated by plain letters (e.g., $C = (C_x, C_y, C_z)^T$). Matrices are denoted by uppercase sans serif font (e.g., $\mathsf{A}$). This notation is also adopted for $n$-dimensional vectors. However, vectors providing a direction in 3D are represented using plain lowercase topped by an arrow (e.g., $\vec{l}$). For convenience, and given two $3 \times 1$ vectors $\vec{l}$ and $\vec{m}$, the dot product is indicated either using $< .,. >$ or using regular matrix/vector multiplication (e.g., $<\vec{l}, \vec{m}> = \vec{l}^{\mathsf{T}} \vec{m}$), and the cross product is carried either using the symbol $\times$ or using the skew symmetric matrix (e.g., $\vec{l} \times \vec{m} = [\vec{l}]_{\times} \vec{m}$). $||.||_2$ denotes the vector 2-norm in any real vector space $\mathbb{R}^n$ of finite dimension $n$. The symbols $\mu m$ and $nm$ designate, respectively, micrometer and nanometer unit distances. The abbreviation "w.r.t." stands for "with respect to."

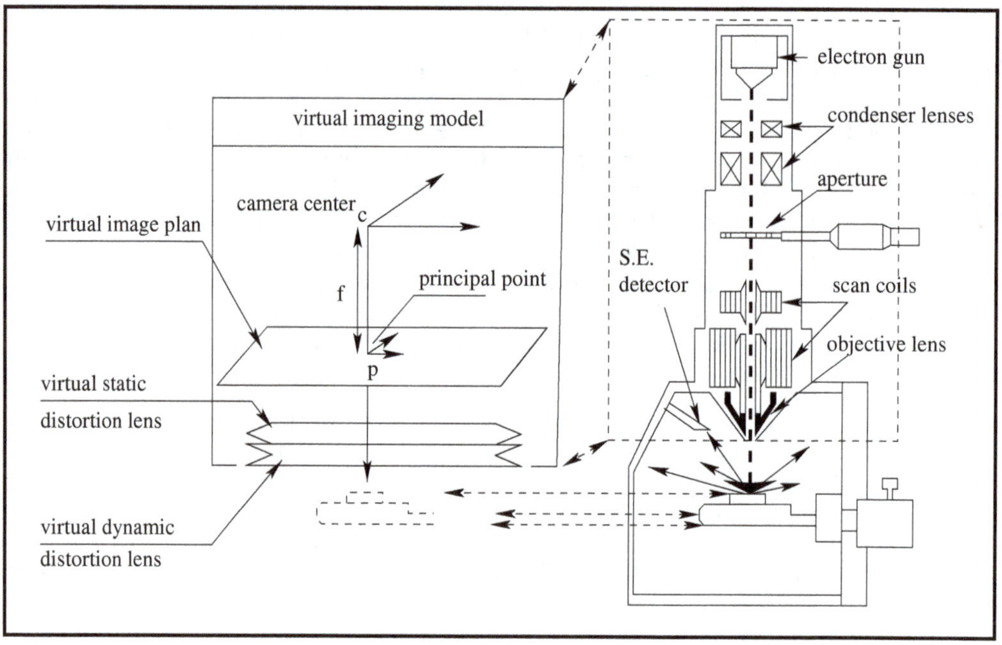


**Figure 1.** Scanning electron microscope with the corresponding imaging model.

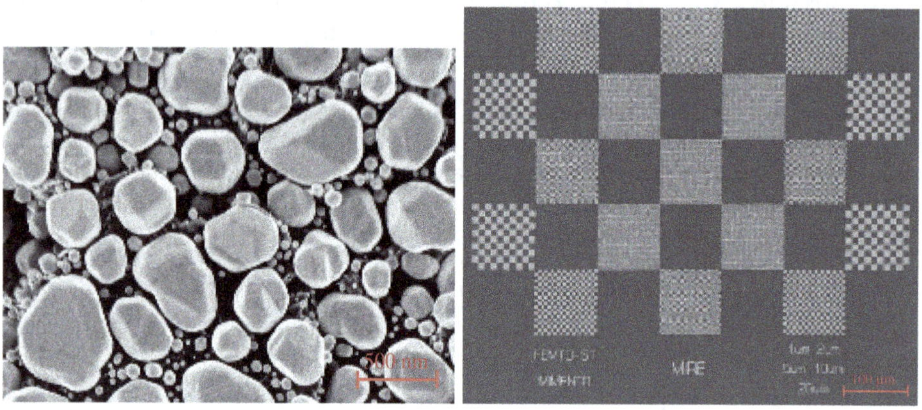

**Figure 2.** Left figure: Gold-on-carbon specimen. Right figure: Multi-scale calibration grid designed at FEMTO-ST lab.

## 2. Formulation of the imaging model

The imaging model of a SEM is the composition of three mappings [19, 20] which are depicted in Figure 1:

1. A pixel-drift mapping from drifted to non-drifted image. It is both magnification and time dependent. To experimentally quantify this drift, we acquire images of a usual specimen (see Figure 2 left) at different times and for different magnifications. Correlation in the

frequency domain of successive pairs of images is used to estimate the drift [13]. The dynamic of the drift with respect to time $t$ and magnification $g$ is modeled by two partial differential equations (PDEs) whose coefficients are estimated by means of the principal differential analysis (PDA) approach [21].

2. A spatial distortion mapping which is magnification dependent. It may concern both radial and tangential distortions [22]. In this chapter, we consider only radial distortions.
3. A 3D-to-2D projection mapping. The type of projection to be applied depends on the magnification. For low magnifications, it can be considered as central perspective projection. But for magnifications above 1000×, it should be interpreted as parallel projective [9]. A specific calibration specimen is used [20]; see Figure 2 right. It contains squares of various sizes, enabling the calibration over a wide range of magnification. Images of this specimen are acquired for various magnifications and poses. An image-centered radial and tangential model that expresses the distorted points w.r.t. the undistorted points is developed. PDEs with respect to magnification of the evolution of the static distortion parameters are established. The estimation of the projection matrix where the magnification factor is embedded is proposed. A bundle-adjustment optimization [23] of the reprojection error between 3D points and their corresponding image pixels allows us to refine the estimated parameters.

Mathematically speaking, the above description can be formalized as:

$$\hat{\mathbf{q}} = \mathrm{T}^d_{t,g}(\mathrm{T}^s_g(\Pi_g\,\mathbf{Q})), \tag{1}$$

where $\mathbf{Q}$ is a 3D point of the observed specimen. $\hat{\mathbf{q}}$ is the corresponding acquired image pixel. $\mathrm{T}^d_{t,g} : \mathbb{R}^2 \to \mathbb{R}^2$ and $\mathrm{T}^s_g : \mathbb{R}^2 \to \mathbb{R}^2$ are two-dimensional mappings (images x-axis and y-axis) which respectively represent the dynamical drift and the static distortion. $\Pi_g$ is the 3D-to-2D projective mapping. To retrieve a 3D ray incident from a 3D point of scanned scene, the corresponding pixel is first corrected from the drift effect; then it is statically undistorted and finally back-projected. In the following, we show how to estimate these mappings. Also, we provide examples of application on a JEOL JSM 820 SEM in secondary electron (SE) imaging mode for time ranging from 0 to 30 min and magnification ranging from 100× to 10 k×.

## 3. Magnification-smooth dynamical drift modeling and calibration

### 3.1. Drift modeling

The dynamical drift of an image frame at a time $t$ can be represented as a trajectory flow $(x(t), y(t))^T$ of pixels with respect to time:

$$\hat{I}_t(x_0 + x(x_0, y_0, t), y_0 + y(x_0, y_0, t)) = I_t(x_0, y_0), \tag{2}$$

where $\hat{I}_t$ is the intensity of the acquired image which is submitted to drift. $I_t$ is the ideal image without drift effects. $(x_0, y_0)^T$ is the non-drifted pixel position and $(x(t), y(t))^T$ is the amount of drift. As described previously, SEM images are produced pixel by pixel with a rastering process. An exact identification of the drift should be to follow the amount of drift flow in a pixel-wise manner $(x(x_0,t), y(y_0,t))^T$. In [10, 19], a full-field correction of the image drift is proposed. For SEMs where the acquisition time $t_F$ of one frame is faster than the pixel-wise dynamic of the drift, it can be assumed as being a global pixel drift between frames especially at low magnification (less than 5000×). This global drift displacement can be easily assessed using image cross-correlation computation between frames. In this case, the first acquired image is considered as the ideal one at $t = 0$. Taking into account this assumption, equation (2) can be reconsidered as

$$\hat{I}_0(x_0, y_0) = I_0(x_0, y_0), \tag{3}$$

$$\hat{I}_t(x_0 + \delta x(t), y_0 + \delta y(t)) = \hat{I}_{t+1}(x_0, y_0). \tag{4}$$

From now on, we drop the hat symbol on the notation of acquired image unless it is worth to mention it. If two successive image frames $I_{t-1}$ and $I_t$ contain the same view at different pixel positions, the cross-correlation integral has a large value at the vector $(\delta x(t), \delta y(t))^T$ which corresponds to the drift of the features.

$$(\delta x(t), \delta y(t)) = \underset{x,y}{\arg\max} \left\{ C_{I_{t-1},I_t}(x,y) \right\}, \tag{5}$$

with

$$C_{I_{t-1},I_t}(x,y) = \sum_i \sum_j I_{t-1}(i,j)\, I_t(i+x, j+y), \tag{6}$$

where $i+x, j+y, x$, and $y$ are pixel coordinates running over the image domain. According to [10], at magnification higher than 5000×, the dynamical drift varies even during the scanning process. It showed that the amount of drift variation within an image ranged from 0.3 to 0.9 pixels at 10 k×. In this chapter, the use of an EKF allows us to face this amount of single pixel drift by assuming a Gaussian noise of 1 pixel in the state model [19].

For a given magnification range from $g_0$ to $g_f$ with a step $G$ and a time interval from $t_0$ to $t_f$ with an image acquisition at each sample time $T$, the image drifts are estimated as follows:

**for** $g = g_0$ *to* $g_f$ *with step* $G$ **do**
  **for** $t = t_0$ *to* $t_f$ *with step* $T$ **do**
    acquire images of the specimen pattern;
    estimate drift elements $\delta x$ and $\delta y$ between frame $t_0$ and current frame;
  **end**
**end**

**Algorithm 1:** Processing of Drift Data.

At the end of this step, two data matrices $\Delta_x(t,g)$ and $\Delta_y(t,g)$ of $\frac{t_f - t_0}{T}$ rows and $\frac{g_f - g_0}{G}$ columns are obtained. The next step is to use this data drift and PDA to evaluate the dynamics of the drift trajectory functions $\Delta_x(t,g)$ and $\Delta_y(t,g)$.

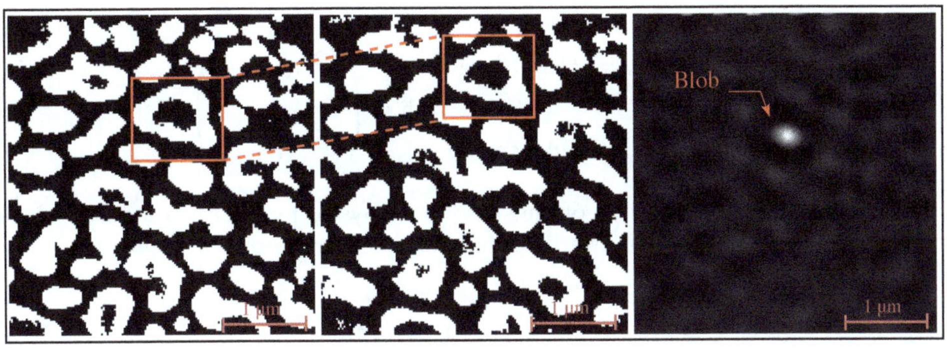


**Figure 3.** An example of drift behavior. The image size is of 512 × 512 pixels. Left figure: The initial figure. Middle figure: After 15 min, the enclosed feature and all the image features have moved. Right figure: The results of the intercorrelation between the two images. The distance between the blob and the center of the image represents the amount of drift.

### 3.2. Estimating PDEs of the dynamical drift

When describing PDA for linear PDE models, Ramsay and Silverman [24] view the system dynamics as a linear differential operator (LDO) acting upon the process variables. For example, let $\Delta_x(t, g_i)$ be the observed data drift which varies w.r.t. the time parameter $t$ at sampled time $T$ from $t = t_0$ until $t = t_f$. Let $D^m x_i$ be the $m^{\text{th}}$ derivative of the function $\Delta_x(t, g)$ w.r.t. $t$. The function $\Delta_x(t, g)$ is assumed to be square integrable. In this chapter, we consider the identification of a second-order LDO which determines the first (speed)- and second (acceleration)-order parameters of the dynamic of the drift [19]:

$$L = w_0^x + w_1^x D + D^2 \tag{7}$$

that comes as close as possible to satisfy the homogeneous linear differential equation:

$$L \, . \, \Delta_x = 0 \tag{8}$$

In other words, if we estimate the first and second derivatives $D\Delta_x$ and $D^2\Delta_x$ of $\Delta_x(t, g_i)$ w.r.t. time using finite differences, we wish the operator $L$ to annihilate the drift function $\Delta_x(t, g)$ as nearly as possible. Thus, we seek a linear differential equation model so that our data satisfies:

$$D^2\Delta_x = -w_0^x(g_i)\,\Delta_x - w_1^x(g_i) D\,\Delta_x \tag{9}$$

to the best possible degree of approximation. To carry out PDA, we adopt a least squares approach to the fitting of the differential equation model. The fitting criterion is to minimize, over $\left(w_0^x(g_i) \;\, w_1^x(g_i)\right)$, the sum of squared norms:

$$J = \left\| \mathsf{A} \begin{pmatrix} w_0^x(g_i) \\ w_1^x(g_i) \end{pmatrix} + \mathsf{b} \right\|^2 \tag{10}$$

where

$$\mathsf{A} = \begin{pmatrix} \Delta_x(t_0, g_i) & D\Delta_x(t_0, g_i) \\ \vdots & \vdots \\ \Delta_x(t_f, g_i) & D\Delta_x(t_f, g_i) \end{pmatrix}, \ \mathsf{b} = \begin{pmatrix} D^2\Delta_x(t_0, g_i) \\ \vdots \\ D^2\Delta_x(t_f, g_i) \end{pmatrix} \tag{11}$$

A is a matrix of $\frac{t_f - t_0}{T}$ rows and 2 columns and is always of rank 2. b is a vector of $\frac{t_f - t_0}{T}$ elements. The solution of such an overdetermined least square problem is given as

$$\begin{pmatrix} w_0^x(g_i) \\ w_1^x(g_i) \end{pmatrix} = (\mathsf{A}^\top \mathsf{A})^{-1}\mathsf{A}\mathsf{b} \tag{12}$$

Solving equation (12) for each $g_i = g_0, g_0 + G, \ldots, g_f$ gives rise to the following weight matrix:

$$\mathsf{W} = \begin{pmatrix} w_0^x(g_0) & w_0^x(g_0 + G) & \cdots & w_0^x(g_f) \\ w_1^x(g_0) & w_1^x(g_0 + G) & \cdots & w_1^x(g_f) \end{pmatrix} \tag{13}$$

W is a matrix of 2 rows and $\frac{g_f - g_0}{G}$ columns. The first row represents a discrete sampling of the function $w_0^x(g)$ and the second row a discrete sampling of $w_1^x(g)$. Once again, computing the first- and second-order derivatives of the first and the second row using finite differences on the variation of the magnification gives rise to two differential equations, driving $w_0^x$ and $w_1^x$ variation w.r.t. magnification:

$$D_g^2 w_i = -\alpha_{0i} w_i - \alpha_{1i} D_g w_i, \ \ i = 0, 1 \tag{14}$$

The two real valued vectors $(\alpha_{0i} \ \alpha_{1i})$, $i = 0, 1$ are estimated as in equation (12) after constructing the corresponding matrix A and vector b by using matrix W, the first and second finite differences of each row w.r.t. $g$. Equation (14) becomes then a PDE of second order that can be easily solved.

Henceforth, the differential equation related to drift function $\Delta_x$ can be expressed w.r.t. smooth magnification-dependent weight functions:

$$D^2\Delta_x = -w_0^x(g)\,\Delta_x - w_1^x(g) D\,\Delta_x, \ \ g \in [g_0, g_f] \tag{15}$$

Finally, equation (15) can be solved as a second-order PDE giving rise to a smooth drift function $\Delta(t, g)$. The development conducted to represent $\Delta_x$ can be easily followed to represent $\Delta_y$:

$$D^2\Delta_y = -w_0^y(g)\,\Delta_y - w_1^y(g) D\,\Delta_y, \ \ g \in [g_0, g_f] \tag{16}$$

In order to take into account the noise in the data estimation and the finite differences computation, we choose to embed the differential equations (15) and (16) related to $\Delta_x$ and $\Delta_y$ in an EKF using a state modeling of the drift function.

### 3.3. Embedding PDEs of the drift in an EKF

If we assume that $x(g,t)$ and $y(g,t)$ are corrupted by a zero mean Gaussian noise $\omega_x(g,t)$ and $\omega_y(g,t)$ with covariance $\mathsf{Q}_x(g)$ and $\mathsf{Q}_y(g)$, then the stochastic state model can be written as

$$\begin{aligned} \mathsf{x}(g,t+T) &= \underbrace{\begin{pmatrix} 0 & 1 \\ -w_0^x(g) & -w_1^x(g) \end{pmatrix}}_{\mathsf{A}_x(g)} \mathsf{x}(g,t) + \omega_x(g,t) \\ x(g,t) &= \underbrace{(1\ 0)}_{\mathsf{C}} \mathsf{x}(g,t) + \gamma_x(g,t) \end{aligned} \tag{17}$$

$$\begin{aligned} \mathsf{y}(g,t+T) &= \underbrace{\begin{pmatrix} 0 & 1 \\ -w_0^y(g) & -w_1^y(g) \end{pmatrix}}_{\mathsf{A}_y(g)} \mathsf{y}(g,t) + \omega_y(g,t) \\ y(g,t) &= \underbrace{(1\ 0)}_{\mathsf{C}} \mathsf{y}(g,t) + \gamma_y(g,t) \end{aligned} \tag{18}$$

where $\mathsf{x}(g,t) = \left(x(g,t), v_x(g,t)\right)^\top$ and $\mathsf{y}(g,t) = \left(y(g,t), v_y(g,t)\right)^\top$ are the state vectors with $v_x$ and $v_y$ the speeds of the drift. The observed variable being the displacement flow $x(g,t)$ and $y(g,t)$, they are assumed to be corrupted by zero mean Gaussian noises $\gamma_x(g,t)$ and $\gamma_y(g,t)$ of covariances $\mathsf{R}_x(g)$ and $\mathsf{R}_y(g)$ . This space model representation allows us to write the EKF time update and prediction equations:

$$\begin{aligned} \hat{\mathsf{x}}(g,t+T|t) &= \mathsf{A}_x(g)\hat{\mathsf{x}}(g,t+T|t) && \textbf{(predicted state drift)}, \\ \mathsf{P}_x(g,t+T|t) &= \mathsf{A}_x(g)\mathsf{P}_x(g,t|t)\mathsf{A}_x^\top(g) + \mathsf{Q}_x(g) && \textbf{(predicted covariance drift)}, \end{aligned} \tag{19}$$

$$\begin{aligned} \mathsf{K}_x(g,t) &= \mathsf{P}_x(g,t|t-T)\mathsf{C}\left(\mathsf{C}\mathsf{P}_x(g,t|t-T)\mathsf{C}^\top + \mathsf{R}_x(g,t)\right)^{-1} && \textbf{(optimal Kalman gain)}, \\ \hat{\mathsf{x}}(g,t|t) &= \hat{\mathsf{x}}(g,t|t-T) + \mathsf{K}_x(t)\left(\tilde{x}(g,t) - \mathsf{C}\hat{\mathsf{x}}(g,t|t-T)\right) && \textbf{(updated state drift)}, \\ \mathsf{P}_x(g,t|t) &= (\mathsf{I} - \mathsf{K}_x(t)\mathsf{C})\mathsf{P}_x(g,t|t-T) && \textbf{(updated covariance drift)}, \end{aligned} \tag{20}$$

The same EKF equations can be stated for the drift flow $y(g,t)$.

### 3.4. The multi-scale drift calibration algorithm

In summary, the multi-scale drift flow is characterized in both magnification axis and time axis. Therefore, using the PDA approach, the PDEs w.r.t. time and the PDEs w.r.t. magnification are assessed. Assume a calibration through the range $[g_0, g_f]$ of magnifications. A calibration pattern with random shapes is positioned upon the stage inside the chamber.

The scene is static, and a set of images are taken at each sample time $T$. The different steps of the dynamical drift calibration can be summarized in Algorithm 2.

**Data**: $\Delta_x(t, g)$, $\Delta_y(t, g)$
**for** $g = g_0$ *to* $g_f$ *with step* $G$ **do**
  estimate $w_0^x(g_i)$, $w_1^x(g_i)$ with equation (10);
  estimate $w_0^y(g_i)$, $w_1^y(g_i)$ with equation (10);
**end**
estimate $\alpha_{00}$ and $\alpha_{10}$ as in equation (10);
estimate $\alpha_{01}$ and $\alpha_{11}$ as in equation (10);
embed equations (13) and (14) in an EKF as explained in Section 3.3;

**Algorithm 2:** Dynamical drift calibration

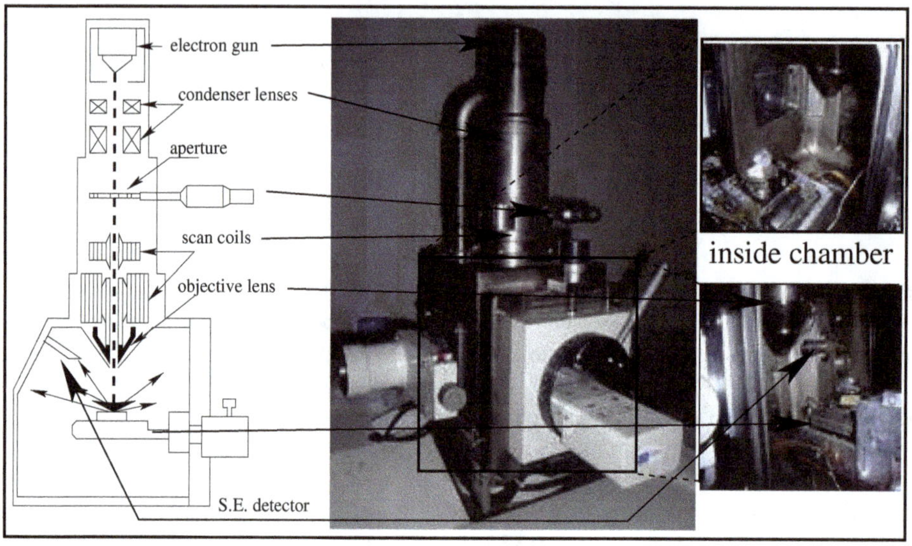


**Figure 4.** The *JSM 820* SEM manufactured by *JEOL*.

## 3.5. An example of dynamical drift calibration of the JEOL JSM 820

In this paragraph, we show how to use the method developed so far to calibrate the dynamical drift of the *JSM 820*, a SEM manufactured by *JEOL* (see Figure 4 for an illustration). The electron gun is equipped with a tungsten filament that can support from 0.3 kV up to 30 kV of acceleration voltage. The acquired images have a size of $512 \times 512$ pixels. The acceleration voltage is 15 kV, the scan rate is 15 frames per second, and the number of scans average is 8. The calibration is done for magnifications from $g_0 = 100\times$ up to $g_f = 30$ k$\times$.

### *3.5.1. Data drift estimation*

To assess the pixel displacement between two frames, a specimen of particles of gold deposed above a layer of carbon is used (Figure 2 left). The particles are randomly positioned and have random shapes of different sizes so that the maximum of cross-correlation images of such a sample can be calculated with less errors [10]. According to the cross-correlation theorem [25], the cross-correlation can be calculated using the Fourier transform. The widely used FFTW3 [26] library is applied for Fourier transform calculations. In order to improve the accuracy of the peak calculation, the cross-correlation is combined with the frequency filtering. A set of 55 images per magnification scale are taken every 30 s. The magnification scale is tuned from 100× until 10 k×. Some of these images are shown in Figure 5 with an illustration of the cross-correlation, resulting peaks between the first frame and the following frames. The pixel displacement vector of all images across time and magnification is depicted in Figure 6. It can be shown that at higher magnification, the drift is more important and can reach up to $(20, 90)^T$ pixel after 20 min of image acquisition. Now, the multi-scale calibration method of Section 3.4 can be applied. The first 30 images of each scale are used to construct the PDA model, and the 25 last images are used as validation data.

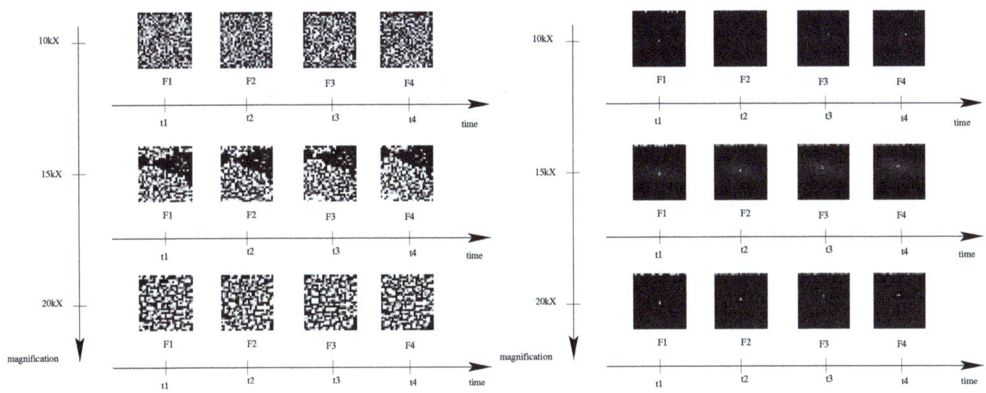

**Figure 5.** Left figure: A sample of images at four successive times along three magnification scales. Right figure: The bright point shows the maximum of the cross-correlation function between the first frame and the following frames. It can be seen that it moves after some acquisition time.

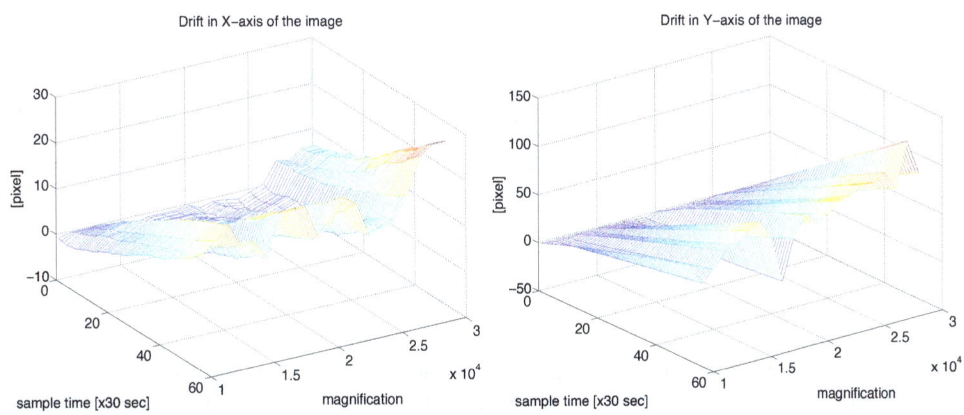


**Figure 6.** The pixel drift across time and magnification (left, image x-axis; right, image y-axis).

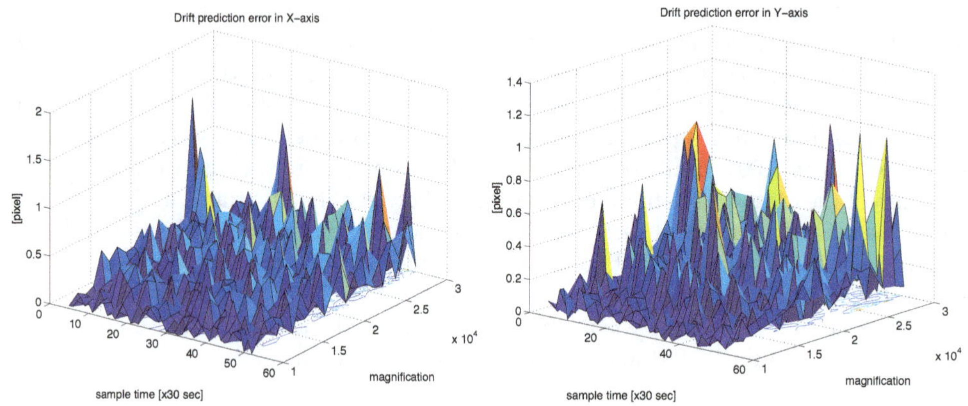


**Figure 7.** The Kalman filter prediction absolute error across time and magnification (left, image x-axis; right, image y-axis). Less than 0.9 pixel error in the x-axis and less than 1 pixel error in the y-axis. 0.28 pixel of global RMS error in the x-axis and 0.23 pixel of RMS error in the y-axis can be observed.

### *3.5.2. PDE estimation of the dynamical drift and EKF embedding*

Using PDA, the estimated differential equations associated to $w_0^x$ and $w_1^x$ are

$$\begin{aligned} D_g^2 w_0^x &= -169.31\, w_0^x + 0.55 \cdot 10^{-2} D_g w_0^x \\ w_0^x(g_0) &= -1.27 \cdot 10^{-7}, \quad w_0^x(g_f) = -1.91 \cdot 10^{-5}, \end{aligned} \tag{21}$$

$$\begin{aligned} D_g^2 w_1^x &= 9.61\, 10^4\, w_1^x - 0.17 D_g w_1^x \\ w_1^x(g_0) &= 3.01 \cdot 10^{-4}, \quad w_1^x(g_f) = 0.04, \end{aligned} \tag{22}$$

Similarly, the weights $w_0^y(g)$ and $w_1^y(g)$ describing the ODE of $\Delta_y(t,g)$ are the solution of the two following differential equations:

$$\begin{aligned} D_g^2 w_0^y &= -169.31\, w_0^y + 0.0055 D_g w_0^y \\ w_0^y(g_0) &= -1.20 \cdot 10^{-6}, \quad w_0^y(g_f) = -1.44 \cdot 10^{-6}, \end{aligned} \tag{23}$$

$$\begin{aligned} D_g^2 w_1^y &= 9.61\, 10^4\, w_1^y - 0.17 D_g w_1^y \\ w_1^y(g_0) &= 0.16 \cdot 10^{-2}, \quad w_1^y(g_f) = 0.14 \cdot 10^{-2}, \end{aligned} \tag{24}$$

The PDA study of the collected data shows that the time dependence is more likely to be a second-order differential equation and so is for the magnification dependence. This order may not be accurate because of the noise in the data. Thus, to take into account this noise, the two dynamical models of equations (15) and (16) are embedded in an EKF. The Gaussian noise associated to the state model and observation of $\Delta_x$ and $\Delta_y$ has a zero mean and 0.25 pixels of standard deviation. The plots of the prediction error are shown in Figure 7. Less than 0.9 pixel error in the x-axis and less than 1 pixel error in the y-axis can be observed. Also, 0.28 pixel of global RMS error in the x-axis and 0.23 pixel of RMS error in the y-axis can be observed.

## 4. Multi-scale static calibration

The multi-scale static calibration concerns the estimation of the static distortion parameters as well as the projection parameters. They are independent of time, but they are magnification dependent.

### 4.1. Static distortion calibration

In contrast with the dynamical drift, at low magnification, the static distortion is much more significant than at high magnification. This is due to the fact that at high magnification, the scanned area is much smaller than at low ones.

#### *4.1.1. The static distortion model*

The most commonly used distortion model represents this physical phenomenon as a decentered distortion which has both a radial and tangential component [27]:

$$\mathbf{q}^d - \mathsf{e} = \underbrace{\frac{2}{1+\sqrt{1-4\xi_r r_u^2}}(\mathbf{q}^u - \mathsf{e})}_{\text{radial}} + \underbrace{\xi_t(r_u^2 + 2(x^d - x^u)^2)}_{\text{tangential}}, \tag{25}$$

where $\xi_r$, $\xi_t$ are, respectively, the radial and tangential distortion parameters. $\mathsf{e}$ is the center of distortion and
$\lambda = 2/1 + \sqrt{1 - 4\xi_r r_u^2}$ is called the factor of distortion.

#### *4.1.2. Estimation of the center of distortion*

The estimation of the center of distortion requires the use of a geometrically structured calibration pattern as the one shown in Figure 8. It consists on a planar grid of vertices $\{\mathbf{q}_i^c\}_{i\in\mathbb{N}}$. The positions of the points $\mathbf{q}_i^c$ are assumed to be known in a Euclidean coordinate frame attached to the grid. If $\mathbf{q}_i^d$ are the corresponding points in the distorted image, then each pair of points $(\mathbf{q}_i^c, \mathbf{q}_i^d)$ is linked by the epipolar relation proposed in [28] and stated as

$$\mathbf{q}_i^d \underbrace{[\mathsf{e}]_\times \mathsf{H}}_{\mathsf{F}} \mathbf{q}_i^c = 0. \tag{26}$$

$[\mathsf{e}]_\times$ is the skew-symmetric $3 \times 3$ matrix representing the cross product. $\mathsf{H}$ is the homography between the planar grid and the image plane. The matrix $\mathsf{F}$ may be called the fundamental matrix for radial distortion. It may be estimated using state-of-the-art methods [15], and the center of radial distortion can be estimated as the left epipole. In the case of no presence of radial distortion, the estimation of the fundamental matrix is unstable, and the value of

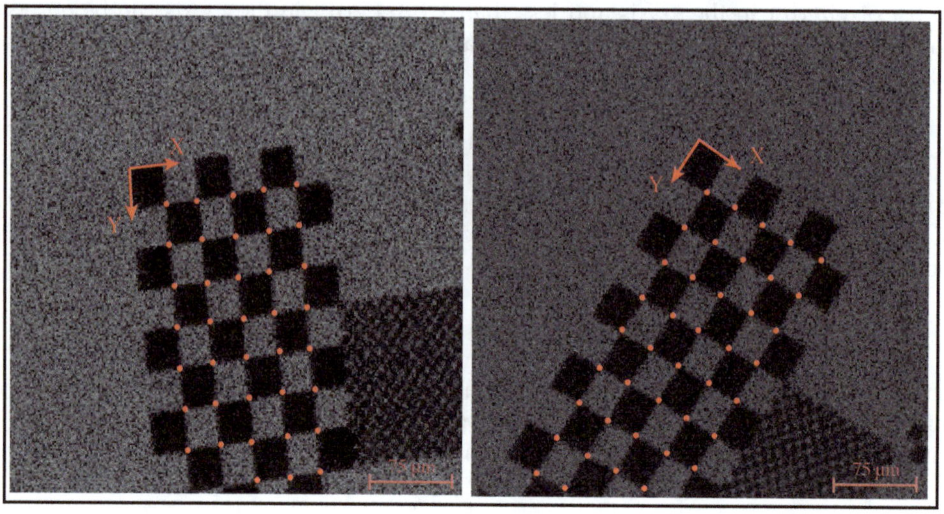


**Figure 8.** An example of static distortion at 400× of magnification. The image size is of $512 \times 512$ pixels, and the size of the squares is of 25 μm per side. The estimated center of distortion is at $\sim (290, 300)^T$ pixels.

e is meaningless. This situation can be detected during the estimation of the fundamental matrix.

The estimation of the distortion parameters $\xi_r$ and $\xi_t$ is processed iteratively. It is first assumed that $\xi_t = 0$ and estimate $e$. Then $\xi_t$ and $\xi_r$ are initialized to zero and estimated using bundle adjustment methods [23].

## 4.2. Projection model

The 3D-to-2D projection mapping can vary from perspective to orthographic. State-of-the-art works [18, 29] use either a perspective model at low magnification or an orthographic projection for high magnification with projection switch at the magnification of transition which is experimentally determined (usually 5 k×). In this chapter, a magnification-dependent projection model which smoothly switches from a perspective projection to an orthographic projection is detailed [20]. A perspective camera model can be written as

$$\mathsf{P}_0 = \mathsf{KR}[\mathsf{I} \ -\mathbf{C}] = K \begin{pmatrix} \vec{r_1}^T & -\vec{r_1}^T\mathbf{C} \\ \vec{r_2}^T & -\vec{r_2}^T\mathbf{C} \\ \vec{r_3}^T & -\vec{r_3}^T\mathbf{C} \end{pmatrix}, \tag{27}$$

where $\mathbf{C} \in \mathbb{R}^3$ is the position of the projection center. $\mathsf{R} \in \mathbb{SO}(3)$ is the orientation of the projection frame. $\vec{r_i}$ is the $i$-th row of R, and K is the matrix of intrinsics of the form

$$\mathsf{K} = \begin{pmatrix} af & s & p_x \\ 0 & a^{-1}f & p_y \\ 0 & 0 & 1 \end{pmatrix}. \tag{28}$$

$f, a,$ and $s$ are, respectively, the focal length, the ratio factor, and the skew parameter. $(p_x, p_y)^T$ are the coordinates of the principal point $e_0$. The principal ray of the imaging system is in the direction of the vector $\vec{r_3}$, and the value $d_0 = -\vec{r_3}^T\mathbf{C}$ is the distance of the planar grid origin from the camera center in the direction of the principal ray.

Multiplying the magnification by a scale factor $g$ acts similarly as moving the camera center backward along the principal ray. The center of the camera is then moved to $\mathbf{C} - g\,\vec{r_3}$. Replacing $\mathbf{C}$ by $\mathbf{C} - g\,\vec{r_3}$ in equation (27) gives the projection matrix at magnification $g$:

$$P_g = \mathsf{K}_g \begin{pmatrix} \vec{r_1}^T & -\vec{r_1}^T(\mathbf{C} - g\,\vec{r_3}) \\ \vec{r_2}^T & -\vec{r_2}^T(\mathbf{C} - g\,\vec{r_3}) \\ \vec{r_3}^T & -\vec{r_3}^T(\mathbf{C} - g\,\vec{r_3}) \end{pmatrix} = \mathsf{K}_g \begin{pmatrix} \vec{r_1}^T & -\vec{r_1}^T\mathbf{C} \\ \vec{r_2}^T & -\vec{r_2}^T\mathbf{C} \\ \vec{r_3}^T & d_g \end{pmatrix}, \tag{29}$$

where the terms $\vec{r_i}^T\vec{r_3}$ are zeros for $i = 1,2$ because $\mathsf{R}$ is a rotation matrix. The scalar $d_g = -\vec{r_3}^T\mathbf{C} + g$ is the depth of the world origin with respect to the imaging system center in the direction of the principal ray $\vec{r_3}$. The effect of zooming by a factor $g$ is to move an image point $\mathbf{q}^u$ on a line radiating from the principal point $\mathsf{e}_0$ to the point $\mathbf{q}'^u = g\mathbf{q}^u + (1-g)e_0$. From similar triangles, we obtain that

$$g = \frac{f_g}{f_0} = \frac{d_g}{d_0} \tag{30}$$

The resulting projection matrix at a magnification $g$ is

$$\begin{aligned} P_g &= \mathsf{K} \begin{pmatrix} g & 0 & 0 \\ 0 & g & 0 \\ 0 & 0 & 1 \end{pmatrix} \begin{pmatrix} \vec{r_1}^T & -\vec{r_1}^T\mathbf{C} \\ \vec{r_2}^T & -\vec{r_2}^T\mathbf{C} \\ \vec{r_3}^T & d_g \end{pmatrix} \\ &= g\mathsf{K} \begin{pmatrix} \vec{r_1}^T & -\vec{r_1}^T\mathbf{C} \\ \vec{r_2}^T & -\vec{r_2}^T\mathbf{C} \\ \vec{r_3}^T g^{-1} & d_0 \end{pmatrix} \end{aligned} \tag{31}$$

When $g \to \infty$, the projection mapping tends to an orthographic projection.

### 4.3. The static calibration method

Let us consider a calibration at the interval of magnification $[g_0, g_t]$. This range is uniformly discretized at a sampling step of $\delta g$. The different steps of the static calibration can be outlined by Algorithm 3.

To estimate $f(g)$, $a(g)$, and $s(g)$, we use second-order PDEs and we drive the same reasoning as the computation of drift PDEs in Section 3.4. Since $f$, $a$, and $s$ are time independent, only one PDE of second-order is estimated for each parameter.

**for** $g = g_0$ *to* $g_f$ *with step* $G$ **do**
  **Data**: A set of $N$ image with grid points
  calibrate the projection matrix assuming $\xi_r(g_i) = 0$ and $\xi_t(g_i) = 0$ [15];
  comptue $f(g_i)$, $s(g_i)$ and $a(g_i)$. estimate distortion center $e(g_i)$ as explained in Section 4.1.2;
  **if** $e(g_i)$ *is not degenerate* **then**
    initialize $\xi_t$ and $\xi_r$ to zero and estimate them using bundle adjustment method[23];
  **else**
    set $e(g_i)$ to the principal point and $\xi_r = 0$;
    initialize $\xi_t$ to zero and estimate it using bundle adjustment method[23];
  **end**
**end**
estimate $f(g)$, $a(g)$ and $s(g)$ using PDA;

**Algorithm 3:** Static calibration

### 4.4. Static calibration of the JEOL JSM 820

This calibration is done with a multi-scale planar grid; see Figure 2 right [20]. This grid contains multiple chessboards of different sizes: 25 μm, 10 μm, 5 μm, 2 μm, and 1 μm; see Figure 9. The grid calibration points $\{\mathbf{Q}_i^c\}_{i \in \mathbb{N}}$ are selected, and three images are taken for each magnification scale $g$. This calibration is done for magnifications from $g_0 = 100\times$ up to $g_f = 10\,\text{k}\times$ with a step of $500\times$. The static calibration follows the procedure explained in Section 4.3. The following results were obtained [20]:

**Figure 9.** From left to right: Square side sizes are, respectively, of 25 μm, 10 μm, 5 μm, and 1 μm.

| | PDE | $g_1 = 100\times$ | $g_2 = 10\,\text{k}\times$ | solution |
|---|---|---|---|---|
| $e_x$ | $\ddot{e}_x = -0.02e_x$ | 230.21 | 225.08 | $e_x(g) = 230 + 10(1 - \exp(2\frac{g}{10^6}))$ |
| $e_y$ | $\ddot{e}_y = -0.02e_y$ | 270.34 | 265.10 | $e_y(g) = 270 + 10(1 - \exp(2\frac{g}{10^6}))$ |

| | PDE | $g_1 = 100\times$ | $g_2 = 10\,\text{k}\times$ |
|---|---|---|---|
| $\xi_r$ | $\ddot{\xi}_r = -0.0086\dot{\xi}_r + 0.015\xi_r$ | 0.003 | 0.002 |
| $\xi_t$ | $\ddot{\xi}_t = -0.0042\dot{\xi}_r + 0.0007\xi_r$ | $-7 \cdot 10^{-7}$ | $-59 \cdot 10^{-8}$ |

and solutions are

$$\begin{aligned} \xi_r(g) &= 0.003 - 13 \cdot 10^{-12} g - 6 \cdot 10^{-9} g^2 \\ \xi_t(g) &= -7 \cdot 10^{-7} - 13 \cdot 10^{-13} g + 6 \cdot 10^{-13} g^2 \end{aligned} \tag{32}$$

| | PDE | $g_1 = 100\times$ | $g_2 = 10\,\text{k}\times$ |
|---|---|---|---|
| focal length | $\dot{f} - 0.35 \cdot 10^{-4} = 0$ | $0.35 \cdot 10^4$ | |
| aspect ratio | $10^{13}\ddot{a} = -0.2225\,\dot{a} + 0.0017\,a$ | 1 | 1.1 |
| skew | $10^{13}\ddot{s} = -0.2059\,\dot{s} + 0.0016\,s$ | $10^{-3}$ | $1.9 \cdot 10^{-3}$ |

and solutions are

$$
\begin{aligned}
f(g) &= 0.35 \cdot 10^{-4}\, g \\
a(g) &= 0.05(g \cdot 10^{-4}) + 0.95(g \cdot 10^{-4})^2 \\
s(g) &= \frac{-0.05 + 0.65(g \cdot 10^{-4}) + 0.40(g \cdot 10^{-4})^2}{10^3}
\end{aligned}
\tag{33}
$$

The calibration results show variations of 10 pixels at most around a median principal point of $(245, 260)^T$ pixel. Finally, Figure 10 shows the RMS reprojection error between image points and back-projected planar grid points. It corresponds to a median value of 7 pixels without any correction and to 3 pixels after drift compensation, and without distortion correction, it decreases down to 1.45 pixel with both drift and distortion correction.

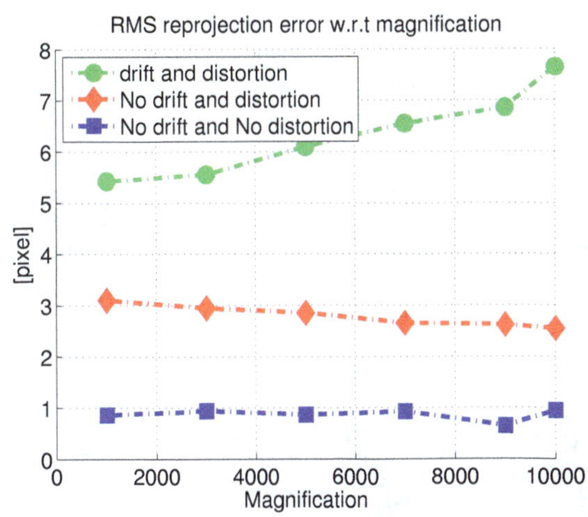


**Figure 10.** The RMS reprojection error through magnification scales. The average RMS error is about 0.9 pixel.

## 5. A small toy application: Cantilever deformation measurement

As a simple application to conclude this chapter, we propose to quantify the deformations of a cantilever that we deform with a micromanipulator. The cantilever is 35 µm long, 3.5 µm wide, and 300 nm thick. The micromanipulator is the *Kleindiek MM3A-EM* with a rigid tool mounted on the tip. The cantilever is fixed within a holder and is deformed by the moving of the tool tip; see Figure 11. Such an experiment may have several applications in the mechanical characterization of cantilevers [30], of biological deformable objects [31], etc. To check the accuracy of the calibration process, a simple test can be made. We estimate the repeatability of the measures for three different magnifications: 190×, 230×, and 260× [20]. The deformation measures which evaluate our calibration method use the magnification-smooth calibration parameters described in the previous section. It is worth to notice that these parameters were not estimated directly at these three magnification scales, but the estimated magnification-smooth functions allow us to find the calibration parameters at any magnification factor in the range [100× , 10 k×]. A set of 11 configurations are taken for each magnification factor (11 × 3 images). Initially the cantilever is straight and free from any contact with the planar surface. Then, it comes close to the tip of the cantilever which is progressively pushed forward by the *MM3A-EM*. After 7 × 3 acquired images at different configurations of the deformation, the cantilever is progressively dragged backward to the initial contact-free configuration; see Figure 12. Through the whole experiment, a time tracking frame acquisition is automatically processed using the PC processor's clock trigger. This time acquisition is important to retrieve the amount of pixel by which the acquired images have drifted. The acquired images are undrifted and undistorted with the estimated calibration parameters. We assume that during the deformation, the cantilever sweeps a virtual plane. The affine homography between image pixels and this deformation plane is estimated [16] by taking into account the estimated parameters of the projection model. The Euclidean stratification is done using the length of the cantilever provided by the manufacturer. The stratified homography gives us the mapping between distances in the image and their corresponding metric values in the deformation plane. The amount of deformation is measured as the distance between the tip of the cantilever at rest and its position after deformation. The measured deformation reaches a maximum of 250 nm. After drift and distortion correction, the standard deviation of the error among the three scales is of about 10 nm which is an acceptable amount of error at this scale of magnification.

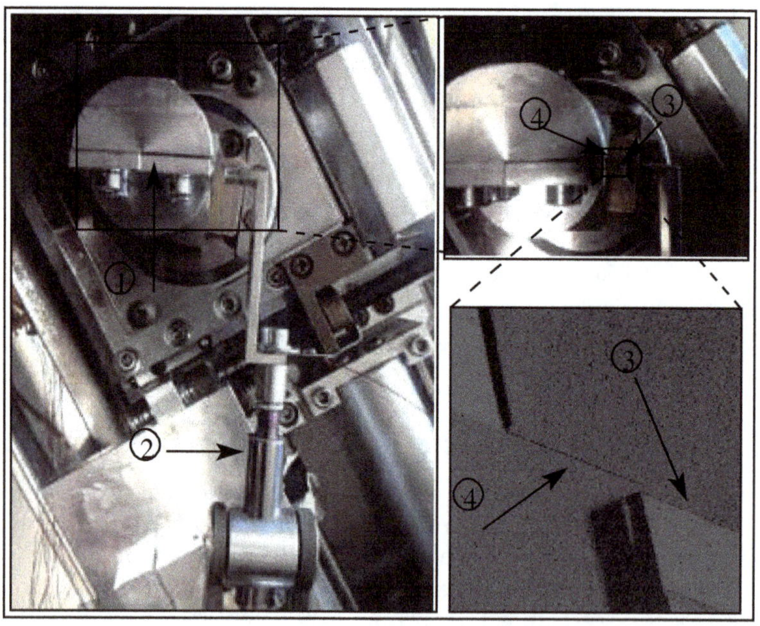


**Figure 11.** The setup of the example: 1. Holder of the cantilever; 2. *Kleindiek MM3A-EM*; 3. deforming tool; and 4. cantilever.

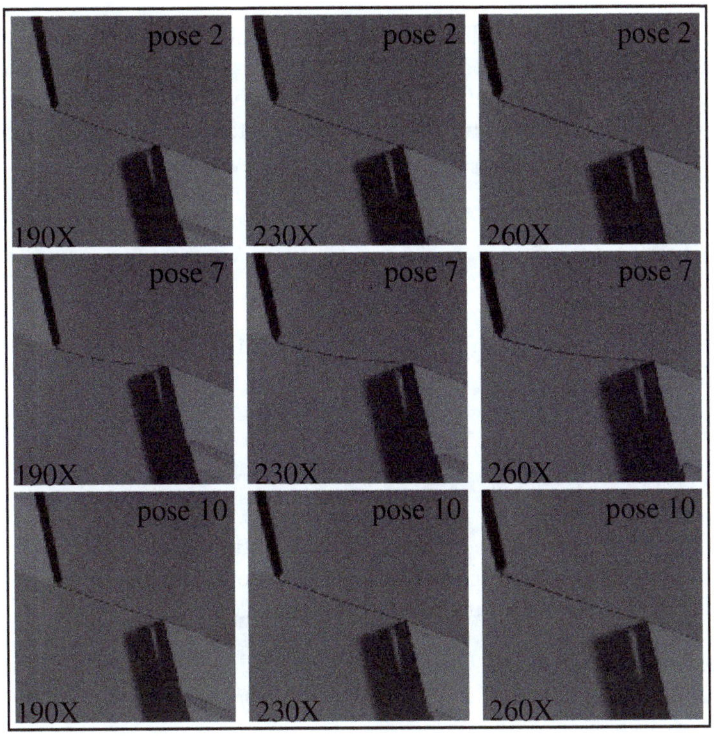


**Figure 12.** Three configurations of deformation (pose: 2, 7 and 10) at the three different magnification factors. The maximum of the deformation is at pose 7. Poses: 1 and 11 are contact-free between the cantilever and the plan.

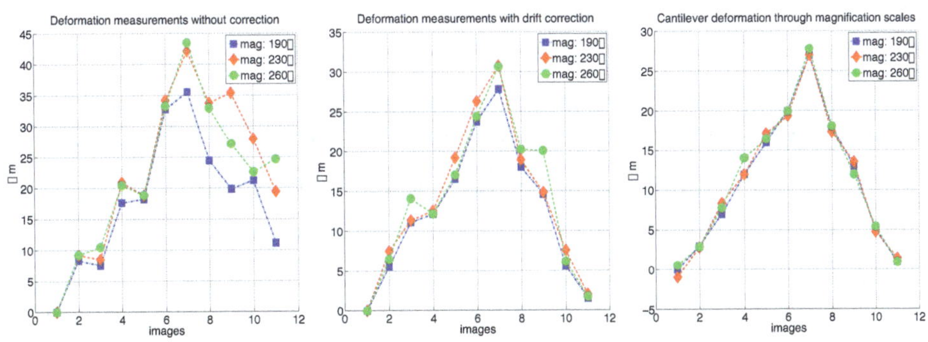


**Figure 13.** Left figure: Measured deformation without neither drift compensation nor static distortion correction. Middle figure: Measured deformation after drift compensation but without static distortion correction. Right figure: Deformation measures for the three magnification scales with drift and static distortion correction. The repeatability 10 nm. It can be seen that the drift and distortion correction improve considerably the repeatability of the measurements for the three scales.

## 6. Conclusion

In this chapter, a spatiotemporal calibration model for SEM imaging systems was presented. This model has the advantage of being smooth with respect to magnification scales. Both dynamic (temporal) and static (spatial) mappings are treated. The evolution over time and magnification of the pixel drifts and of the spatial distortion and projection matrices are modeled by mean of PDA.

## Author details

Abed Malti*

*Address all correspondence to: abed.malti@gmail.com

INRIA, France

## References

[1] F. Vignon, G. Le Besnerais, D. Boivin, J.L. Pouchou, and L. Quan. 3D reconstruction from scanning electron microscopy using stereovision and self-calibration. In *Physics in Signal and Image Processing*, Marseille, pages 23–24, 2001.

[2] M. Jaehnisch and S. Fatikow. 3-d vision feedback for nanohandling monitoring in a scanning electron microscope. *International Journal of Optomechatronics*, 1:4–26, 2007.

[3] S. Fatikow, C. Dahmen, T. Wortmann, and R. Tunnel. Visual feedback methods for nanohandling automation. *International Journal of Information Acquisition*, 6(3):159–169, 2009.

[4] M.A. Sutton, N. Li, D. Garcia, N. Cornille, J.J. Orteu, S.R. McNeill, H.W. Schreier, X. Li, and A.P. Reynolds. Scanning electron microscopy for quantitative small and large deformation measurements part II: Experimental validation for magnifications from 200 to 10,000. *Experimental Mechanics*, 47:789–804, 2007.

[5] T. Sievers and S. Fatikow. Real-time object tracking for the robot-based nanohandling in a scanning electron microscope. *Journal of Micromechatronics*, 3(3–4):267–284, 2006.

[6] B.E. Kratochvil, L. Dong, and B.J. Nelson. Real-time rigid-body visual tracking in a scanning electron microscope. *International Journal of Robotics Research*, 28:498–511, 2009.

[7] T. Kasaya, H.-T. Miyazaki, S. Saito, K. Koyano, T. Yamaura, and T. Sato. Image-based autonomous micromanipulation system for arrangement of spheres in a scanning electron microscope. *Review of scientific instruments*, 75(6):2033–2042, 2004.

[8] T. Sievers and S. Fatikow. Visual servoing of a mobile microrobot inside a scanning electron microscope. In *2005 IEEE/RSJ International Conference on Intelligent Robots and Systems*, Edmonton, 2005.

[9] M.A. Sutton, N. Li, D. Garcia, N. Cornille, J.J. Orteu, S.R. McNeill, H.W. Schreier, and X.D. Li. Metrology in a scanning electron microscope: Theoretical developments and experimental validation. *Measurement Science and Technology*, 17:2613–2622, 2006.

[10] M.A. Sutton, N. Li, D. Garcia, N. Cornille, J.J. Orteu, S.R. McNeill, H.W. Schreier, X. Li, and A.P. Reynolds. Scanning electron microscopy for quantitative small and large deformation measurements part I: Sem imaging at magnification from 200 to 10,000. *Experimental Mechanics*, 47:755–788, 2007.

[11] N. Li, M.A. Sutton, X. Li, and H.W. Schreier. Full-field thermal deformation measurements in a scanning electron microscope by 2d digital image correlation. *Experimental Mechanics*, 48(5):635–646, 2008.

[12] T. Zhu, M.A. Sutton, N. Li, J.-J. Orteu, N. Cornille, X. Li, and A.P. Reynolds. Quantitative stereovision in a scanning electron microscope. *Experimental Mechanics*, 51(1):97–109, 2011.

[13] P. Cizmar, A.E. Vladar, and M.T. Postek. *Real-Time Image Composition with Correction of Drift Distortion*. Technical Report, arXiv:0910.0213, Oct 2009.

[14] B. Mokaberi and A.A.G. Requicha. Towards automatic nanomanipulation: Drift compensation in scanning probe microscopes. In *Proceedings of the 2004 IEEE International Conference on Robotics and Automation (ICRA 2004)*, New Orleans, pages 416–421, 2004.

[15] R.I. Hartley and A. Zisserman. *Multiple View Geometry in Computer Vision*, 2nd ed. Cambridge: Cambridge University Press, 2004.

[16] Y. Ma, S. Soatto, J. Kosecka, and S.S. Sastry. *An Invitation to 3-D Vision: From Images to Geometric Models*. New York: Springer, 2003.

[17] O. Sinram, M. Ritter, S. Kleindiek, A. Schertel, H. Hohenberg, and J. Albertz. Calibration of an sem, using a nano positioning tilting table and a microscopic calibration pyramid. In *ISPRS Commission V Symposium*, Corfu, pages 210–215, 2002.

[18] N. Cornille. *Accurate 3D Shape and Displacement Measurement using a Scanning Electron Microscope*. PhD thesis, University of South Carolina and Institut National des Sciences Appliquées, 2005.

[19] A. Malti, S. Dembélé, N. Piat, C. Arnoult, and N. Marturi. Toward fast calibration of global drift in scanning electron microscopes with respect to time and magnification. *International Journal of Optomechatronics*, 6(1):1–16, 2012.

[20] A. Malti, S. Dembélé, N.L. Fort-Piat, P. Rougeot, and R. Salut. Magnification-continuous static calibration model of a scanning-electron microscope. *Journal of Electronic Imaging*, 21:1–16, 2012.

[21] J.O. Ramsay. Principal differential analysis: Data reduction by differential operators. *Journal of the Royal Statistical Society, Series B*, 58:495–508, 1996.

[22] A.W. Fitzgibbon. Simultaneous linear estimation of multiple view geometry and lens distortion. In *2001 IEEE Computer Society Conference on Computer Vision and Pattern Recognition (CVPR 2001)*, Kauai, pages 125–132, 2001.

[23] I. Lourakis and A. Argyros. Sba: A software package for generic sparse bundle adjustment. *ACM Transactions on Mathematical Software*, 36:2:1–2:30, 2009.

[24] J.O. Ramsay and B.W. Silverman. *Functional data analysis*, 2nd ed. New York: Springer, 2005.

[25] G.B. Folland. *Fourier Analysis and Its Applications*. The Wadsworth and Brooks/Cole Mathematics Series. Pacific Grove: Wadsworth & Brooks/Cole Advanced Books & Software, 1992.

[26] M. Frigo and S.G. Johnson. The design and implementation of FFTW3. *Proceedings of the IEEE*, 93(2):216–231, 2005. Special issue on "Program Generation, Optimization, and Platform Adaptation".

[27] J. Heikkila and O. Silven. A four-step camera calibration procedure with implicit image correction. In *Proceedings of the 1997 Conference on Computer Vision and Pattern Recognition (CVPR '97)*, Washington, DC, page 1106. IEEE Computer Society, 1997.

[28] R.I. Hartley and S.B. Kang. Parameter-free radial distortion correction with center of distortion estimation. *Pattern Analysis and Machine Intelligence*, 29(8):1309–1321, Aug 2007.

[29] M. Ritter, M. Hemmleb, P. Faber, B. Lich, and H. Hohenberg. Sem/fib stage calibration with photogrammetric methods. In *ISPRS Image Engineering and Vision Metrology*, Dresden, 2006.

[30] G. Ionascu, E. Manea, C. D. Comeaga, N. Alexandrescu, I. Cernica, and L. Bogatu. Silicon cantilever beam micromachining and structure geometry characterization. In *International Conference on Semiconductor*, Kobe, 2009.

[31] M. Boukallel, M. Gauthier, M. Dauge, E. Piat, and J. Abadie. Smart microrobots for mechanical cell characterization and cell convoying. *IEEE Transactions on Biomedical Engineering*, 54(8):1536–1540, 2007.

# 8

# Focused Ion Beams (FIB) — Novel Methodologies and Recent Applications for Multidisciplinary Sciences

Meltem Sezen

**Abstract**

Considered as the newest field of electron microscopy, focused ion beam (FIB) technologies are used in many fields of science for site-specific analysis, imaging, milling, deposition, micromachining, and manipulation. Dual-beam platforms, combining a high-resolution scanning electron microscope (HR-SEM) and an FIB column, additionally equipped with precursor-based gas injection systems (GIS), micromanipulators, and chemical analysis tools (such as energy-dispersive spectra (EDS) or wavelength-dispersive spectra (WDS)), serve as multifunctional tools for direct lithography in terms of nano-machining and nano-prototyping, while advanced specimen preparation for transmission electron microscopy (TEM) can practically be carried out with ultrahigh precision. Especially, when hard materials and material systems with hard substrates are concerned, FIB is the only technique for site-specific micro- and nanostructuring. Moreover, FIB sectioning and sampling techniques are frequently used for revealing the structural and morphological distribution of material systems with three-dimensional (3D) network at micro-/nanoscale.This book chapter includes many examples on conventional and novel processes of FIB technologies, ranging from analysis of semiconductors to electron tomography-based imaging of hard materials such as nanoporous ceramics and composites. In addition, recent studies concerning the active use of dual-beam platforms are mentioned

**Keywords:** Focused Ion Beams, Electron Microscopy, Dual-Beam Platforms, Nanostructuring, Nanoanalysis

## 1. Introduction

The miniaturization of novel materials, structures, and systems down to the atomic scale has assigned *electron microscopy*, a complementary branch of *nanotechnology*, for multidisciplinary sciences. In particular, transmission electron microscopy (TEM), scanning electron microscopy

(SEM), focused ion beams (FIB), and atomic force microscopy (AFM) can be considered as the most comprehensive techniques for advanced and precise analysis on different material species. Among all, *focused ion beam* microscopes are becoming more popular due to their versatility and configurational flexibility, as numerous tasks can practically be carried out with a single tool. Currently, many cutting-edge integrated microscope systems allow for nanosurgery applications inside a pressurized microscope chamber using in situ applications by means of using the capabilities of related tools and attachments that are coupled to related equipment. The best candidate is the dual-beam (also called two-beam or multi-beam) platforms.

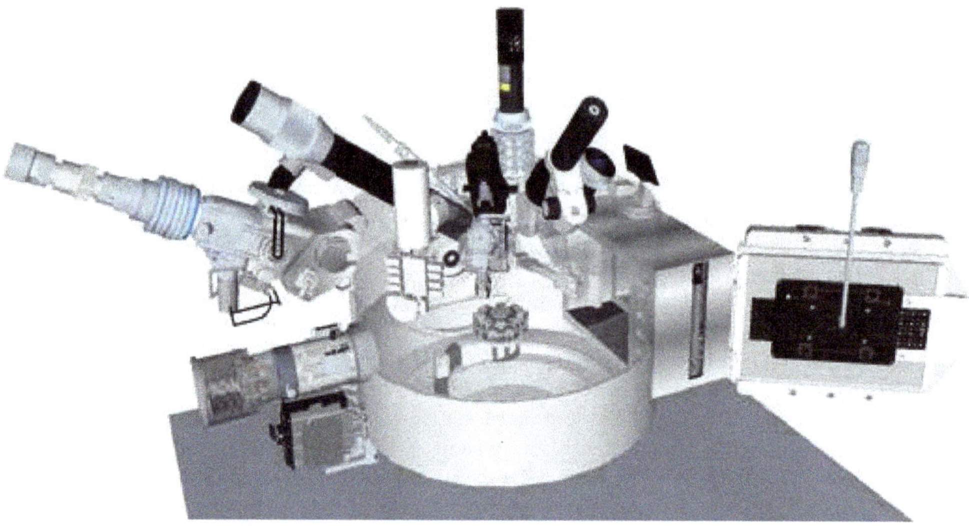

**Figure 1.** Illustration of a dual-beam platform with several attachments

Dual-beam platforms, consisting of an HR-SEM and an FIB column and additionally equipped with gas injection systems (GIS), micromanipulators, and detectors, serve as multifunctional tools for direct lithography in terms of nano-machining and nano-prototyping. A commercial dual-beam platform is shown in Figure 1. In the dual-beam platforms, in addition to imaging and chemical analysis of matter at ultrahigh resolution, many more processes can successfully be carried out from the micro- down to nanoscale, such as fabrication, structuring, deposition, prototyping, machining, 3D sculpturing, and manipulation. Using electron and ion beams in an integrated dual-beam platform, a wide variety of applications can be performed for multidisciplinary fields of nanotechnology: from material sciences and semiconductor technologies to biosciences. Whereas the initial development of focused ion beam (FIB) instruments was driven by their unique capabilities for computer chip repair and circuit modification in semiconductor technology, present FIB applications support a much broader range of scientific and technological disciplines [1,2]. The versatility and multifunctionality of these platforms give users inspiration and *enthusiasm* for developing new methodologies of nanostructuring and nanoanalysis upon the needs and the properties of diverse materials and systems.

**Figure 2.** A commercial FIB/SEM workstation equipped with GIS, micromanipulator, and EDS detector (JEOL JIB 4601F MultiBeam System at Sabanci University Nanotechnology Research and Application Center-SUNUM)

Fundamentally, an FIB column produces and directs a stream of high-energy ionized atoms of a relatively massive element, focusing them onto the sample both for the purpose of etching or milling the surface and as a method of imaging. The ions' greater mass allows them to easily expel surface atoms from their positions and produces secondary electrons (SE) from the surface, allowing the ion beam to image the sample before, during, and after the lithography process. The ion beam has a number of other uses as well, including the deposition of material from a gaseous layer above the sample [3]. Most of the recent commercial FIB systems and dual-beam platforms use gallium ($Ga^+$) as the ion source. The reason for preferring gallium element as the ion source is explained in the subsequent sections.

Concerning micro- and nano-sectioning processes, rapid FIB slicing techniques are frequently preferred for especially hard materials (e.g., metals, glass, ceramics) or layered structures (e.g., semiconductors) with hard substrates (silicon, glass, ceramics, etc.) because corresponding materials cannot be mechanically sliced by cutting tools. It has been always reported that ion milling is not a proper method for soft materials due to their sensitivity to beam damage and heat-dependent shape distortion. Therefore, for soft materials, ultramicrotomy has so far been accepted as the most convenient preparation technique, which is a mechanical serial sectioning process that uses diamond knives or blades for cutting cross sections from the materials that were embedded in epoxy resin. However, currently, due to the progress in the FIB technologies and due to advanced procedures and novel recipes developed by experienced users, same processes can also be allowed for soft materials as FIB applications still have many advantages over mechanical sectioning methods. Most of the newest procedures include the application of low voltage, low current, and gentle instrumental parameters, and bringing the microscope chamber to low temperatures and to even cryogenic conditions might also be necessary for specific experimental studies particularly when soft cells and organic materials are to be examined and structured.

FIB is recently considered to be the best technique for site-specific TEM specimen preparation. For an efficient TEM tomography analysis, the specimen has to be precisely structured so that it represents all its properties originally in 3D at the nanometer scale. Such samples can be successfully structured using the capabilities of FIB-SEM systems, e.g., ion milling, micro-/nanostructuring, deposition, manipulation, and polishing. Another application of dual-beam instruments is the 3D FIB tomography, where the material is cross-sectioned by ion milling sequentially, and from each sliced surface, an electron image (e.g., SE) or elemental information by energy-dispersive spectra (EDS) is acquired. Consequently, the collected serial two-dimensional (2D) data can be stacked and reconstructed to form ultimate comprehensive 3D data. Using this procedure, different material systems can be characterized for their morphological and elemental distributions in 3D.

Three-dimensional electron tomography is extensively used for revealing the structural and morphological distribution of material systems with 3D network at micro-/nanoscale. Accordingly, specific sample preparation techniques that keep the original structures to be investigated at the TEM are often required. FIB technologies, which are gaining high impact for their applications on diverse fields of science, provide material-dependent and TEM analysis spectrum-based solutions that are highly rapid, practical, creative, and reliable. The main approach is to overcome the limitations and difficulties (e.g., to be able to acquire only 2D data instead of 3D, projection problem that occurs at TEMs), which can be faced during electron microscopy-based characterization. In order to investigate the structural properties of the related materials in detail, practical, rapid, and reliable electron microscopy techniques and characterization methodologies have to be determined. This is very supportive in the direction of improving the quality of the production and the final product. TEM tomography enables to image the samples for different angles of tilting, which can be performed by 2D tilt series, and consequently, 3D reconstruction and segmentation can be carried out. Eventually, the resolving power of electron tomography (3D atomic resolution achieved) can be accompanied with analytical possibilities such as electron energy loss spectroscopy (EELS) and EDS. For performing TEM tomography work, special FIB samples (e.g., pin-like) that enable tilt series without projection problems are also needed to be prepared by experts.

Nevertheless, *irradiation damage*, caused by the use of beams in the electron microscopes, which leads to undesired physical/chemical material property changes or uncontrollable modification of structures that are being processed, should not be underestimated. Especially, soft matter such as polymers or biological materials is highly susceptible and very much prone to react on electron/ion beam irradiation. The effect is even higher when the ions are used as incident particles, and the end effect might even be the total loss of the material properties. The reason for that is, focused ions (in case of FIB, this is usually $Ga^+$) are energetic species with a high momentum and relatively low mean path, due to their mass. Therefore, they strongly affect surface composition, leading to extensive chemical modification and sometimes resulting in graphization. Nonetheless, it is well possible to turn degradation-dependent physical/chemical changes from negative to positive use when materials are intentionally exposed to beams. Especially, controllable surface modification allows tuning of surface properties in intended directions and thus provides the use of the ultimate materials and their systems toward desired and predefined concepts. Moreover, FIB is capable of performing maskless site-specific structuring, which are considered to be the major advantages of FIB over e-beam lithography. In particular, surface modification processes can be carried out through gas-assisted etching (GAE) in an FIB-SEM dual-beam instrument equipped with gas injections systems (GIS).

In the following sections, fundamentals of dual-beam technologies are revealed and examples of conventional and novel dual-beam applications and processes as well as structuring and modification of structures and surfaces are explained in detail.

## 2. Fundamentals of FIB technologies

### 2.1. Focused ion beams and FIB/SEM platforms

FIB systems are very similar to SEM, while the only difference is the use of an ion beam for scanning the sample surfaces, instead of an electron beam. In the FIB systems, a focused beam of metal ions is generated by a liquid metal ion source (LMIS). The LMIS is able to provide a source of ions of ≈5 nm in diameter, and a typical LMIS contains a tungsten (W) needle attached to a reservoir that holds the metal source material. There are several metallic elements or alloy sources that can be used in LMIS [4]. Among all, gallium ($Ga^+$) is commonly preferred in commercial FIB instruments owing to its advantages that are summarized as follows [2]:

- **i.** Gallium has low melting point ($T_m$ = 29.8 °C), minimizing reaction or inter-diffusion probabilities between the liquid and the tungsten needle, from where the ions are emitted.
- **ii.** The low volatility of gallium at the melting point protects the supply of metal, and this yields a long source life.
- **iii.** The low surface free energy leads to viscous behavior of the liquid on the substrate.
- **iv.** The low vapor pressure allows gallium to be used in its pure form rather than in the form of an alloy source. This yields a long lifetime of the source as the liquid will not evaporate.
- **v.** Gallium has excellent mechanical, electrical, and vacuum properties.
- **vi.** Gallium's emission characteristics provide high angular intensity with a small energy spread.

Once the $Ga^+$ ions are extracted from the LMIS, they are accelerated down to the ion column up to 30 keV and subsequently focused onto the sample using electrostatic lenses. The ion column typically has two lenses, a *condenser lens* and an *objective lens*. The condenser lens is the probe-forming lens and the objective lens is used to focus the beam of ions at the sample surface. Beam currents from a few picoamperes up to 60 nA can be obtained.

An optimum ion probe can be achieved by adjusting apertures, tuning all lenses, and doing final settings of the beam, such as stigmator and focus corrections. Cylindrical octopole lenses placed in an FIB system have multiple uses, which include beam deflection, alignment, and stigmation correction [2]. It should be reminded that, in both SEM and TEM systems, *magnetic lenses* are used for focusing the beam. Because ions are massive and they travel at much lower velocities, their *Lorentz force* is lower and magnetic lenses are less effective on ions than they would be on electrons at the same accelerating voltages. As a result, FIB columns are equipped with electrostatic lenses rather than magnetic lenses.

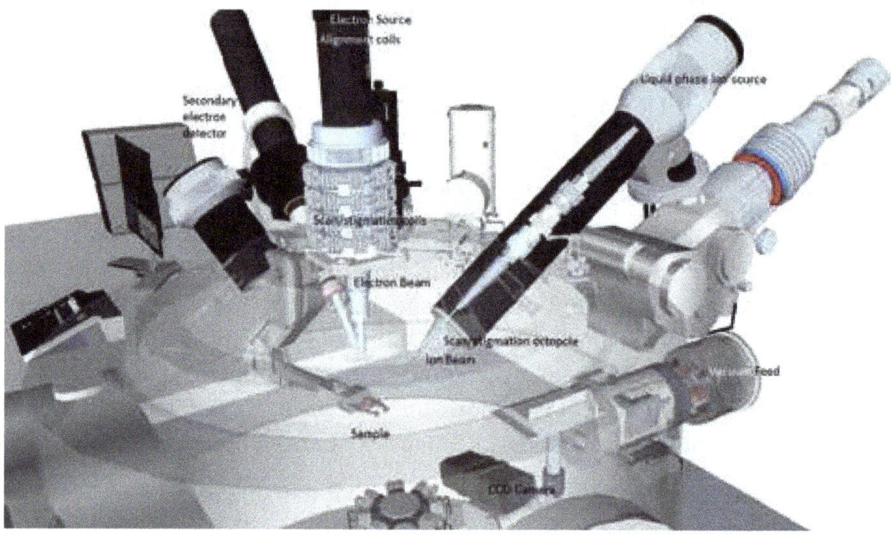

**Figure 3.** Construction of a dual-beam chamber

Commercial dual-beam platforms incorporate both a FIB column and a SEM column in a single system. This combination is especially useful for sample preparation of cross-sections using the electron beam to view this cross-sectional face as the ion beam mills normal to the sample surface. This monitoring allows the milling to be stopped precisely when the task is completed. The typical dual-beam column configuration is a vertical electron column with a tilted ion column. In dual-beam platforms, the tilt angle of the FIB column can vary from manufacturer to the model, but usually it is between 52° and 55° tilt to the vertical.

Both SEM and FIB can be used to acquire high-resolution images by collecting the secondary electrons (SE) that are emitted from the interactions between the beam and the surface atoms, although backscattered electrons (BSE) and/or secondary ions (SI) can contribute to form images. For secondary electron detection, Everhart–Thornley electron multiplier detector is the most common design used recently [3]. The main difference between scanning and/or transmission electron microscopy and focused ion beams is the use of ions as the beam that is

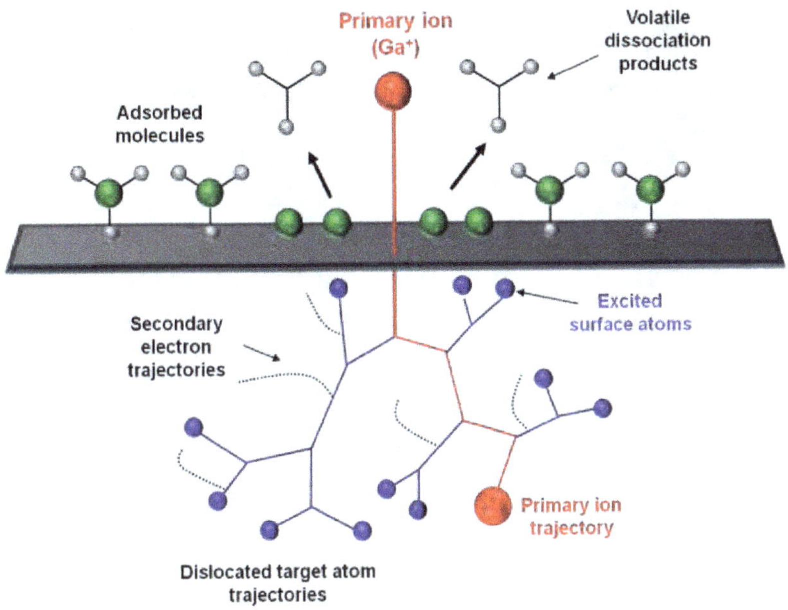


**Figure 4.** The interactions of primary ions generating secondary electrons and the collision cascade. The non-sputtered target atoms remain as excited surface atoms contributing to the molecule dissociation (modified from [5])

also responsible for many interactions occurring at the sample surface. Because ions are much larger in size than electrons, they are not able to penetrate within specimens' individual atoms, and outer shell interaction results in atomic ionization and breaking of chemical bonds of the target material. Secondary electrons are formed and alterations in the chemical stability are generated by this means. The interactions of primary ions and the target surface are responsible for the generation of secondary electrons, while the non-sputtered target atoms remain as excited surface atoms and contribute to dissociation of molecules [5].

### 2.2. Beam–specimen interactions

Dual-beam platforms allow the use of electron and ion beams for several applications through particle–sample interactions and/or reactions. As already very well known by electron microscope users, electrons may adversely affect organic or inorganic samples when they are placed in an electron microscope (e.g., SEM or TEM) for different purposes and may cause temporary or permanent changes within the specimen. The main effects might be in the form of electrostatic charging, ionization damage (radiolysis), displacement damage, sputtering heating, and hydrocarbon contamination. Typically, the amount of radiation damage is proportional to the electron dose, and the extent of damage depends on the amount of energy deposited in the specimen [6].

When ions are accelerated from a source and hit on the specimen surface in an ion column, they enter the target material as they might interact with specimen in various ways within a penetration volume. This interaction, depending on the ion energy, can be in the form of

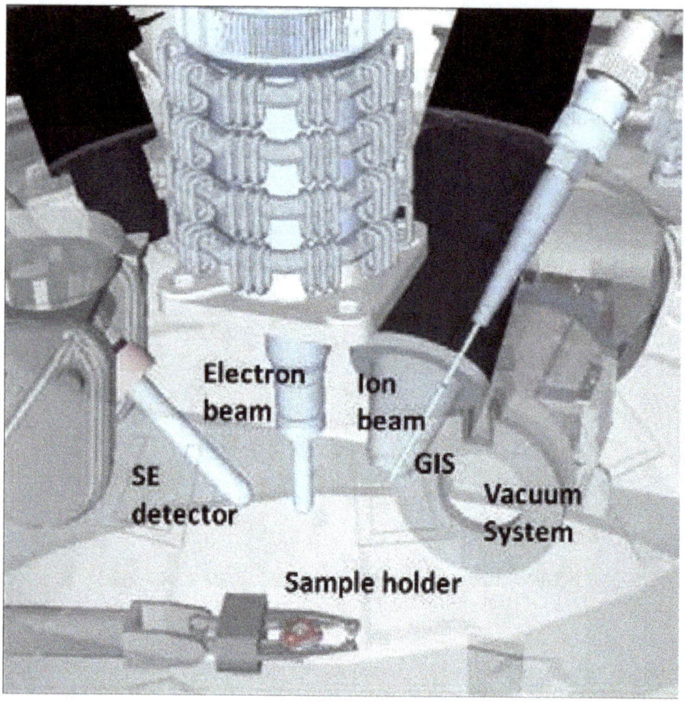


**Figure 5.** Illustration of the specimen chamber of a dual-beam platform

sputtering, amorphization, swelling, deposition, redeposition, implantation, backscattering, or nuclear reaction. However, many of the interactions are not completely separable and occur simultaneously, and it is often not very well understood which mechanism is dominating in the degradation process of specimens being irradiated by ions. Radiation damage not only leads to morphological changes but also alters intrinsic physical properties (crystallinity, elasticity, conductivity, electrostatic charge), as well as chemical characteristics (hydrophilicity, surface composition) of the surface.

The most dominating mechanisms for ion irradiation are discussed below:

i. **Sputtering:** Sputtering is actually the same as ion milling process and is the major mechanism for material removal. Usually, sputtering range increases with increased ion energies, while heavier ion sources or lower surface binding energies of target materials can contribute to higher sputtering regimes.

A software package called SRIM (Stopping and Range of Ions in Matter)/TRIM (TRansport of Ions in Matter) has been widely used for predicting the range of sputtering for many different ions at a wide energy range. SRIM/TRIM uses a Monte Carlo treatment of ion–atom collisions to calculate the stopping range of ions into the matter. SRIM/TRIM calculations agree very well with the experimental data for the cases considered, and the sputter yield is dependent not only on the material but also on many processing parameters, including the ion energy, angle of incidence, and scanning procedures.

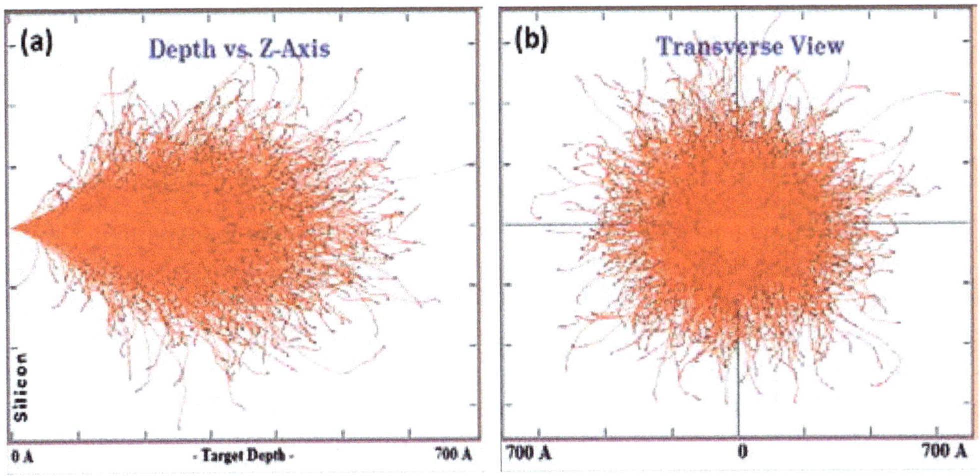


**Figure 6.** TRIM calculations for gallium penetration on silicon (for 30 keV ion energy, 0° angle and 5000 ions): (a) through z-axis, (b) transverse projection

ii. **Gallium Implantation:** Sputtering mechanism may cause gallium ions to be implanted and mixed into the specimen. This might be responsible for the alteration of the specimen's local composition within the interaction volume. Gallium implantation may lead to structural changes, as well as alteration in, e.g., thermal, electrical, optical, and mechanical properties.

iii. **Redeposition:** Sputtered particles leaving material's surface in gas phase are very prone to condense back into the solid phase upon collision with solid surfaces, as they are not thermodynamically in equilibrium. As a result, a portion of ejected atoms tend to stick back into the sputtered surfaces and redeposit. Redeposition can be minimized via using low ion doses (e.g., currents), deposition of protective layers using low ion energies, and optimizing ion milling geometries.

iv. **Amorphization:** Amorphization of the materials that are processed with focused ions may occur in the bombarded area of a crystalline substrate and may induce the substrate to swell. This mechanism can be attributed to sufficient atom displacement within the collision cascade, resulting in the loss of crystalline orientation. Amorphization is usually faced during TEM specimen preparation and might be a serious problem for the crystalline structures to be investigated in TEM. Hence, the use of low ion energies during the polishing step of the preparation process can drastically help in minimizing amorphization effects.

v. **Swelling:** Swelling of the target material due to ion bombardment during FIB processing is dependent on two major mechanisms: amorphization and ion implantation. Swelling is mostly attributed to material amorphization, while ion implantation does not seem to remarkably contribute to volume expansion. Distortion of the crystalline orientation to amorphous structures leads to volumetric alterations and hence swelling of the material. Also implanting of gallium ions into the target material is found to be another reason for swelling mechanism during FIB processing.

### 2.3. Basic applications of FIB

#### *2.3.1. Ion milling*

Ion milling, as the fundamental application of FIB systems, is a continuous sputtering process that occurs during ion beam exposure on the sample. Milling is actually an atomic collision process that ends up in the removal of the material from the ion–sample interaction volume, mainly depending on the beam current and voltage used in the process. Ion milling can be used to create both simple structures, such as lines, rectangles, or circles in the material, and complex patterns, bitmaps, and streamline files. In addition, specific singular or serial patterns can be created by internal patterning engines or external lithography software and can be easily imported to FIB user interfaces.

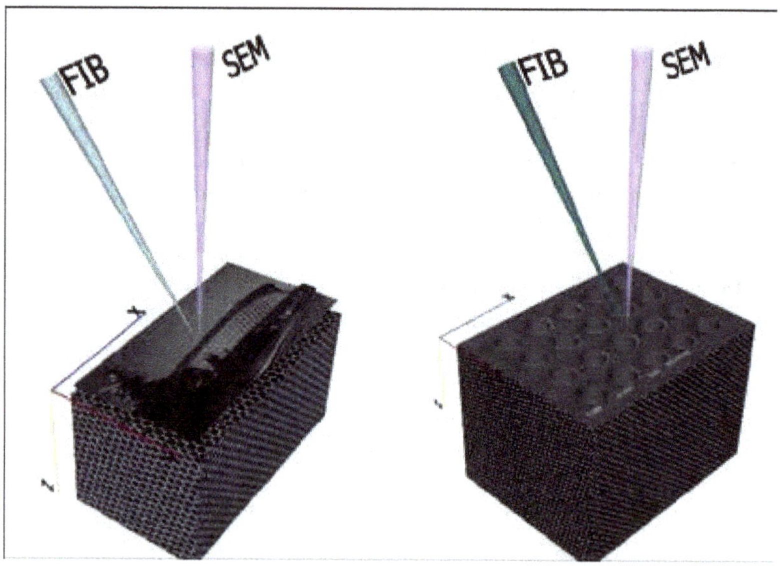


**Figure 7.** Representation of ion milling: maskless lithography and nanostructuring

Milling allows for creating cross sections or developing structures with desired geometries to control not only the lateral position but also local depth. Ion milling can also be described as a "direct writing technique" via etching the material surfaces with the exposure of ion beams. This process is similar to lithography; however, the advantage here is that it does not require the use of masks. Local assisting gases exposed by gas injection systems (GIS) that are integrated into dual-beam platforms can help in enhancing the removal of atoms from the material surfaces. Local gas delivery systems can change the oxidation state of released particles and speed up the milling process, while exposure of gases into the region of interest may also reduce the local redeposition of atoms released from the surface. $XeF_2$ and $I_2$ are the most common gases that are used in FIB to enhance ion milling processes, and their supply is usually dependent on the manufacturer.

### *2.3.2. Deposition*

Deposition is the second most powerful feature of FIB technologies, as the ion or electron beams can be used in a deposition system, allowing the addition of material instead of removing the material. Deposit materials are often supplied by an internal gas delivery system that locally exposes a chemical compound close to the surface impact point via gas injection systems (GIS) that can be incorporated into dual-beam platforms. The chemical gas compound is usually in the precursor form and consists of organometallic molecules. When this compound is exposed to the region of interest, beams decompose the molecules locally and deposit almost-pure material onto the surface. In other words, gas compound is exposed onto the target specimen and adsorbed on its surface, and this is followed by bombardment of focused beams on the adsorbed molecules within predefined patterns. Finally, as a result of complex beam-induced reactions, gas molecules dissociate into deposits and volatile fragments, while dissociated molecules are adsorbed and deposited into desired structures. This procedure is demonstrated in a scheme in Figure 8.

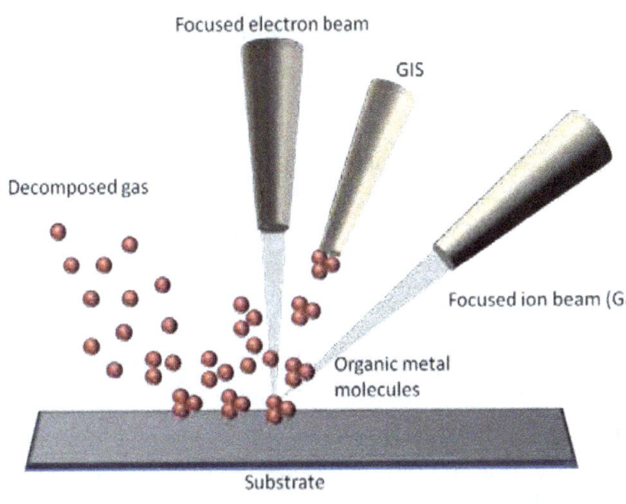


**Figure 8.** Illustration of deposition process in dual-beam instruments via electron beam-induced deposition (EBID) and ion beam-assisted deposition (IBAD)

Usually the decomposition of the precursor molecules is not hundred percent, and therefore some additional matrix molecules (e.g., organic residues) are also deposited together with the diverted material. For this reason, the purity of the deposits is lower when compared to other deposition techniques such as chemical vapor deposition (CVD) or physical vapor deposition (PVD). The materials used for beam-induced deposition in the dual-beam platforms are determined by their different gas chemistries, and several precursor gases are commercially available for the deposition of Pt, W, $SiO_2$, and C provided by different manufacturers and suppliers. There has been already a COST Action (CM 1301 – CELINA) ongoing that aims to develop novel gas chemistries for beam-induced deposition techniques and to perform applications for multidisciplinary sciences using deposition capabilities of precursor materials.

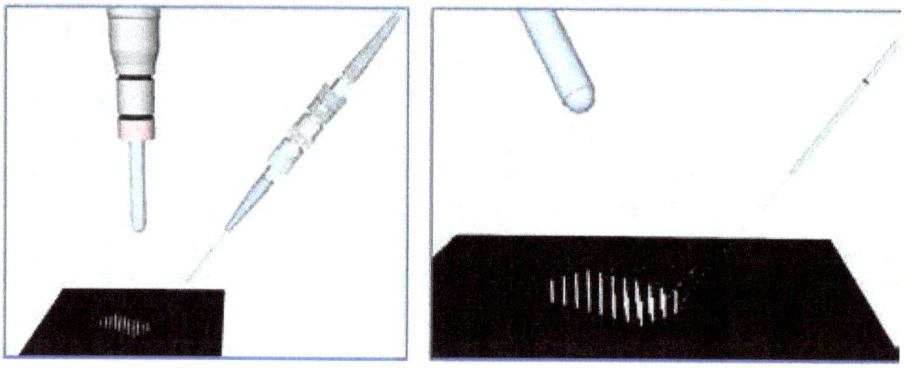

**Figure 9.** Illustration of gas-assisted deposition process

## 3. FIB-based conventional and novel processes

### 3.1. TEM specimen preparation

One of the most important applications of dual-beam instruments is preparing samples for transmission electron microscopy, as an important capability owing to controlled ion milling abilities is the production of ultrathin and uniform lamellae that are electron-transparent and hence can serve as TEM samples. Advantages of using an FIB for TEM specimen preparation are listed below:

- TEM lamellae can be prepared site-specifically with a spatial accuracy as fine as ≈30 nm from any region of interest.
- When compared to other techniques (microtomy, low-energy ion milling, dimpling, etc.), the duration for site-specific and ultrathin specimens' preparation process is considerably short, varying from less than 1 hour for noncomplex structures to 4–5 hours for challenging specimens.
- FIB allows for the preparation of TEM samples from both hard and soft materials, regardless of how brittle, ductile, or mechanically sensitive the material is. Even FIB serves for the TEM investigation of samples of life sciences studies when proper conditions of dual-beam systems are maintained (e.g., using cooling stage and cryo-conditions; low ion energy and ion current options), and the experience of the FIB user is sufficient.
- It is possible to develop new methodologies and specific geometries according to the nature and properties of the materials from which TEM specimens are to be prepared. The main

approach in special TEM sample designs is to overcome the limitations and difficulties which can be faced during TEM investigations.

The most common route for TEM specimen preparation is the "in situ lift-out technique," for which the dual-beam instrument has to be equipped with a micromanipulator unit for allowing the transfer of the lamella to the TEM grid, when both are placed in the microscope chamber at the same time. In this procedure, initially a metal protection layer is coated on the region of interest via beam-induced deposition and two opposing trenches are milled away, leaving behind a 1–2-μm-thin section on the block sample. This section can be named as "pre-lamella." The next step involves cutting the bottom part and the side trenches away, until the section is held by the bulk sample from its shoulders. Then, this pre-lamella can be welded to the micromanipulator using ion beam-assisted platinum deposition (IBAD) by simultaneously cutting away the shoulders, and can be lifted out from the sample, transferred, and welded to a TEM grid. Afterwards, final thinning and polishing down to a thickness of < 100 nm is achieved using low incident angles and low ion currents. Finally, the sample is ready for TEM analysis. This type of TEM specimen preparation method is demonstrated in Figure 10, giving an example of preparing a cross-sectional lamella from an organic thin film transistor.

In addition to in situ lift-out route, there are many more procedures in the literature that have been developed for more than a decade. Specimen preparation by FIB is often preferred for hard materials (e.g., metals, glass, ceramics) or layered structures (e.g., semiconductors) with hard substrates (silicon, glass, ceramics, etc.) since ion milling is not a proper method for soft materials due to their sensitivity to beam damage and heat-dependent shape distortion. For soft materials, ultramicrotomy is considered to be the most convenient preparation technique, which is a mechanical sectioning process using a diamond knife. As far as TEM investigations for organic and inorganic electronics are concerned, dual-beam instruments are effective for preparation of cross-sections from multilayer integrated devices with hard substrates as the active layer is often very thin (in nanometers) and sandwiched between supporting materials. However, it should be reminded that, for all conditions ion milling and deposition parameters have to be optimized in order to avoid potential ion irradiation damage. Hence, special care should be taken during the entire FIB-based TEM specimen preparation process for minimizing the radiation effects triggered by damage mechanisms mentioned in the previous section (2.2), such as amorphization, gallium implantation, and swelling.

Moreover, when three-dimensional investigations of materials are of interest, TEM tomography-based analysis can be performed. For this type of characterization, the specimen has to be precisely structured so that it has to be representing all its properties of the original material in 3D at the nanometer scale. Such samples can be successfully structured using the capabilities of dual-beam systems, as the details of the corresponding technique are given in the following parts of this section.

### 3.2. Serial slicing and imaging

Dual-beam platforms provide the use of electron and ion beams simultaneously, which opens a way to perform cross sectioning by means of sequential ion milling and monitoring and/or

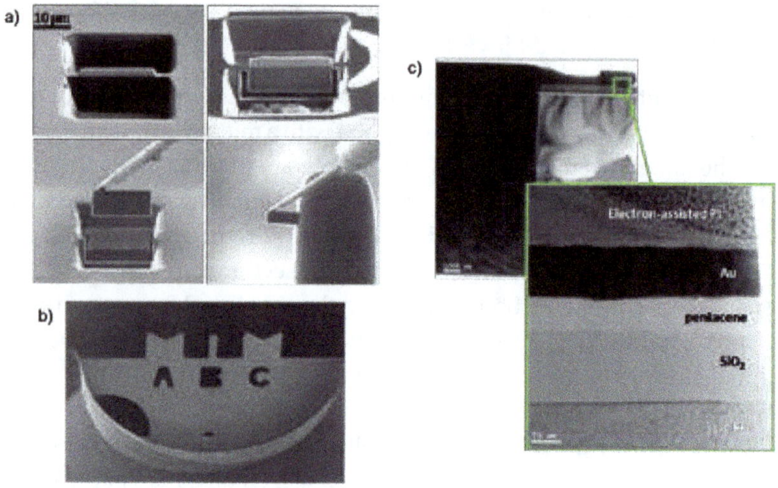


**Figure 10.** TEM specimen preparation of an organic thin film transistor (O-TFT) structure using a dual-beam tool: (a) coarse milling, lift-out, and mounting steps, (b) lamella on a TEM semi-grid, (c) thinned and polished cross-sections investigated by TEM in bright field (BF) mode (images taken at Felmi-Zfe, TU Graz, Austria)

acquiring images of the corresponding cross section of the specimen at the same time using the electron beam. There are mainly two different ways to collect a stack of SEM images of sectioned surfaces: static mode and dynamic mode. In dynamic image acquisition mode, SEM images are acquired in real time during the FIB milling process. In static image acquisition mode, after each slicing, the process is either paused or stopped and therefore slow-scan high-resolution SEM images are acquired.

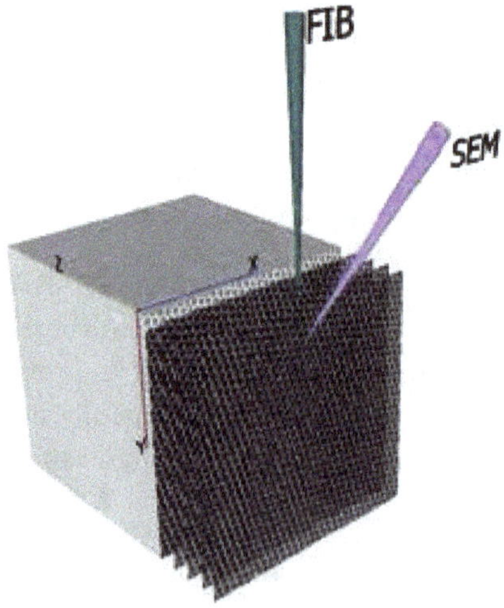


**Figure 11.** Illustration of serial slicing and imaging application of dual-beam platforms. Ion beam is used for creating cross sections, while electron beam allows for monitoring and imaging of the sliced regions.

In particular, the application of this process is very useful for failure analysis of semiconductor devices because it is much faster than TEM specimen preparation. Also rapid monitoring of inner structures of several materials and gaining information of the features down to a few nanometers possible with serial slicing and imaging techniques.

### 3.3. 3D microstructural characterization and FIB tomography

Dual-beam platform also enables three-dimensional information methodologies, especially for quantitative characterization of materials, while the measurement of a number of important geometric properties that cannot be obtained using a 2D analysis can be performed using FIB tomography methods. These are, for instance, the number of features per unit volume, feature connectivity, real feature shapes and sizes, and spatial distribution information.

One step further of serial slicing and imaging application is the 3D FIB tomography, which is based on the principle that continuous 2D data are collected from the surface of the bulk material by serial-sectioning and are stacked together to form reconstructed data, giving information in 3D. By the removal of each section, SE images, BSE images, EBSD maps, and/or EDS data can be acquired from the specimen surface, collected for 3D reconstruction and processed for 3D material characterization. Consequently, dual-beam microscopes are capable of high-fidelity characterization of the morphology, crystallography, and chemistry of micron- and submicron-sized features in 3D. The FIB slicing and HR-SEM imaging-based reconstruction showing the distribution of dentinal tubules in human tooth is shown in Figure 12.

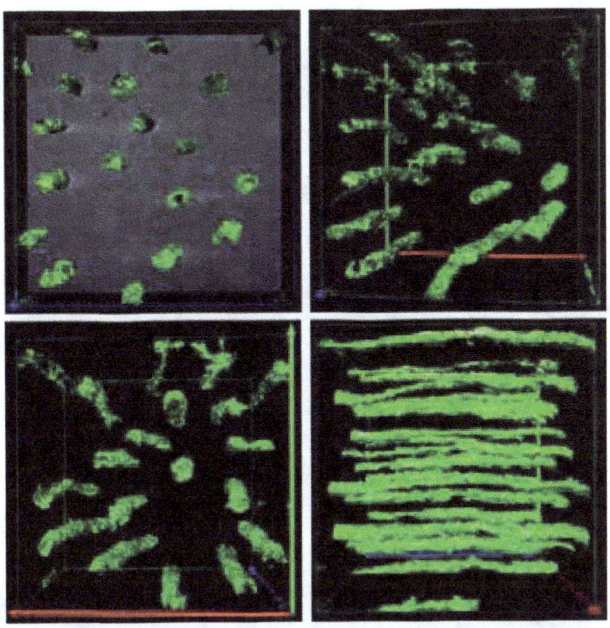

**Figure 12.** The 3D reconstruction of dentin showing the tubule distribution: red arrow corresponds to $x$-axis, while blue arrow represents the $y$-axis and green arrow the $z$-axis. For the reconstruction, Stack N-Viz software was used (reprinted by permission from [7])

### 3.4. Micro-/nano-fabrication, micro-/nano-modification, and other applications

In addition to applications mentioned above, structuring capabilities of dual-beam platforms in small scales fall into two major categories: one is *fabrication* and *machining*, while the other is *rapid prototyping* or *modification* abilities of structures and devices using both ion and electron beam-based processing. For the former, FIB is used for preparing structures that are difficult to form using conventional processes due to material or geometry constraints. The latter, when FIB is utilized, processes can be carried out in more practical and less time-consuming steps than conventional routes. Within user-defined patterns, it is possible to mill away lines down to 10 nm in width, and deposit materials as small as 30 nm by the breakdown of organometallic precursors. The finest ion beam spot size is approximately 5–10 nm, enabling small features to be patterned, while the shape of an FIB cut is dependent on many factors, such as its geometry, milled depth, ion beam profile, and the redeposition of sputtered material.

As already mentioned, FIB provides a very convenient technique for material removal using gallium ions. The advantage of the instrument is that the structure that is to be milled can be predefined as patterns and the process can be performed in an automated way. Milling patterns may be defined in different forms such as scripts, stream files, or image files. However, as already given in the previous section concerning platinum deposition, the quality and the efficiency of the ion milling process are dependent on the instrumental and process parameters, and those have to be optimized for the achievement of the desired structures.

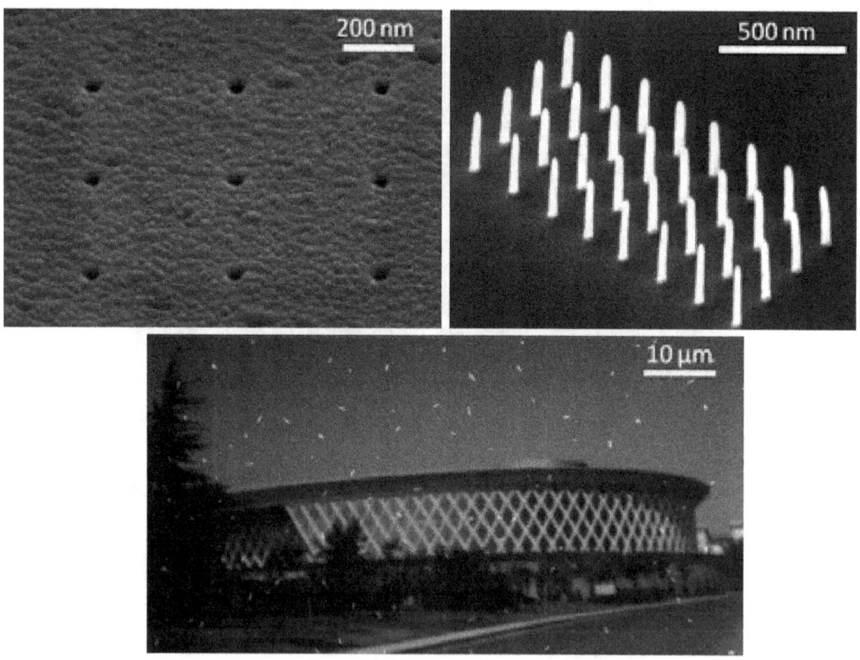


**Figure 13.** Nanostructuring, nano-fabrication and maskless ion lithography examples performed by dual-beam instruments

Recently, FIB technologies are becoming more popular for machining miniaturized samples to investigate the influence of sample dimensions on mechanical properties, in terms of determining size-dependent effects, particularly in metals, alloys, and ceramic composites. Sometimes dual-beam platforms enable not only the fabrication of the test structures but also the application of in situ examination in micro-/nano-size when they are coupled to proper mechanical test stages and nano-indentation devices. The most popular mechanical behavior tests that are recently in application and development include tensile strength and yield strength measurements, giving out the data for deformation in the form of stress–strain curves. The acquired data are very important to get compared to the data in macroscale, for the evaluation of size-dependent effects. Also in situ hardness tests can be performed on the coatings that include applications on several material systems used for multidisciplinary sciences and many fields of industry.

### 3.5. Special sample designs for TEM tomography

Electron tomography applications are recently becoming more popular for imaging of three-dimensional material systems such as alloys, composites, or samples having spatial features, such as porous network or multicomponents. Especially, in order to reveal the geometric and elemental distribution of material systems that are based on nanostructures, electron tomography applications are being widely used. For this reason, specific sample preparation techniques that keep the original structures of the sections to be investigated at the TEM are often required. Dual-beam technologies provide material-dependent and TEM analysis spectrum-based solutions that are highly rapid, practical, creative, and reliable.

The main approach for novel specimen designs is to overcome the limitations and difficulties (e.g., to be able to acquire only 2D data instead of 3D, projection problem that occurs at TEMs, etc.) that can be faced during electron microcopy-based characterization. In order to investigate the structural properties of these materials in detail, unique electron microscopy techniques and characterization methodologies have to be developed according to the needs and what is expected to be analyzed.

When TEM tomography investigation of a sample is coupled to its FIB slicing and imaging and FIB tomography work, while TEM will be able to provide 3D information at the nanometer scale and below, via FIB tomography sectioning, the information in the scale ranging from micrometers to tens of nanometers will be collected from the identical sample and the data can be comparatively and complementarily evaluated.

One of the solutions for the problems faced during TEM tilt series in tomography applications is the preparation of samples with special geometries that do not cause any projection, thickness variation, and shadowing problems during tilting. This might be a problem for lamellar samples; however, pin-like TEM structures are proper for tilting without any projection or shadowing, as well as this type of sample can represent the three-dimensional nature for the nanocomposite systems or porous materials. Especially, nanoparticles can be kicked out of the matrix during final thinning of ultrathin lamellae and the pores in nano-size can be enlarged with respect to ion bombardment via FIB processing in thin samples. Hence, pin-like structures are found to be more stable both physically and chemically when compared to lamellar samples. An example for pin-like sample preparation of human dentin using dual-beam instruments is given in Figure 14.

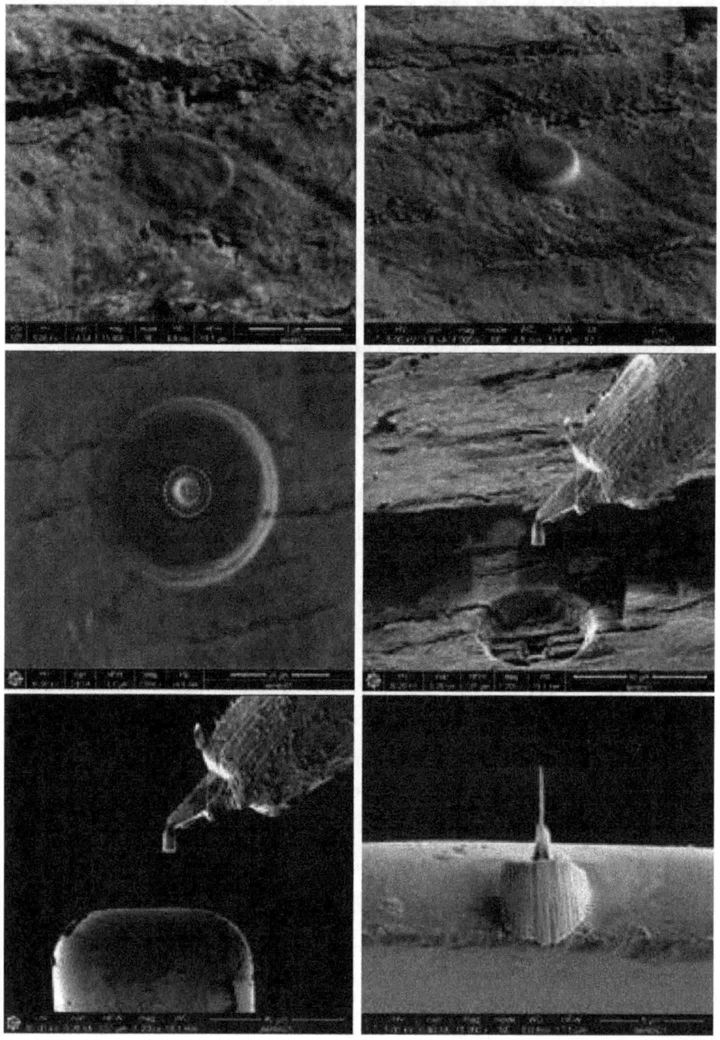

**Figure 14.** The steps for preparation of pin-like TEM sample using the dual-beam instruments: (a) deposition of electron beam-assisted Pt layer, (b) deposition of ion beam-assisted Pt layer, (c) ion beam milling via annular patterns, (d) lift-out of the pre-section, (e) mounting of the pre-section onto the grid; (f) final thinning and polishing (reprinted from [7] by permission)

Pin-like structures investigated by TEM allow for revealing structures in 3D at the nanoscale and below, and when combined to FIB tomography data, the overall results help in observing materials' features at different scales. An example again on human dentin is given in Figure 15 for tracking micro-sized dentinal tubules, and nano-sized collagen fibrils are investigated using FIB slicing/SEM imaging and TEM bright field imaging, respectively.

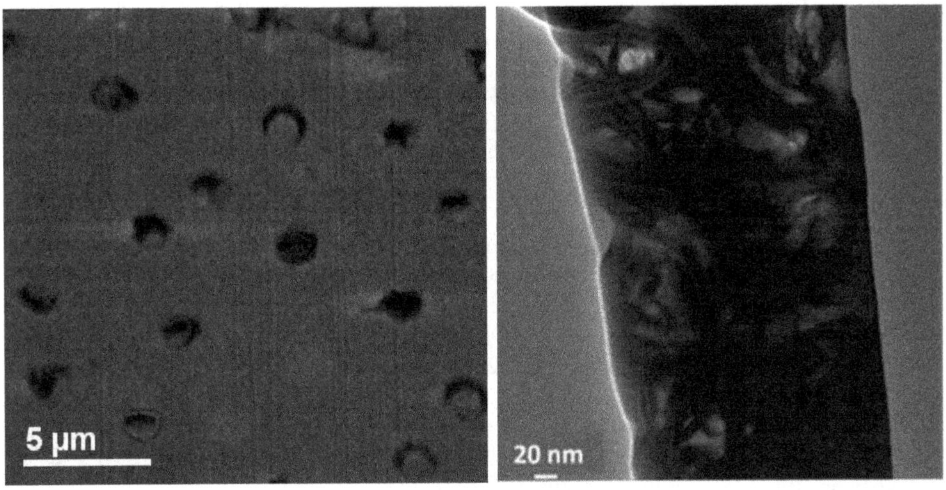


**Figure 15.** FIB slice showing the cross sections of tubules elongated in $y$-axis in human dentin and the TEM bight field images showing the 3D distribution of collagen fibrils within the human dentin. The micrographs show the nanofeatures within the dentin structure [7].

## Acknowledgements

I would like to cordially thank Dr. Feray Bakan, Sina Sadighikia, and Ertuğrul Bülbül for their valuable help with this book chapter. Also support by Dr. Volkan Özgüz at Sabanci University and my former colleagues at Felmi-Zfe, TU Graz, is gratefully acknowledged. In addition, I owe many thanks to the national TUBITAK projects 112M195 and 114M040, as well as COST Actions MP1103 and CM1301 (CELINA) for providing financial support and collaboration opportunities.

## Author details

Meltem Sezen*

Sabanci University Nanotechnology Research and Application Center (SUNUM), Orhanlı-Tuzla, Istanbul, Turkey

## References

[1] Mayer J, Giannuzzi L.A, Kamino T, Michael J. TEM sample preparation and FIB-induced damage. MRS Bulletin. 2007; 32: 400–407. DOI: 10.1557/mrs2007.63

[2] Giannuzzi L.A, Stevie F.A. Introduction to Focused Ion Beams: Instrumentation, Theory, Techniques, and Practice. New York: Springer; 2005. 357 p. DOI: 10.1007/b101190

[3] Yao N, editor. Focused Ion Beam Systems: Basics and Applications. Cambridge University Press; 2011. 395 p. DOI: 10.1017/CBO9780511600302

[4] Orloff J, Utlaut M, Swanson L. High Resolution Focused Ion Beams. New York: Kluwer Academic/Plenum Publishers, Springer US; 2002. 303 p. DOI: 10.1007/978-1-4615-0765-9

[5] Utke I, Hoffmann P, Melngailis J. Gas-assisted focused electron beam and ion beam processing. Journal of Vacuum Science and Technology. B. 2008; 26, 4: 1197–1276. DOI: 10.1116/1.2955728

[6] Egerton R.F, Li P, Malac M. Radiation damage in the TEM and SEM. Micron. 2004; 35, 6: 399–409. DOI: 10.1016/j.micron.2004.02.003

[7] Sezen M, Sadighikia S. 3D electron microscopy investigations of human dentin at the micro/nanoscale using focused ion beams based nanostructuring. RSC Advances. 2015; 5: 7196–7199. DOI: 10.1039/C4RA14364G

# 9

# Microstructure Evolution in Ultrafine-grained Magnesium Alloy AZ31 Processed by Severe Plastic Deformation

Jitka Stráská, Josef Stráský, Peter Minárik, Miloš Janeček and Robert Král

**Abstract**

Commercial MgAlZn alloy AZ31 was processed by two techniques of severe plastic deformation (SPD): equal channel angular pressing and high pressure torsion. Several microscopic techniques, namely light, scanning and transmission electron microscopy, electron backscatter diffraction, and automated crystallographic orientation mapping were employed to characterize the details of microstructure evolution and grain fragmentation of the alloy as a function of strain imposed to the material using these SPD techniques. The advantages and drawbacks of these techniques, as well as the limits of their resolution, are discussed in detail. The results of microstructure observations indicate the effectiveness of grain refinement by severe plastic deformation in this alloy. The thermal stability of ultrafine-grained structure that is important for practical applications is also discussed.

**Keywords:** Magnesium, ECAP, HPT, grain refinement, thermal stability

## 1. Introduction

Due to its low density, magnesium alloys are very attractive materials for structural components in automotive, aerospace and other transport industries with the potential to replace steel or aluminium alloys in various applications [1]. Nevertheless, the use of magnesium alloys in more complex applications is limited because of the problems associated with poor corrosion and creep resistance and above all the low ductility. The limited ductility is a consequence of the hexagonal structure providing the lack of independent slip systems and the large difference in the values of the critical resolved shear stress in different slip systems.

Moreover, the strong deformation textures and stress anisotropy in magnesium alloys reduces significantly the variety of possible industrial applications.

The mechanical and other essential properties determining the application of magnesium alloys may be improved by refining the grain size to the submicrocrystalline or even nano-crystalline level. In the last three decades, a variety of new techniques have been proposed for the production of the ultra-fine grain (UFG) structures in materials. The common feature of all these techniques is the imposition of large straining and consequent introduction of very high density of lattice defects in the material resulting in exceptional grain refinement. Since these procedures introduce severe plastic deformation (SPD) to bulk solids, it became convenient to describe all of them as SPD processing. Several processes of SPD are now available but only two of them receiving the most attention at present time, in particular, equal channel angular pressing (ECAP) and high-pressure torsion (HPT) [2–5].

The practical applications of the UFG materials are limited due to low microstructure stability at elevated temperatures that complicates the processing of final products. Thermal stability depends on many variables, such as stacking fault energy of the material, processing, or properties of grain boundaries [6]. Microstructure stability can be improved by various alloying elements or composite reinforcements.

In order to understand the microscopic mechanisms of grain fragmentation during SPD processing, detailed characterization of the microstructure is needed. The objective of this review paper is to employ a wide variety of standard and enhanced microscopic techniques to characterize the microstructure evolution of the UFG magnesium alloy AZ31 by employing several microscopic techniques, in particular, light microscopy, scanning electron microscopy (SEM), transmission electron microscopy (TEM), electron backscatter diffraction (EBSD), and a new sophisticated microscopic method—automated crystallographic orientation mapping in a TEM (ACOM-TEM).

## 2. Experiment

The commercial magnesium alloy AZ31 (3 wt. % of Al, 0.8 wt. % of Zn, and 0.2 wt. % of Mn) was used in this investigation. Two most popular techniques of severe plastic deformation were employed for material processing and microstructure refinement, namely equal-channel angular pressing preceded by hot extrusion (EX-ECAP) and high-pressure torsion (HPT).

### 2.1. Equal-channel angular pressing

Equal-channel angular pressing (ECAP) is one of the most developed and also commercially used SPD procedures. ECAP was invented by Segal et al. in the 1970s and 1980s in the former Soviet Union [7]. Since its invention, ECAP has become well-known and widely used technique. Nowadays, ECAP is a commonly used SPD method applicable in many branches of industries. The main advantage of the ECAP technique is the same cross-sectional dimensions of the specimens before and after pressing. This fact allows repetitive pressing and the accumulation of strain in the specimen. Furthermore, the ECAP is a quite simple process commonly used worldwide.

The ECAP die consists of two channels with the same cross-section. Generally, the ECAP die is characterized by two angles: $\Phi$, which is formed by these two channels, and the angle $\Psi$ that indicates the outer arc of curvature at the intersection of the two channels (see Figure 1).

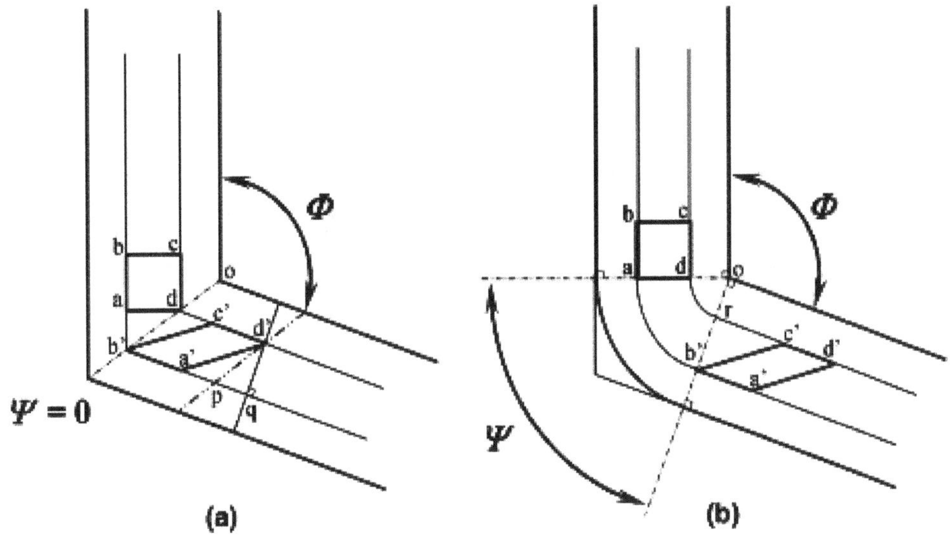


**Figure 1.** The scheme of pressing of the rectangular sample with the square cross-section through the ECAP die a) with $\Psi = 0$ and b) $\Psi \neq 0$ [8].

The specimen pressed through the ECAP die is deformed by a simple shear in the intersection point of the channels. The imposed strain in each ECAP pass is dependent primarily upon the angle $\Phi$ and, to a lesser extent, on the angle $\Psi$. It can be shown from the first principles that the imposed strain $\varepsilon_N$ after N passes is given by a relation in the form [8]:

$$\epsilon_N = \frac{N}{\sqrt{3}}\left[2cotg\left(\frac{\Phi}{2}+\frac{\Psi}{2}\right)+\Psi cosec\left(\frac{\Phi}{2}+\frac{\Psi}{2}\right)\right]. \tag{1}$$

## 2.2. High-pressure torsion

High-pressure torsion (HPT) is another well-known SPD technique. In this method, the mechanical properties of a material are improved by a high pressure and concurrent torsional straining [9]. HPT was also for the first time applied to metals in the former Soviet Union in the 1980s [10]. The experimental setup of HPT is schematically illustrated in Figure 2.

The typical size of the disk-shaped sample varies from 10 to 20 mm in diameter and about 1 mm in thickness. A disk sample is placed between two anvils where it is subjected to a compressive pressure of several GPa. Simultaneously, one of the anvils rotates and the torsional strain is imposed to the sample.

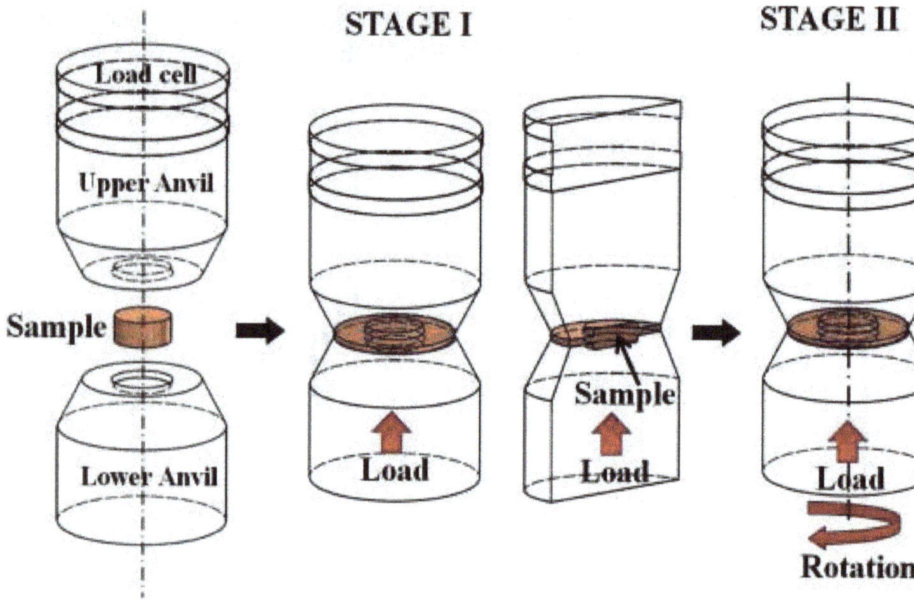


**Figure 2.** Schematic of the HPT device showing the set-up, compression stage (stage I), and compression-torsion stage (stage II) [11].

The total strain imposed by HPT in the sample can be calculated by different approaches. The first widely used approach (Hencky/Eichinger) can be expressed by the logarithmic relation [12]:

$$\epsilon = ln\left(\frac{2\pi N r h_0}{h^2}\right), \tag{2}$$

where $N$ is the number of rotations, $r$ represents the radius of the sample, and $h_0$ and $h$ its initial and final thickness, respectively. The second possible approach (von Mises) using a model of simple torsion could be expressed by the following linear relation [13]:

$$\epsilon = \frac{\gamma}{\sqrt{3}} = \frac{r\theta}{\sqrt{3}h} = \frac{2\pi N r}{\sqrt{3}h}, \tag{3}$$

where $\gamma$ is the shear strain and $\theta$ is the rotation angle.

However, the real strain generated in the workpiece during HPT may be different depending mainly on the die geometry and other factors. No unambiguous conclusion was adopted whether the Hencky or von Mises strain describes better the total strain imposed by HPT. Finite element method (FEM) may be employed to analyze the plastic deformation distribution in specimens processed by HPT. Several authors [14, 15] have used the FEM technique to determine the stress distribution and other parameters, e.g., the influence of the coefficient of friction, torque, etc. Recently, Lee et al. [13] employed a combined analysis of FEM and the model of dislocation cell evolution to describe the local deformation in specimen processed by HPT. The results of this study are shown in Figure 3.

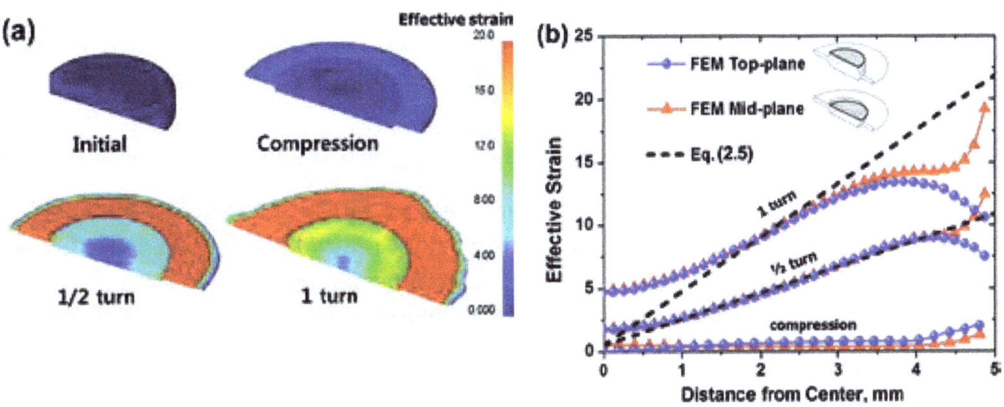


**Figure 3.** FEM simulation of effective strain imposed by HPT. a) The effective strain distribution of the as compressed specimen (N=0), the specimens after 1/2 (N=1/2) and 1 turn (N=1). b) The effective strain distribution throughout the thickness of the specimens N=0, N=1/2, and N=1 in the top and mid-plane showing the comparison of the FEM simulations [13] with the theoretical strain (Eq. 2.5).

## 2.3. Material processing

Prior to ECAP, the material was extruded at 350°C with an extrusion ratio of 22. ECAP pressing was performed at 180°C following route $B_c$, i.e., rotating the sample 90° between the individual passes, with the velocity of 50 mm/min. The angle $\Phi$ between two intersecting channels and the corner angle $\Psi$ were 90° and 0°, respectively. Both channels have a square cross section of 10 mm × 10 mm. A series of specimens after $N$ = 1, 2, 4, 8, and 12 passes was processed by ECAP. The as-extruded specimen was also included in the investigation as the reference material (marked $N$ = 0).

Prior to HPT, the material was homogenized at 390°C for 12 hours. After homogenization the disk specimens of the diameter of 19 mm and the thickness of 1-2 mm were cut from the billet. These specimens were processed by quasi-constrained HPT at room temperature for $N$ = ¼, ½, 1, 5, and 15 rotations. In addition to samples subjected to various number $N$ of HPT rotations, a sample that was pressed only between the anvils but not subjected to any HPT rotation ($N$ = 0) was investigated as well.

### 2.4. Experimental techniques of microstructure investigation

#### *2.4.1. Light microscopy*

The specimen surface for light microscopy observation needs to be plain, without impurities or scratches. Polypropylene was used for mounting the specimens for better manipulation. Mounted specimens were prepared by three consequent procedures: mechanical grinding, polishing, and etching. In the first step, the specimens were mechanically grinded on watered abrasive papers. Then the specimens were polished on a polishing disc with polishing suspension of grade 3 and 1 μm. Using this procedure, flat specimens with minimum surface scratches were obtained. Finally, the polished samples were etched by a diluted solution of picric acid. The time period of etching has to be optimized to attain the best possible visibility and no artefacts on the surface [16].

#### *2.4.2. Electron Backscatter Diffraction (EBSD)*

Electron backscatter diffraction (EBSD) is a microstructural crystallographic technique for determining crystallographic orientation. Texture, grain size and distribution, the misorientation and shape of individual grains, the types of grain boundaries, and many other microstructural features may be obtained from EBSD results.

The principle of the EBSD method is that the accelerated high-energy electrons interact with the atomic lattice of the crystalline material and form the so-called Kikuchi lines on the screen. The measurement is done automatically with a defined step. The formation of Kikuchi lines is caused by inelastic electron scattering. The inelastically scattered electron wavelength is slightly higher than the wavelength of elastically scattered electrons and the inelastic scattered electron intensity decreases with increasing scattering angle. In certain crystal orientations, some planes satisfy the Bragg condition (with diffraction angle $\theta$), and the inelastically scattered electrons are Bragg diffracted. These electrons are called Kikuchi electrons. Kikuchi electrons move along conical surfaces whose top angle is equal ($\pi - 2\theta$) and the axis of diffraction planes is the normal line. Two hyperbolas are formed by the intersection of the conical surfaces and a screen. The hyperbolas seem like straight lines in the central part of the electron diffractogram and the distance between lines corresponds to the angle $2\theta$ [16].

The SEM FEI Quanta 200 FX equipped with EDAX EBSD camera and OIM software was utilized for EBSD observations. A field emission gun (FEG) of a Schottky type was used as a source of electrons in the microscope. Beam voltage in the range from 500 V to 30 kV is applicable. The acceleration voltage employed was 10 kV.

The samples for EBSD investigation were first mechanically grinded on watered abrasive papers of grade 800, 1200, 2400, and 4000, respectively. Then the specimens were mechanically polished with diamond suspensions for water-sensitive materials of grade 3, 1, and ¼ μm, respectively. The final ion-polishing using a Gatan PIPS device was used to remove the surface layer influenced by polishing.

### *2.4.3. Transmission Electron Microscopy (TEM)*

Transmission electron microscopy (TEM) is a technique that studies the local microstructure of the material using a beam of high-energy electrons. The specimens for TEM have to be very thin (≈ 100 nm depending on the acceleration voltage). An image is formed from the interaction of the electrons transmitted through the specimen and focused on a fluorescent screen or CCD camera.

The sample (thin foil) preparation consists of two steps: mechanical thinning and polishing. The type of polishing depends on the physical and chemical properties of the material.

The specimens of AZ31 alloy were first cut from the billet plane X (perpendicular to the extrusion direction) into slices of the thickness of approximately 1 mm. The slices were mechanically grinded from both sides to the thickness of 100 μm. The 3 mm diameter disks were cut out of the thin slices and dimpled from one side to the thickness of approximately 30 μm using the Gatan dimple grinder. This dimpling technique reduces electrolytic or ion polishing times and ensures that the ultra-thin area is done in the central part of the foil.

Ion polishing is used for materials that cannot be polished electrolytically. It is the case of many Mg alloys including AZ31. Ion polishing is a method of removing very fine quantities of the material. It uses an inert gas (argon) to generate a plasma stream that is emitted to the thinned area of the sample and removes the individual layers of the material. Acceleration energies vary usually from 2 to 4 keV. Ion beam always enters the sample. The penetration depth depends on the angle between the direction of the ion beam and the sample surface. The optimum conditions found for AZ31 were the following: acceleration voltage of 4 kV, angle of incidence of 4°, and room temperature. In the final stage the acceleration, voltage was reduced to 2 kV and the angle of incidence to 2°. The specimen preparation for TEM is finished when a small hole is formed in the foil and the surrounding area is thin enough to allow high-energy electrons to pass through the specimen. The schematic of the Precision Ion-Polishing System (PIPS) used for ion polishing is displayed in Figure 4 [17].

The microstructure observations were made with the TEM JEOL 2000FX at Charles University in Prague. The applied accelerated voltage was 200 kV.

### *2.4.4. Automated Crystallographic Orientation Mapping in a TEM (ACOM-TEM)*

Automated crystallographic orientation mapping in a TEM (ACOM-TEM) is an effective technique for mapping phase and crystal orientation and an alternative to the well-known EBSD attachment SEM based on the Kikuchi lines. ACOM-TEM was first introduced by Schwarzer et al. [18]. This method of microstructural characterization represents a powerful tool especially for UFG or nano-grained materials where the limited resolution of SEM disallows EBSD scans. Its basic principle is similar to EBDS mapping—a selected area is scanned with defined step size and the electron diffraction patterns are collected using an external CCD camera. Off-line, every diffraction pattern is compared to the pre-calculated template and the best match is selected. The main difference with respect to EBSD is that point diffraction patterns are analyzed instead of Kikuchi lines, which significantly increases the

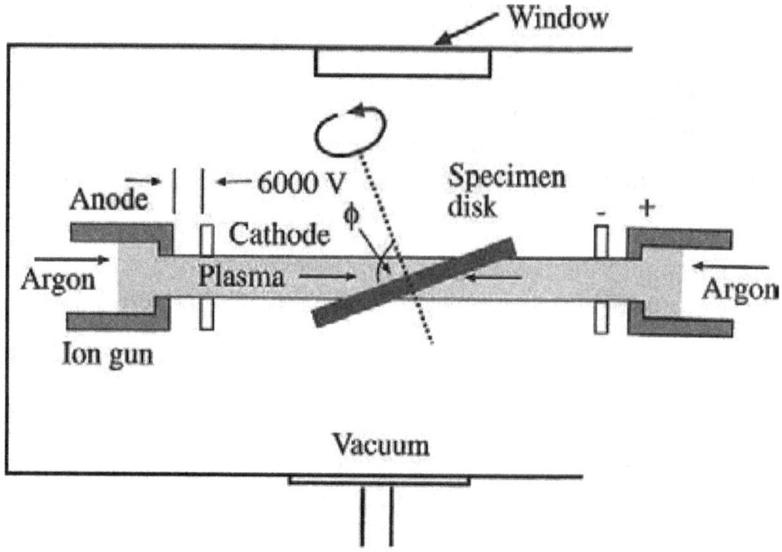


**Figure 4.** Schematic illustration of Precision Ion-Polishing System (PIPS).

precision of orientation determination. This type of ACOM-TEM device was developed in Grenoble by E. Rauch et al. [19, 20].

The experimental measurements were carried out using a TEM Philips CM200 equipped with a $LaB_6$ gun at 200 kV at Université Libre de Bruxelles in Belgium. For the data treatment, TSL OIM Analysis 5 was used.

## 3. Results and discussion

### 3.1. Microstructure evolution of AZ31 processed by ECAP

#### *3.1.1. Light microscopy*

The limited resolution of light microscopy allows to perform only overview images with no structure details. Typical examples of microstructure evolution of extruded specimen and the specimens processed by different number of ECAP passes (N) are shown in Figure 5. The extruded sample microstructure shown in Figure 5a consists of two types of grains—several large grains surrounded by many small grains. The microstructure developed only slightly after the first and the second ECAP pass, the bimodal structure is still observed (see Figure 5b). The microstructure changes begin to be significant in light microscopy after 4 ECAP passes. One can see in Figure 5c that the large grains become smaller. However, the bimodal structure is not observed after 8 and 12 ECAP passes. The homogeneous microstructure of sample after 12 ECAP passes is shown in Figure 5d.

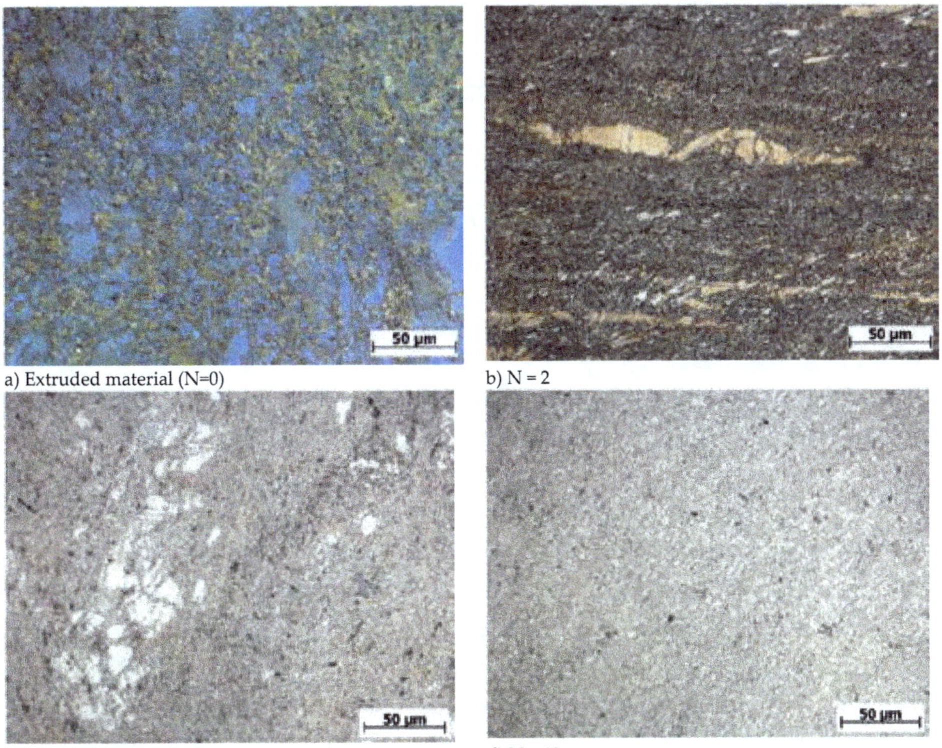


a) Extruded material (N=0) b) N = 2 c) N = 4 d) N = 12

**Figure 5.** The microstructure of the extruded and ECAPed AZ31 alloy.

### *3.1.2. EBSD*

Figure 6 shows the microstructure and texture of the extruded AZ31 alloy. The EBSD inverse pole figure map is in the extrusion direction (ED). The microstructure of the initial extruded bar (0P) consists of large grains of 50–100 ≈m mixed with relatively fine grains of 2–5 μm. Most grains have their crystallographic c-axis perpendicular to the extrusion direction (ED), i.e., <10.0> axes parallel to the ED, which is typically found after extrusion of Mg alloys [21].

Figure 7 presents the microstructure and texture of the sample after 1 ECAP pass (1P). EBSD measurements were carried out at the mid-part of the cross-section of the billet. Sample coordinate system, i.e., X–Y–Z directions, which is used for representing the texture are illustrated in Figure 7d.

As shown in Figure 7a, the bimodal distribution of grain sizes is still observed in the 1P sample. A new texture component that corresponds to the basal poles rotated about 40° from the initial orientation towards the pressing direction is visible in the 1P sample, Figure 7b. The mentioned orientation change, however, is observed mainly in large grains (grain size> 10 μm). Figure 7c presents the (0001) pole figure of grains smaller than 3 μm in the 1P sample. The intensity of the tilted basal poles is rather weak compared to the pole figure constructed using all grains.

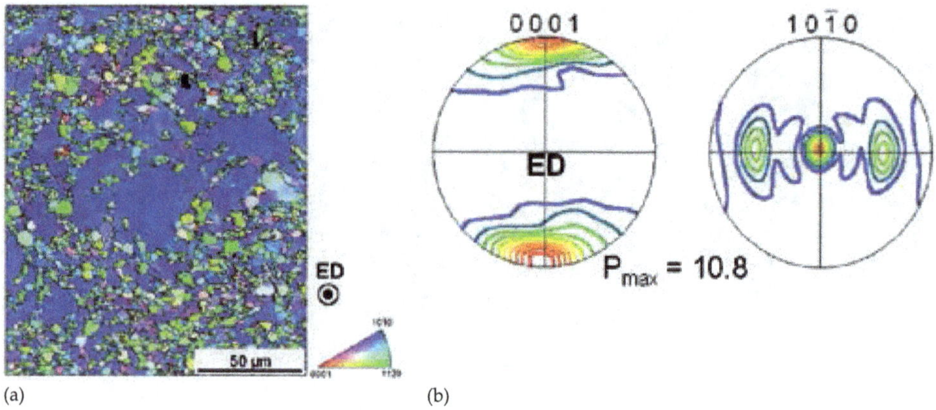


**Figure 6.** (a) EBSD orientation map and (b) recalculated pole figures of the extruded bar (0P), measured at the cross-section transverse to the extrusion direction (contour level = 1, 2,...10).

Moreover, the small grains (< 3 μm) with the tilted basal pole are found mainly in neighboring areas around large grains.

These results regarding distinct textures depending on the grain sizes indicate that the shear strain by the first ECAP pass is mainly accommodated within the large grains in accordance with the Hall–Petch relation. The discontinuity of material flow caused by the inhomogeneous deformation seems to be compensated by the occurrence of the dynamic recrystallization in the vicinity of grain boundaries of large grains, such that the sample could be deformed without failure.

Microstructural features of the sample after 2 ECAP passes (not shown here) are very similar to those after 1 ECAP pass, i.e., the bimodal distribution of grain sizes and the orientation change in large grains remain almost unchanged.

The amount of large grains decreases significantly after 4 ECAP passes (4P), and their size becomes smaller when compared to the initial and the 1P samples (see Figure 8). As shown in Figure 8a with dashed circles, the fine grains usually surround large grains. Unlike the 1P sample, the fine grains (<3 μm) have mainly the orientation of the rotated basal poles in the 4P sample (Figure 8c). It is clear that the texture heterogeneity depending on the grain size disappears after 4 ECAP passes, compared pole figures evaluated from the whole area of the EBSD measurement to that from fine grains, Figure 8b and c, respectively.

The microstructure and the texture of the sample after 12 ECAP passes (12P) are presented in Figure 9. The 12P sample shows the homogeneous distribution of fine, almost equiaxed grains, i.e., no large grains are visible. Figure 9a shows grain agglomerates with different colors distributed along a diagonal line. This indicates the heterogeneity in texture depending on the locations. This texture inhomogeneity can be understood as a result of non-uniform deformation along the ECAP billet after multiple passes [22]. Though the heterogeneous texture is visible, the fraction of the grains relating to the inhomogeneity is small such that the main texture component is found at the rotated basal pole, Figure 9b.

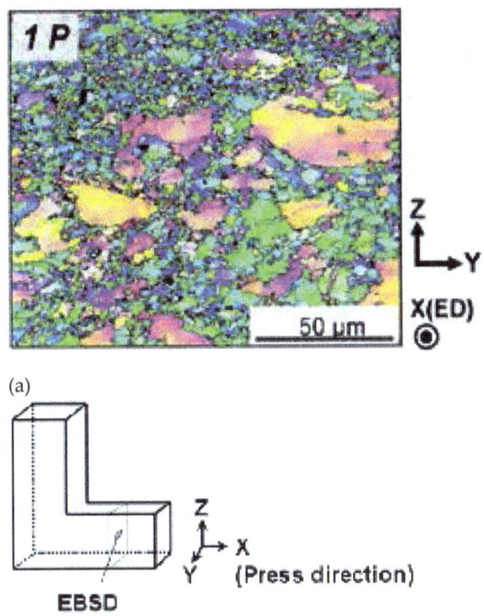


(a) (b) (c)

(d)

**Figure 7.** (a) EBSD orientation map of the sample after 1 ECAP pass, (b) (0001) pole figure calculated using all grains, (c) (0001) pole figure of grains smaller than 3 μm, and (d) the scheme of the geometry of specimens for EBSD measurements (contour level = 1, 2,... 10).

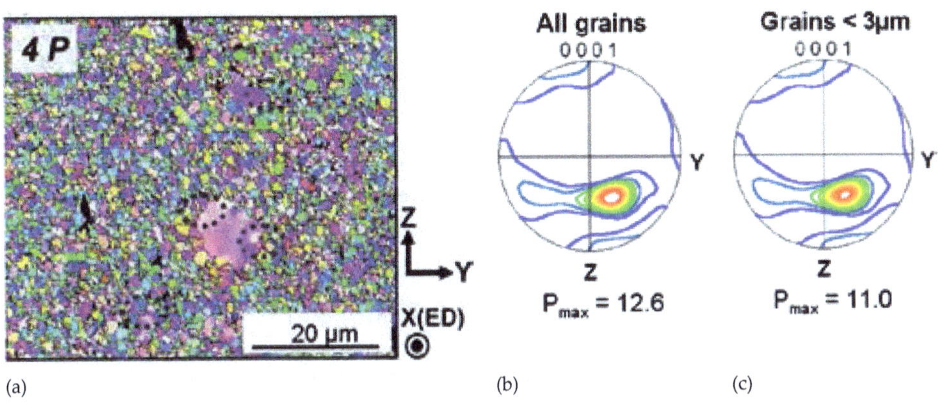


(a) (b) (c)

**Figure 8.** (a) EBSD orientation map of the sample after 4 ECAP passes, (b)(0001) pole figure of all grains, and (c) grains smaller than 3 μm (contour level = 1, 2,... 10).

Figure 10 displays size distributions of fine grains (< 10 μm) after different numbers of ECAP passes. The variation in the area fraction of large grains (>10 μm) is shown in Table 1. Grains smaller than 3 μm present in the extruded sample (0P) become finer (μ 1 lm) after the first ECAP pass. In subsequent ECAP passes no significant change in the size distribution in the range of the relatively fine grains is observed, while the area fraction of grains larger than 10 ≈m decreases gradually with ECAP passes, see Table 1.

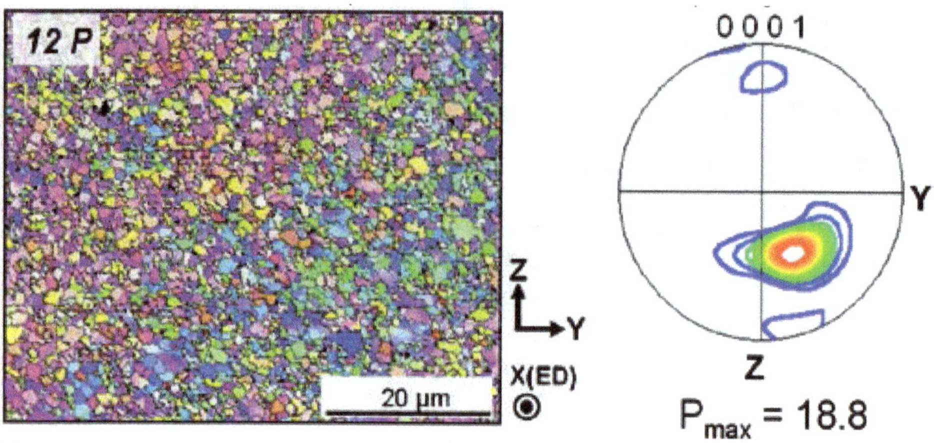


(a) (b)

**Figure 9.** (a) EBSD orientation map and (b) (0001) pole figure of the sample after 12 ECAP passes (contour level = 1, 2,... 15).

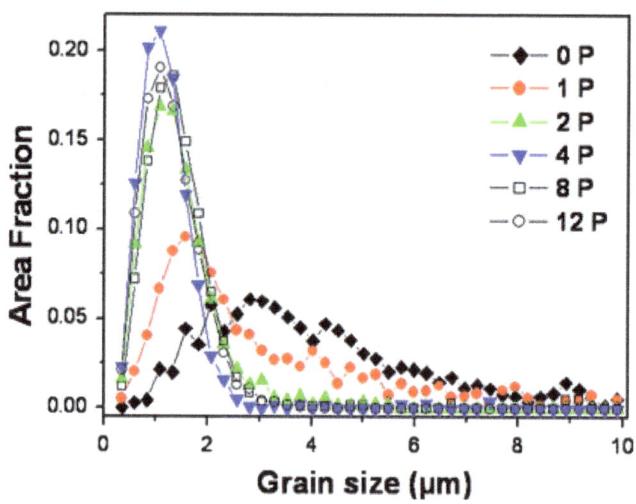


**Figure 10.** Grain size distributions within the range up to 10 μm and the area fraction of the grains larger than 10 μm as a function of the ECAP pass number.

| No. ECAP Passes | 0 | 1 | 2 | 4 | 8 |
|---|---|---|---|---|---|
| Area fraction of grains > 10μm | 17.9% | 18.1% | 7.9% | 2.8% | None |

**Table 1.** Area fraction of large grains as a function of ECAP pass number.

The variation in the fraction of low-angle boundaries (misorientation angle < 15°, LAGB) and high-angle boundaries (misorientation angle > 15°, HAGB) as a function of the ECAP pass number is shown in Figure 11. The LAGB fraction increases after 1 ECAP pass to μ 50%, and is significantly reduced (≈20%) after two ECAP passes and remains almost unchanged as the number of passes increases above 2. This stabilization of the HAGB formed by 2 ECAP passes indicates that the dislocations generated by ECAP did not reach the level necessary for the formation of dislocation in the cell-structure and further development of HAGB. Since the ECAP in this study was carried out at moderate temperature, the dislocation annihilation process by dynamic recovery occurred fast enough such that the grains could not be further refined after 2 ECAP passes.

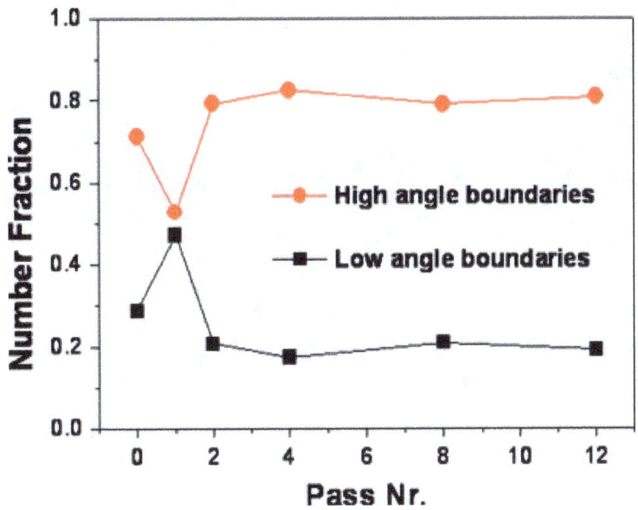


**Figure 11.** Variation of fraction of the HAGB and LAGB as functions of the ECAP pass number.

### *3.1.3. TEM*

Additional detailed observations using TEM broadened the EBSD research. The samples for TEM observations were cut perpendicular to the pressing direction from the EX-ECAPed billet and prepared by mechanical grinding, dimpling, and ion polishing. The microstructure of extruded samples is bimodal—consists of large grains of 50–100 μm mixed with relatively fine grains of 2–5 μm. Figure 12 shows the area with smaller grains of the average size of few μm. Typical twins are clearly seen in several grains. The (0001) basal texture of individual grains in extruded magnesium alloy observed in EBSD measurements was confirmed by electron diffraction analysis. The contrast of individual grains on this micrograph with typical low-angle grain boundaries confirmed the analysis of diffraction patterns. The area of larger grains had a typical heavily deformed structure with high density of tangled dislocations (not shown here).

Figure 13 presents the microstructure of the specimen after 1 EX-ECAP pass. The bimodal character of the microstructure did not change after the first ECAP pass; the average grain size is only slightly smaller in comparison with the extruded material, see Figure 13a. However,

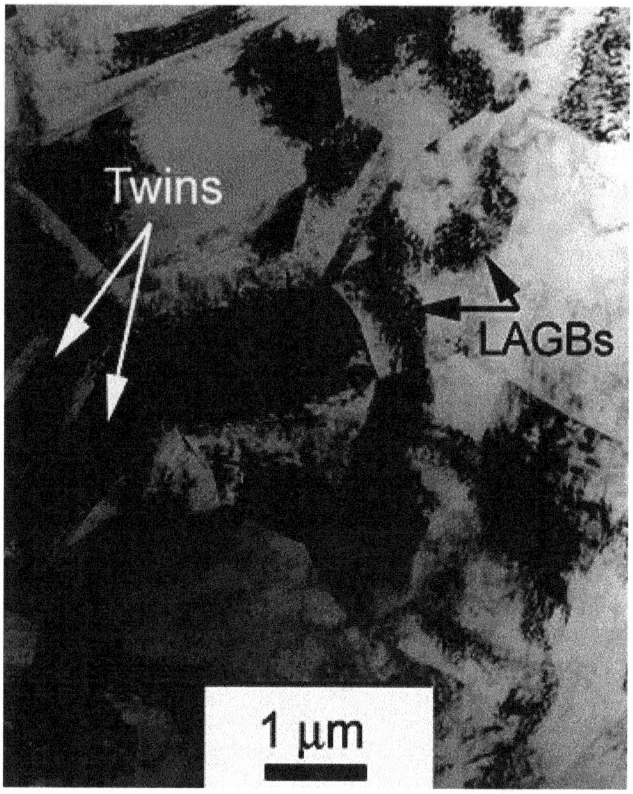


**Figure 12.** Transmission electron micrograph of extruded AZ31 alloy.

one significant difference between these two micrographs is observed, namely the character of grain boundaries. One can see many high-angle grain boundaries with typical thickness fringe band contrast confirming their almost equilibrium state in this micrograph. On the other hand, several grain boundaries remained in a non-equilibrium state with diffuse fuzzy contrast and many dislocations lying in a grain boundary plane were also observed. Several areas with high density of tangled dislocations with no or only exceptional signs of substructure formation were found in this specimen, see Figure 13b.

Subsequent TEM observation confirmed that the microstructure of EX-ECAPed samples changed only partly with increasing number of ECAP passes—fine grains were refined only slightly while the large grain zones were refined significantly. Typical microstructure of the sample processed by 2 EX-ECAP passes is shown in Figure 14.

Only fine grains of the average grain size in the submicrometer range were observed in the micrographs of the specimens after 8 and 12 ECAP passes. Figure 15a presents the typical microstructure of the sample after 8 passes. One can see that the grains in this micrograph are equiaxed, and the grain size is approximately 800 nm. This material has significantly lower density of dislocations and equilibrium grains boundaries. A few newly recrystallized very small grains with no dislocations and sharp equilibrium boundaries were also found in the

a) b)

**Figure 13.** Transmission electron micrographs of AZ31 alloy after extrusion and 1 pass of ECAP, (a) equilibrium and HAGB region and (b) nonequilibrium GB region.

**Figure 14.** Transmission electron micrograph of AZ31 alloy after extrusion and 2 passes of ECAP.

microstructure after 8 and 12 passes of ECAP as seen in Figure 15b. TEM observations confirm the process of microstructure evolution and its fragmentation as observed by EBSD. In particular, large grains were continuously refined during subsequent ECAP pressing while the fine grains remained almost unchanged.

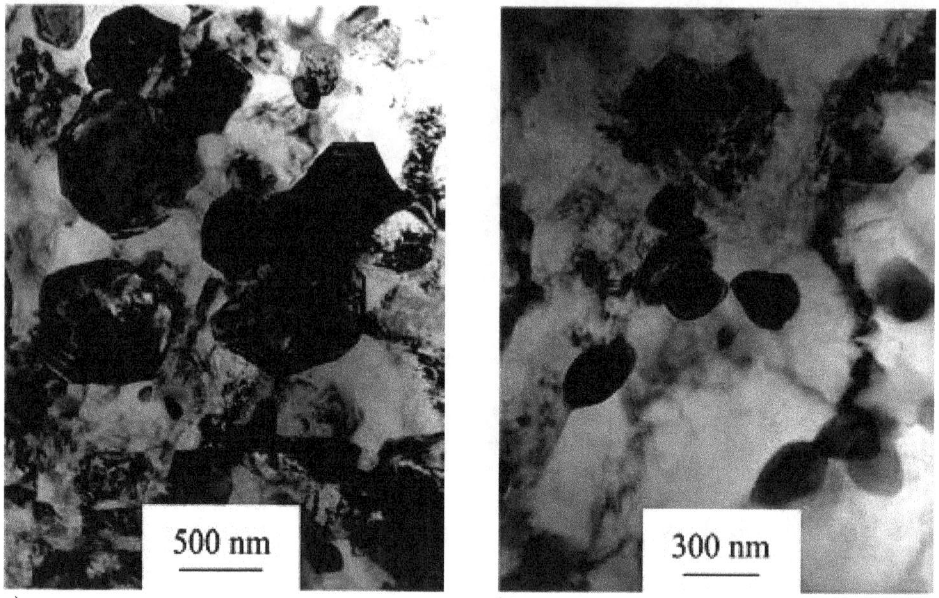


a) b)

**Figure 15.** Transmission electron micrograph of the AZ31 alloy (a) after 8 passes of EX-ECAP and (b) newly recrystallized small grains after 12 passes of EX-ECAP.

### 3.2. Thermal stability of the UFG structure investigated by EBSD

Microscopic techniques, in particular EBSD, can also contribute to the characterization of thermal stability of the UFG structure. The maintenance of the fine-grained structure is of particular importance for many physical properties, e.g., the superplasticity, etc. [23, 24].

Series of specimens for thermal stability investigation were prepared by isochronal annealing at the temperatures 150–500°C for 1 h followed by water-quench. The samples after extrusion and 4 passes of ECAP were chosen as suitable materials because of their UFG and homogeneous microstructure with relatively high dislocation density (see also Figure 8a).

The microstructure and grain size distribution of the specimen in the initial non-annealed condition after extrusion and 4 passes of ECAP are shown in Figure 16a and 16b, respectively. The microstructure is homogeneous comprising of very fine grains with an average size of 0.9 μm. The microstructure and grain sizes of the samples after 1 h of isochronal annealing at 150°C and 170°C (not shown here) are very similar to the non-annealed specimen.

Inhomogeneous grain growth is observed at higher annealing temperatures as displayed at Figure 17. Some grains start to grow at temperatures of 190°C (Figure 17a) and 210°C (not shown here). The fraction of coarse grains increases with increasing annealing temperature. At the temperature of 250°C, some areas with original fine grains are still observed, see Figure 17b. However, the small grains are continuously disappearing at higher annealing temperatures and nearly no small grains are observed after annealing at 400°C, see Figure 17e. Please

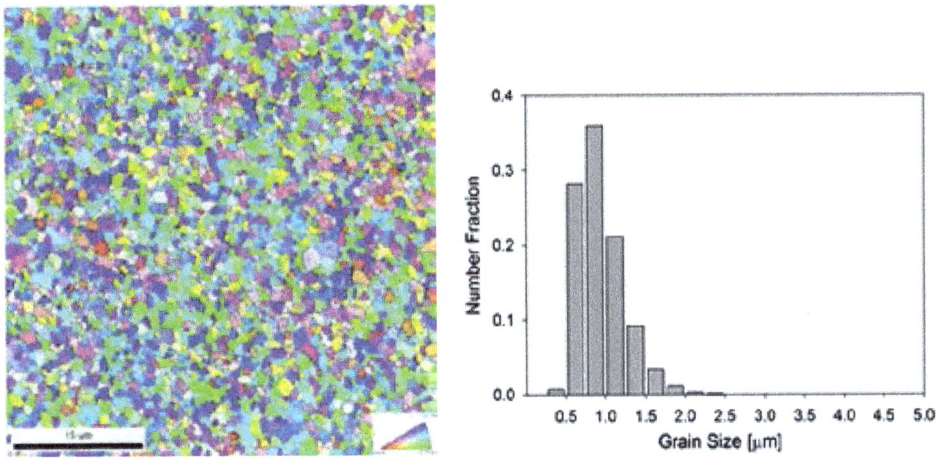


a) b)

**Figure 16.** Microstructure of the AZ31 sample after extrusion and 4 passes of ECAP, (a) inverse pole figure map and (b) grain size distribution.

note that magnification of Figure 17d and 17e is two times smaller than the magnification of the previous inverse pole figure maps; simultaneously, four times bigger area of the sample (100 × 100 μm) is depicted in Figure 17d and 17e. Microstructure of the specimens annealed at 450°C and 500°C was observed by light microscope and is displayed in Figure 18.

Grain size distribution and average grain size of the samples studied by EBSD were calculated from data measured in the area 100 × 100 μm for all annealing temperatures to achieve better statistics. Statistical values from the samples studied by light microscopy were calculated from two or more images in order to get more than 1000 grains for each sample. Grains intersecting the edge of EBSD maps and light microscopy images were not included in the analysis. The dependence of average grain size (number average) on annealing temperature is plotted in Figure 19. In samples annealed at 250°C and 300°C, the average values were calculated from the bimodal grain size distribution. The values of the average grain sizes are summarized in Table 2.

| **Annealing temperature [°C]** | - | 170 | 190 | 210 | 250 | 300 | 350 | 400 | 450 | 500 |
|---|---|---|---|---|---|---|---|---|---|---|
| **Average grain size d [μm]** | 0.94 | 0.99 | 1.05 | 1.48 | 1.83 | 2.06 | 3.04 | 3.79 | 10.09 | 24.53 |

**Table 2.** The average grain sizes at different annealing temperatures.

Annealing twins observed after annealing at 250–400°C (see Figures 17b and 17e) were excluded from grain size calculations to achieve true grain size values (twin boundaries were ignored in grain size calculations, i.e., the twin is considered to be a part of the grain). All these

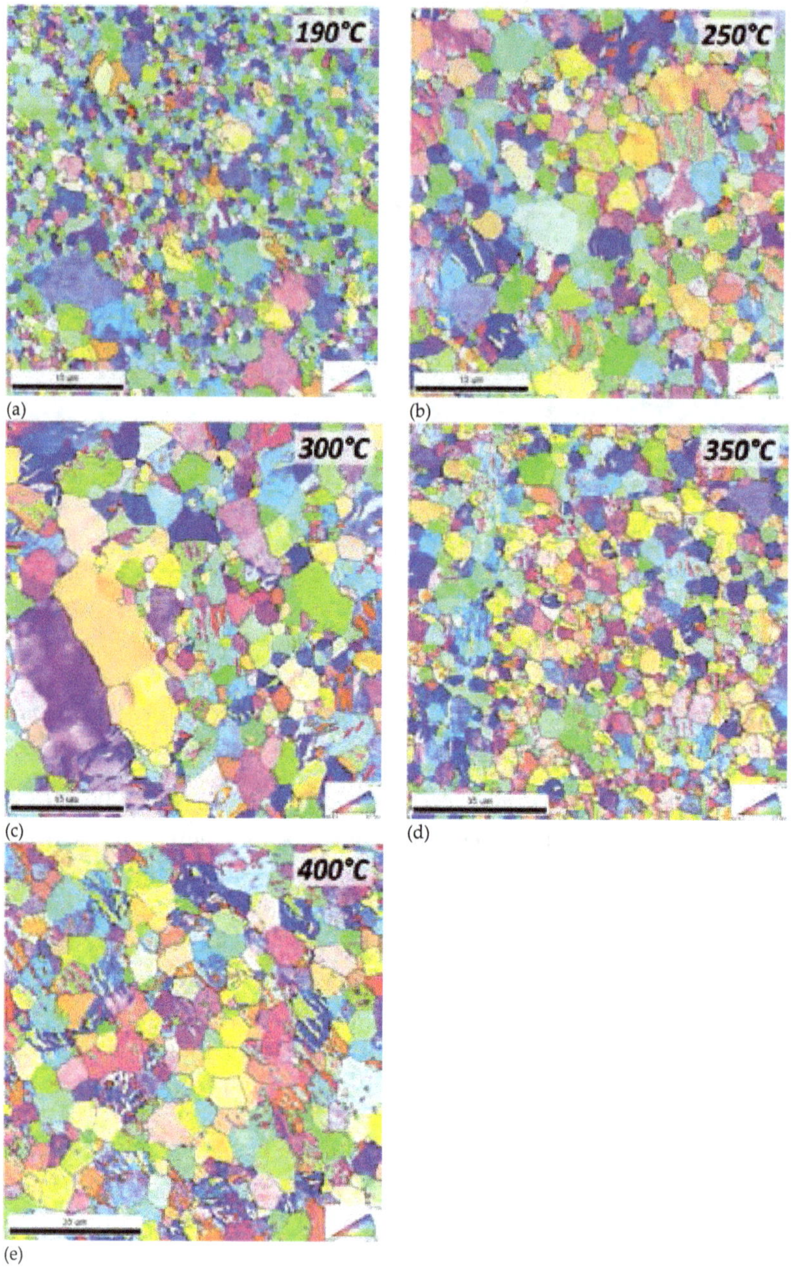


Figure 17. Microstructure of the AZ31 EX-ECAP specimen after 1 h of isochronal annealing at (a) 190°C, (b) 250°C, (c) 300°C, (d) 350°C, and (e) 400°C (EBSD inverse pole figure maps).

twins were determined as the tensile twins with misorientation angle of 86° [25]. Twin fractions, fraction of twinned grains and the area fraction of twinned grains in samples after annealing at $T \geq 250$ °C are shown in Table 3.

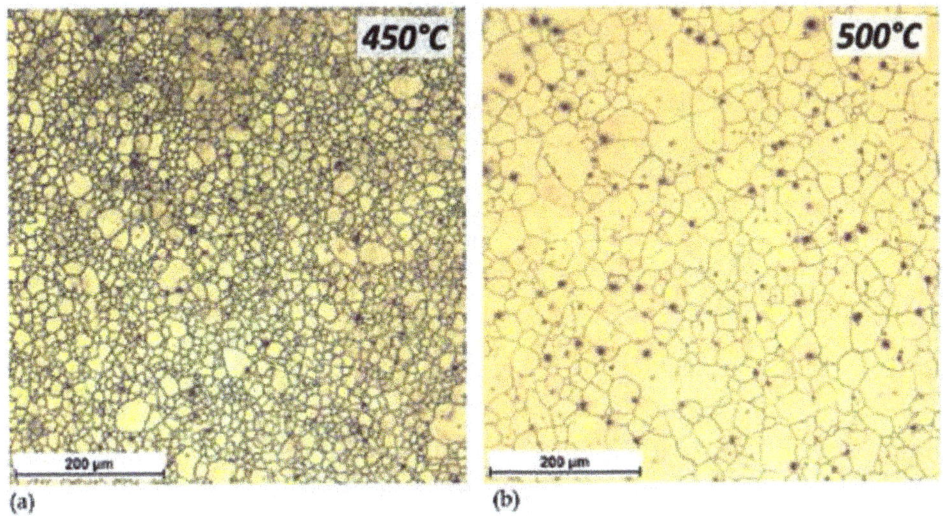


**Figure 18.** Microstructure of the AZ31 EX-ECAP specimen after 1 h of isochronal annealing at (a) 450°C and (b) 500 °C (light microscope images).

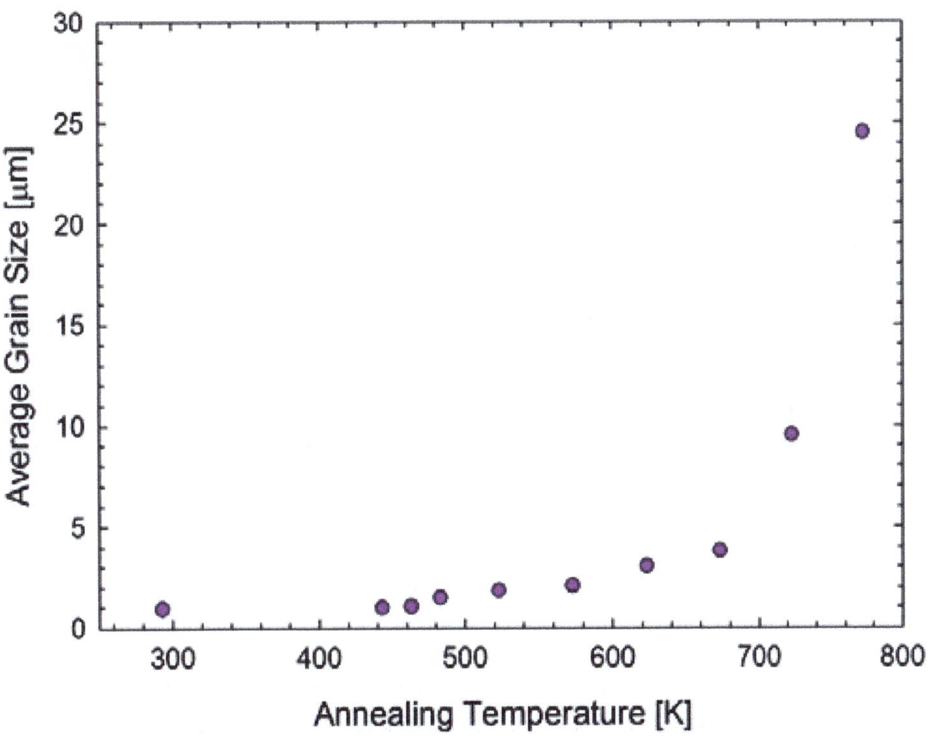


**Figure 19.** Dependence of the average grain size (number average, excluding twins) of the EX-ECAP AZ31 alloy on annealing temperature after 1 h of isochronal annealing process.

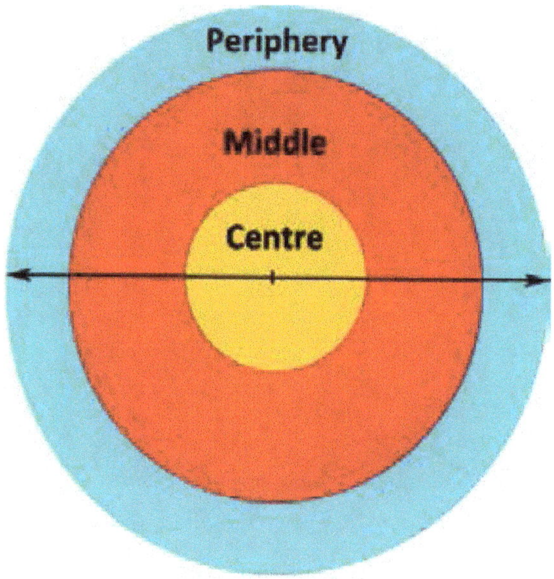


**Figure 20.** Schematic illustration of three different parts of the disks processed by HPT.

| Annealing Temperature | 250°C | 300°C | 350°C | 400°C |
|---|---|---|---|---|
| Twin Fraction | 0.13 | 0.12 | 0.16 | 0.14 |
| Fraction of Twinned Grains | 0.26 | 0.26 | 0.37 | 0.40 |
| Area Fraction of Twinned Grains | 0.55 | 0.52 | 0.60 | 0.71 |

**Table 3.** Twinning in isochronally annealed EX-ECAP AZ31 alloy.

### 3.3. Microstructure evolution in UFG AZ31 processed by HPT investigated by ACOM-TEM

The microstructure of the materials after SPD cannot be often observed by light microscopy because of their very small grain sizes. EBSD is usually a much better and suitable method in this case. However, even EBSD has some limitation in resolution or some "more local" method could be required for microstructure observations. This often occurs in specimens processed by high pressure torsion. This method proved to be more effective in grain refinement than the ECAP due to mainly much larger strains imposed on the HPT disk during processing and also lower temperatures of processing as compared to ECAP [2, 9]. TEM is a well-known technique that usually solves the problem. ACOM-TEM is a very sophisticated method, alternative to EBSD, which enables to study extremely fine-grained materials. We used this technique for microstructure observations of the most deformed parts of HPT specimens. Due to the inhomogeneous character of strain imposed to the material by HPT, disks processed by HPT could be divided into three parts: center, middle, and periphery (see Figure 20).

Due to relatively low imposed strain after one HPT rotation, EBSD is sufficient to be employed for microstructure observation. Figures 21a and 21b show deformed microstructures of the central and middle parts of the specimen after 1 turn studied by EBSD. The areas investigated by EBSD could be relatively big—in our case it was the area 100 × 100 μm. The step size of these measurements was 100 nm. Figure 21a displays a heavily deformed microstructure of the central part of the specimen with a lot of multiple twinning. The microstructure of the middle part of the sample after 1 HPT rotation, shown in Figure 21b, consists of several large elongated grains with no twins and a lot of new grains of the average size of few microns.

The same middle part of this specimen was studied using ACOM-TEM and the typical results are displayed in Figure 22a. Measurements were done with step size 20 nm, which is five times smaller than the step size of the EBSD measurements. ACOM-TEM measurements confirmed heterogeneous microstructures with large elongated grains and a lot of small grains. The microstructure of the peripheral part (see Figure 22b) is homogeneous with very small grains only (≈ 100–300 nm).

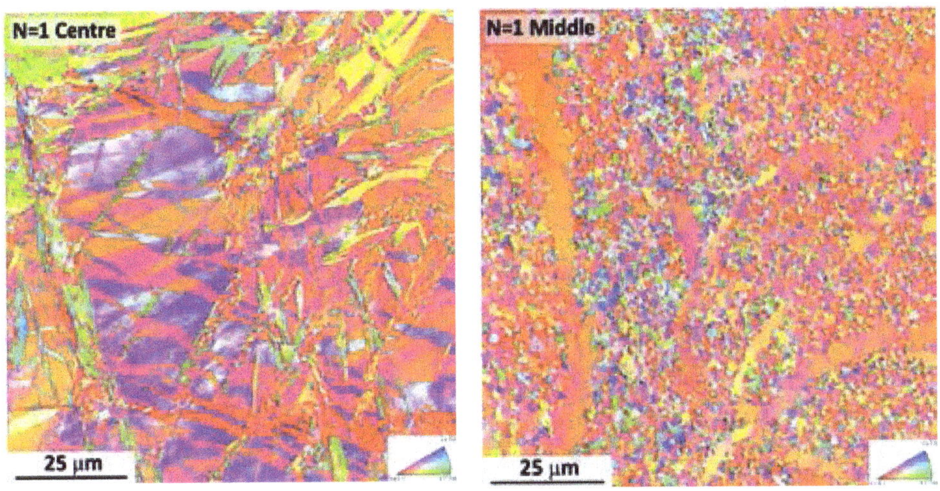


(a) (b)

**Figure 21.** EBSD IPF image of microstructure of the AZ31 sample after 1 turn of HPT, (a) central part and (b) middle part.

The sample, after 5 turns of HPT, was investigated using ACOM-TEM in the middle and the peripheral part of the disk. The microstructure of the middle part (see Figure 23a) is comparable with the middle part of the sample after 1 HPT rotation; it consists of large elongated deformed grains and a lot of small grains surrounding the bigger grains. The microstructure of the peripheral part of the disk after 5 HPT turns, shown in Figure 23b, is composed mainly of very small grains. However, few remaining grains with the size of several microns could be still observed in the peripheral part. The ACOM-TEM is a very local method where the relatively small part of a thin foil could be investigated. Thus, the statistics of measurements of heterogeneous microstructure is not sufficient. Preparation and investigation of more TEM foils is needed to improve the statistics (despite each image consisting of observations at four different zones of a foil).

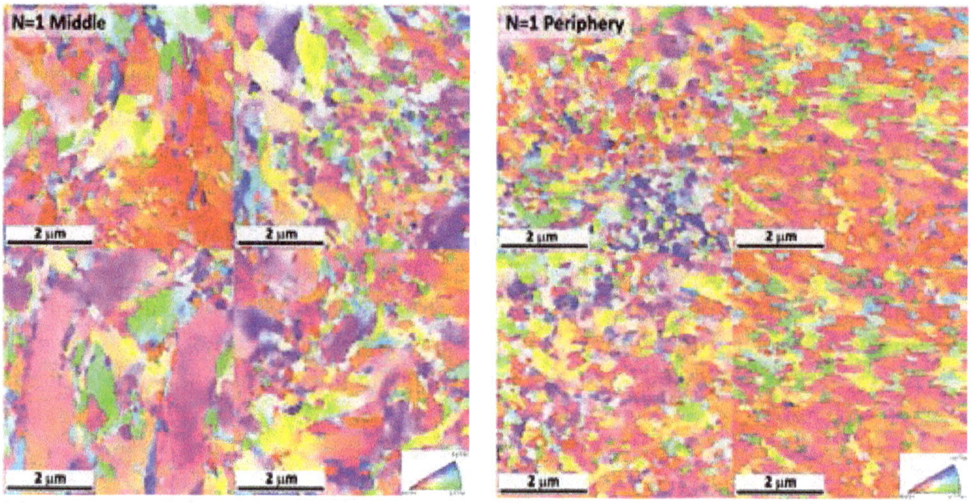


(a) (b)

**Figure 22.** ACOM-TEM image of the microstructure of the AZ31 sample after 1 turn of HPT, (a) middle part and (b) periphery.

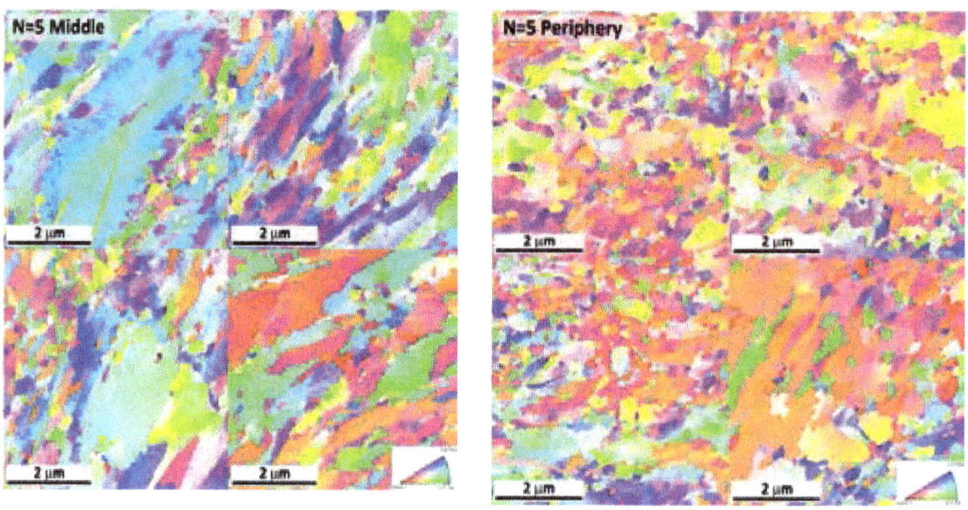


(a) (b)

**Figure 23.** ACOM-TEM image of the microstructure of the AZ31 sample after 5 turns of HPT, (a) middle part and (b) periphery.

Figures 24a and 24b display the microstructure of the middle and peripheral part of the disk after 15 HPT rotations, respectively. The microstructure of both parts is more homogeneous than the microstructure of the same parts of the disk after 5 HPT turns.

Grain size distributions of extremely fine grained microstructure of the peripheral parts of the disks after 1 and 15 HPT rotations (see Figure 25) show that about 60%, and nearly 70%, of

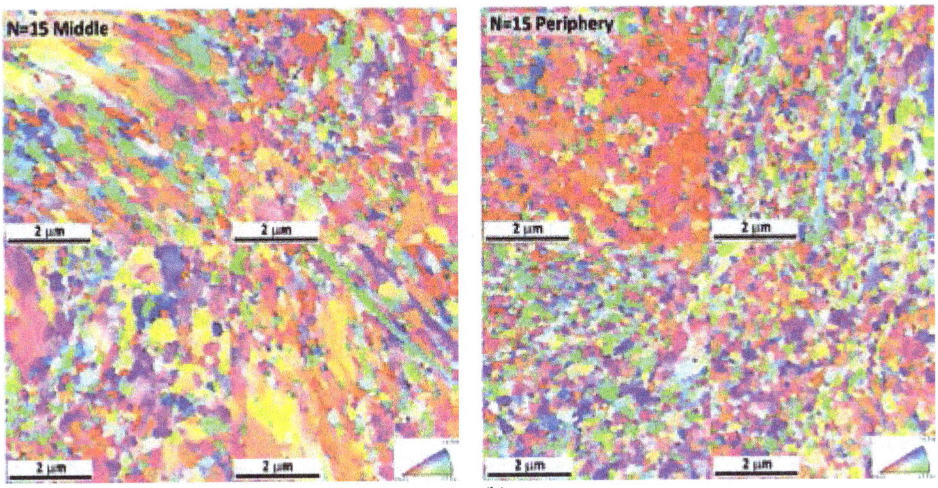


(a) (b)

**Figure 24.** ACOM-TEM image of the microstructure of the AZ31 sample after 15 turns of HPT, (a) middle part and (b) periphery.

grains, respectively, have a diameter of 100–200 nm that proves clearly the efficiency of grain refinement by HPT.

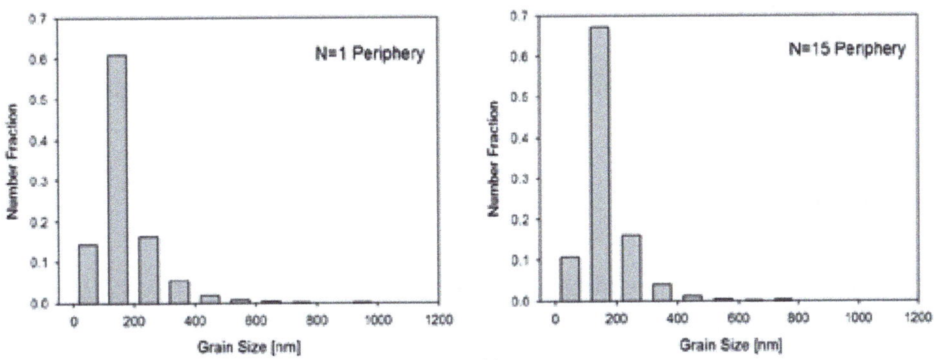


a) b)

**Figure 25.** Grain size distribution of the peripheral part of the disk after (a) 1 and (b) 15 rotations of HPT.

## 4. Conclusions

Microstructure evolution in ultra-fine grained magnesium alloy AZ31 prepared by severe plastic deformation was investigated by several experimental techniques. The following conclusions may be drawn from this investigation:

- Due to limited resolution light microscopy may be employed to investigate the initial stages of grain fragmentation, i.e., the specimens processed by low number of ECAP passes or low number of HPT turns and especially in zones around the disk centers;
- EBSD proved to be a very powerful technique for the investigation of various stages of grain refinement. It allows to characterize not only the grain fragmentation, but also to determine other microstructural features as the grain size distribution and the grain boundary character distribution (the fraction of LAGBS vs. HAGBs), as well as the texture evolution with strain imposed to the material by ECAP and/or HPT;
- TEM allows to characterize the details of the microstructure, namely the dislocation arrangements, grain boundary character (equilibrium vs. nonequilibrium grain boundaries), twins, twin and other special boundaries, etc.;
- The special technique of ACOM-TEM may be used to characterize the terminal stages of grain refinement (saturation) with grain sizes approaching to nanometer range (grain size < 100 nm), which is typical for peripheral zones of specimens processed by multiple turns of HPT;
- Complex microstructure characterization by different techniques of electron microscopy allows to understand the microscopic mechanisms of grain refinement, grain fragmentation, structure stability, as well as other important properties of ultra-fine grained materials processed by severe plastic deformation.

## Acknowledgements

This work was financially supported by the Czech Science Foundation under the project GB14-36566G.

## Author details

Jitka Stráská, Josef Stráský, Peter Minárik, Miloš Janeček* and Robert Král

*Address all correspondence to: janecek@met.mff.cuni.cz

Charles University in Prague, Department of Physics of Materials, Prague, Czech Republic

## References

[1] Gupta M, Sharon NML. Magnesium, Magnesium Alloys, and Magnesium Composites. 1st ed. Wiley; 2011.

[2] Langdon TG. Twenty-five years of ultrafine-grained materials: Achieving exceptional properties through grain refinement. Acta Mater. 2013;61:7035–7059. DOI: 10.1016/j.actamat.2013.08.018.

[3] Estrin Y, Vinogradov A. Extreme grain refinement by severe plastic deformation: A wealth of challenging science. Acta Mater. 2013;61:782–817. DOI: 10.1016/j.actamat.2012.10.038.

[4] Horita Z, Matsubara K, Makii K, Langdon TG. A two-step processing route for achieving a superplastic forming capability in dilute magnesium alloys. Scr. Mater. 2002;47:255–260. DOI: 10.1016/S1359-6462(02)00135-5.

[5] Matsubara K, Miyahara Y, Horita Z, Langdon TG. Developing superplasticity in a magnesium alloy through a combination of extrusion and ECAP. Acta Mater. 2003;51:3073–3084. DOI: 10.1016/S1359-6454(03)00118-6.

[6] Gubicza J, Chinh NQ, Dobatkin SV, Khosravi E, Langdon TG. Stability of Ultrafine-Grained Microstructure in Fcc Metals Processed by Severe Plastic Deformation. Key Eng. Mater. 2011;465:195–198. DOI: 10.4028/www.scientific.net/KEM.465.195.

[7] Segal VM, Reznikov VI, Drobyshevskiy AE, Kopylov VI. Plastic working of metals by simple shear. Russ. Metall. 1981;1:99.

[8] Iwahashi Y, Wang J, Horita Z, Nemoto M, Langdon TG. Principle of equal-channel angular pressing for the processing of ultra-fine grained materials. Scr. Mater. 1996;35:143–146. DOI: 10.1016/1359-6462(96)00107-8.

[9] Zhilyaev AP, Langdon TG. Using high-pressure torsion for metal processing: Fundamentals and applications. Prog. Mater. Sci. 2008;53:893–979. DOI: 10.1016/j.pmatsci.2008.03.002.

[10] Kuznetsov RI, Bykov VI, Chernyshev VP, Pilyugin VP, Yefremov NA, Pasheyev AV. Plastic Deformation of Solid Bodies Under Pressure: 1. Equipment and Methods. Preprint 4/85, Sverdlovsk: IFM UNTs AN SSSR; 1985.

[11] Song Y, Yoon EY, Lee DJ, Lee JH, Kim HS. Mechanical properties of copper after compression stage of high-pressure torsion. Mater. Sci. Eng. A. 2011;528:4840–4844. DOI: 10.1016/j.msea.2011.02.020.

[12] Degtyarev MV, Chashchukhina TI, Voronova LM, Patselov AM, Pilyugin VP. Influence of the relaxation processes on the structure formation in pure metals and alloys under high-pressure torsion. Acta Mater. 2007;55:6039–6050. DOI: 10.1016/j.actamat.2007.04.017.

[13] Lee DJ, Yoon EY, Park LJ, Kim HS. The dead metal zone in high-pressure torsion. Scr. Mater. 2012;67:384–387. DOI: 10.1016/j.scriptamat.2012.05.024.

[14] Kim HS. Finite element analysis of high pressure torsion processing. J. Mater. Process. Technol. 2001;113:617–621. DOI: 10.1016/S0924-0136(01)00709-9.

[15] Yoon SC, Horita Z, Kim HS. Finite element analysis of plastic deformation behavior during high pressure torsion processing. J. Mater. Process. Technol. 2008;201:32–36. DOI: 10.1016/j.jmatprotec.2007.11.204.

[16] Vrátná J. Physical Properties of Ultrafine-grained Polycrystals of Magnesium Based Alloys [thesis]. Prague: Charles University in Prague; 2010.

[17] Williams DB, Carter CB. Transmission Electron Microscopy: A Textbook for Materials Science. 2nd ed. New York: Springer, 2009. ISBN 9780387765020.

[18] Schwarzer RA, Zaefferer S. An Inexpensive CCD Camera System for the Recording and On-Line Interpretation of TEM Kikuchi Patterns. Mater. Sci. Forum. 1994;157-162:189–194. DOI: 10.4028/www.scientific.net/MSF.157-162.189.

[19] Rauch EF, Veron M. Coupled microstructural observations and local texture measurements with an automated crystallographic orientation mapping tool attached to a TEM. Mater. Werkst. 2005;36:552–556. DOI: 10.1002/mawe.200500923.

[20] Rauch EF, Dupuy L. Rapid spot diffraction patterns idendification through template matching. Arch. Metall. Mater. 2005;50:87–99.

[21] Bohlen J, Yi SB, Swiostek J, Letzig D, Brokmeier HG, Kainer KU. Microstructure and texture development during hydrostatic extrusion of magnesium alloy AZ31. Scr. Mater. 2005;53:259–264. DOI: 10.1016/j.scriptamat.2005.03.036.

[22] Agnew SR, Mehrotra P, Lillo TM, Stoica GM and Liaw PK. Crystallographic texture evolution of three wrought magnesium alloys during equal channel angular extrusion. Mater. Sci. Eng. A. 2005;408:72–78. DOI: 10.1016/j.msea.2005.07.052.

[23] Valiev RZ, Islamgaliev RK, Alexandrov IV. Bulk nanostructured materials from severe plastic deformation. Prog. Mater. Sci. 2000;45:103–189. DOI: 10.1016/S0079-6425(99)00007-9.

[24] Valiev RZ, Estrin Y, Horita Z, Langdon TG, Zechetbauer MJ, Zhu YT. Producing bulk ultrafine-grained materials by severe plastic deformation. JOM. 2006;58:33–39. DOI: 10.1007/s11837-006-0213-7.

[25] Aghababaei R, Joshi SP. Micromechanics of tensile twinning in magnesium gleaned from molecular dynamics simulations. Acta Mater. 2014;69:326–342. DOI: 10.1016/j.actamat.2014.01.014.

# 10

# Structure and Dynamics of Macromolecular Assemblies from Electron Microscopy Maps

Xiongwu Wu and Bernard R. Brooks

**Abstract**

Electron microscopy (EM) and electron tomography (ET) have found extensive application to probe structural and dynamic properties of macromolecular assemblies. As an important complementary to X-ray and NMR in atomic structural determination, EM has reached a milestone resolution of 2.2Å, enough for understanding atomic interactions, interpreting mechanism of functions, and for structure-based drug design. This work describes approaches to derive structural information from EM/ET images and methods to study dynamics of macromolecular systems using EM/ET images as starting or ending targets. For low-resolution EM/ET maps, X-ray or NMR atomic structures of molecular components are needed to reduce the number of degrees of uncertainty. Depending on the resolution of EM/ET maps and the conformational differences from the X-ray or NMR structures, either rigid fitting or flexible fitting is used to obtain atomic structures. To illustrate the procedures of the atomic structure derivation, this work describes the core-weighted grid-threading Monte Carlo (CW-GTMC) rigid fitting and the map-restrained self-guided Langevin dynamics (MapSGLD) flexible fitting methods. Their applications are highlighted with four examples: architecture of an icosahedral pyruvate dehydrogenase complex, dynamics of a group II chaperonin, high-resolution structure of the cell-permeant inhibitor phenylethyl β-D-thiogalactopyranoside, and the mechanism of kinesin walking on microtube.

**Keywords:** Electron microscopy, electron tomography, rigid fitting, flexible fitting, molecular simulation, self-guided Langevin dynamics, map restraint

## 1. Introduction

The advance of cryo-electron microscopy (cryo-EM) and electron tomography (ET) opens a new window to the analysis of large biomolecular assemblies under biologically relevant conditions. Figure 1 illustrates the basic procedure from EM/ET experiment to the understanding of molecular complexes. 2D images of samples are the raw data from EM/ET experiment. A 3D reconstruction is needed to build volume maps, which often needs an initial 3D model to start with. In most cases, the resolution of EM/ET maps are not high enough to determine atom positions; therefore, people rely on known high-resolution structures from X-ray, NMR, or homology modelling to derive structures of molecular assemblies through either rigid fitting or flexible fitting. Because molecular environments are different in X-ray crystal, in NMR solution, and on cryo-EM grids, molecular structures in these samples are not identical, especially in the side chains, in loops, and around hinge regions. When the resolution of EM/ET maps is not high enough to show structure difference from X-ray or NMR structures, rigid fitting is often used to produce complex structures [1-8].

When EM/ET maps have high enough resolution or for systems with flexible components, rigid fitting cannot handle structural variation information in EM/ET maps. In different states, proteins and protein assemblies often adopt different conformations, such as in bound and unbound states. Typically, protein side chains have certain conformational flexibilities and can adapt to different environmental conditions. To accommodate conformational variations, a process called flexible fitting is used to change structures from X-ray or NMR experiments to match EM/ET maps. There are a series of methods available to perform flexible fitting [9-14].

Whether to choose rigid fitting or flexible fitting depends on the resolution of EM/ET maps. There is no absolute resolution boundary for choosing rigid fitting from flexible fitting. The rule of thumb is that when the resolution cannot distinguish structure difference of EM/ET maps from fitting structures, then rigid fitting should be used. Flexible fitting becomes necessary when the resolution is high enough to show significant structural variation between the fitting X-ray/NMR structures and the EM/ET maps. For proteins undergoing domain hinge motion, a map with a resolution as low as 50 Å could show the difference, while for side chain rotations, a map needs a resolution of 10Å or higher to identify an alternate orientation.

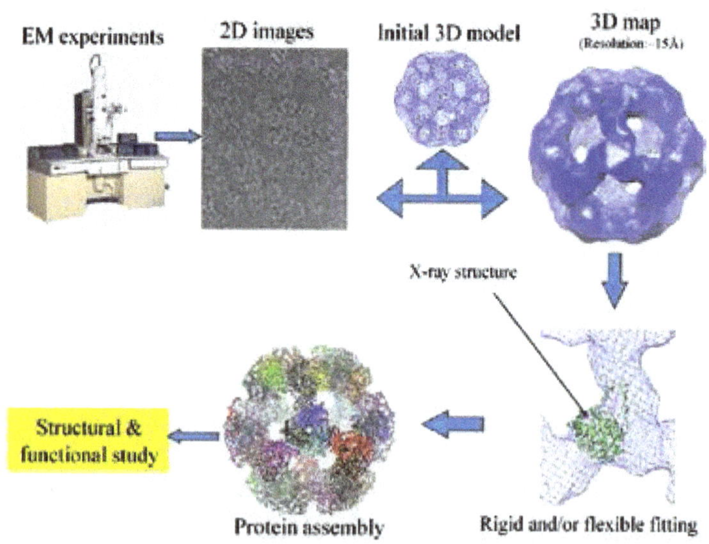


**Figure 1.** Basic procedure from electron microscopy experiment to macromolecular structures.

To illustrate the procedure of atomic structure derivation from EM maps, we describe here the core-weighted grid-threading Monte Carlo (CW-GTMC) rigid fitting method [2] and the map-restrained self-guided Langevin dynamics (MapSGLD) flexible fitting method[14]. Using four examples, we highlight their applications in structural, dynamics, and mechanism studies. The rigid fitting and flexible fitting methods have been implemented in both AMBER [15] and CHARMM [16]. In AMBER, this method has been implemented in Sander, which is part of AmberTools and is freely available at: http://ambermd.org/#AmberTools, as well as in pmemd, which is the main AMBER simulation engine (http://ambermd.org/#Amber14). In CHARMM, it is part of the EMAP module, where a rigid fitting is done through EMAP DOCK command, and a flexible fitting is carried out through DYNA simulation command. Users can refer to the emap test cases in both AMBER and CHARMM as examples of rigid and flexible fitting.

## 2. Methods

### 2.1. Rigid fitting

Rigid fitting of atomic structures to electron microscopy images is a common approach to interpret low-resolution maps and to obtain complex structural information. Figure 2 illustrates the procedure of a rigid fitting to obtain a complex structure. Assume EM/ET experiment produced the complex map em0 and the structures of components, a7na and a7nb are available. By rigid fitting a7na and a7nb into em0, one can obtain the complex structure of a7na+a7nb. A typical rigid fitting starts with blurring the molecular structures, a7na and a7nb, to create maps, ema and emb, which is often done by distributing atomic masses into grid points in the space. The component maps, ema and emb, are translated and rotated to fit the EM/ET map em0. After all component maps are fit, component molecules are positioned accordingly to produce the complex structure.

A map represents a distribution over a spatial region. Typically, a map is described by quantities, e.g., density, on lattice grid points:

$$M(N_x, N_y, N_z) = \bigcup_{i,j,k}^{N_x, N_y, N_z} \rho(x_i, y_j, z_k) \tag{1}$$

Here, the map grids are defined by grid point positions, $(x_i, y_j, z_k)$. The numbers of grids in the $x$, $y$, and $z$ directions are $N_x$, $N_y$, and $N_z$, respectively. The union symbol $\bigcup_{i,j,k}^{N_x,N_y,N_z}$ indicates that all entities with $i \in N_x$, $j \in N_y$, and $k \in N_z$ are included in the map. The $x$, $y$, and $z$ axes can form any angle, but for convenience, in this work, we assume them to be orthogonal to each other. The distribution property $\rho(x_i, y_j, z_k)$ can be any special properties such as charge density, electric potential, solvent accessibility, etc. It is electron density for EM/ET maps.

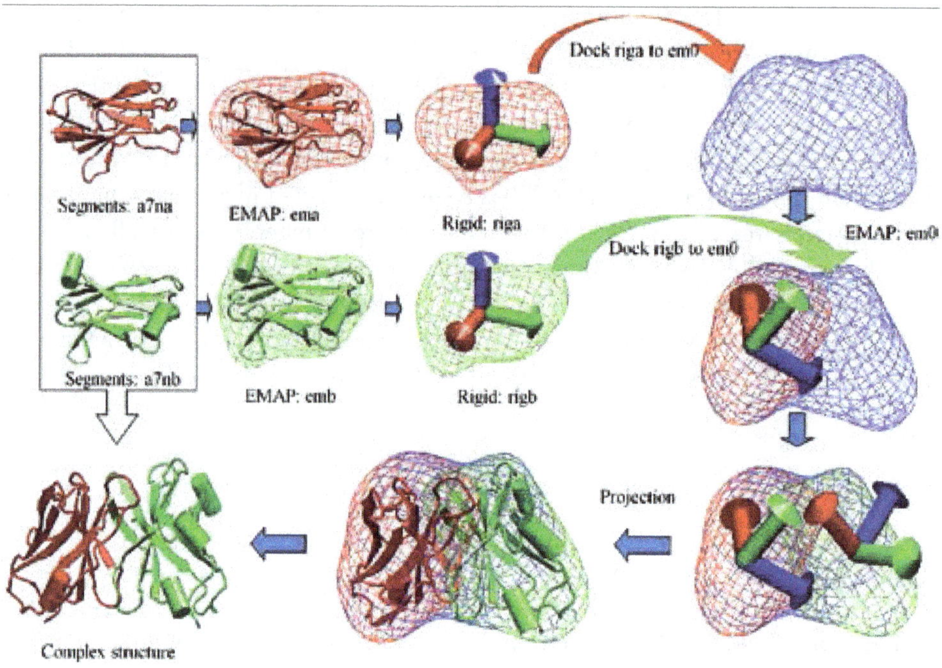


**Figure 2.** Rigid fitting steps to derive atomic structures of complexes.

A major application of EM/ET experiment is to determine structures of molecular assemblies. For low resolution of EM/ET maps, electron densities of structure components spread to a certain range beyond atomic boundaries, which causes significant density overlap between neighboring structures. We define the "core" region of a structure as the part whose density distribution is unlikely to overlap with the density distribution of the other maps and to be altered by the presence of adjacent components. The "surface" region is defined as the part that is accessible or can interact with other components. Therefore, the core region is always enclosed by the surface region.

It is then possible to define a core index, which describes the depth of a grid point located within this core as follows:

$$f_{ijk} = \begin{cases} 0 & \rho_{ijk} \le \rho_c \text{ and } \min[f_{i\pm1jk}, f_{ij\pm1k}, f_{ijk\pm1}] = 0 \\ 0 & \nabla^2 \rho_{ijk} > 0 \text{ and } \min[f_{i\pm1jk}, f_{ij\pm1k}, f_{ijk\pm1}] = 0 \\ \min[f_{i\pm1jk}, f_{ij\pm1k}, f_{ijk\pm1}] + 1 & \text{otherwise} \end{cases} \quad (2)$$

where, $f_{ijk}$ is the core index of grid point $(i, j, k)$, $\rho_c$ is a cut-off density to define the surface position. The core index is zero for grid points outside the surface region and increases progressively as a grid point goes deeper into the core. The core index is calculated with

$\min[f_{i\pm1\,jk}, f_{ij\pm1k}, f_{ijk\pm1}]+1$, that is, the index of a grid point (*i, j, k*) is the minimum core index of the neighboring grid points around this grid point plus 1. The core index has the following features when matching between the map of an individual component and the map of a multicomponent assembly:

1. For a correct fit, a match in the core regions (high core index) corresponds to a high correlation in densities.
2. For a correct fit, a match in the surface regions (low core index) corresponds to a high correlation in densities.
3. For a correct fit, a match between the core region of components and the surface region of assemblies must have low correlation in densities.
4. For a correct fit, a match between a surface (low core index) region and a core region may have low correlation in densities.

For scenarios 1–3, a normal correlation function works fine to distinguish the correct fit from wrong fits. But for scenario 4, the distribution property is altered by the overlap density from neighboring components in a complex map and a regular correlation function is likely to fail. To account for the effect of overlap, and properly recognize the correct fit from misfit, we need to minimize the contribution to correlation function from scenario 4. This can be achieved by "down-weighting" the match between a region with low core index in the map of individual components and a region with high core index in the complex map. We choose the following weighting function to implement this idea:

$$w_{mn} = \frac{f_m^a}{f_m^a + bf_n^a + c} \tag{3}$$

where $w_{mn}$ is the core-weighting function for the individual component, *m*, to the complex, *n*. Three parameters, *a*, *b*, and *c*, control the dependence of the function on the core indices. Typically, we chose $a = 2$ and $b = 1$, and set *c* to be a very small constant (e.g., $10^{-6}$) to ensure $w_{mn}=0$ when $f_m=0$ and $f_n=0$. We call eq. (3) the core-weighting function because it is based on the core index. Using this core-weighting function we calculate the core-weighted correlation function as shown below:

$$CWDC_{mn} = \frac{\overline{(\rho_m\rho_n)_w} - \overline{(\rho_m)_w}\,\overline{(\rho_n)_w}}{\delta_w(\rho_m)\delta_w(\rho_n)}, \tag{4}$$

where $\overline{(X)_w} = \dfrac{\sum_{i,j,k} w_{mn}(i,\ j,\ k)X(i,\ j,\ k)}{\sum_{i,j,k} w_{mn}(i,\ j,\ k)}$ is a core-weighted average and $\delta_w(X)=\sqrt{\overline{(X^2)_w}-\overline{(X)_w}^2}$ is a core-weighted standard deviation of property *X*.

Figure 3 illustrates the definition of core indices for the maps of two individual proteins, A and B, and their complex. In each map, the core index is zero outside the density regions, 1 at the outer edge and becomes larger for the grid points that locate more deeply in the core region. Please note that the core region does not necessarily correspond to the region with high density. For internal cavities that are buried well below the surface of a structure (e.g., the cavity in protein B), it is possible that the core index can have high values while densities are low. When proteins A and B form complex, the core indices of their interaction surfaces dramatically increase in regions where the surfaces become deeply buried in the AB complex.

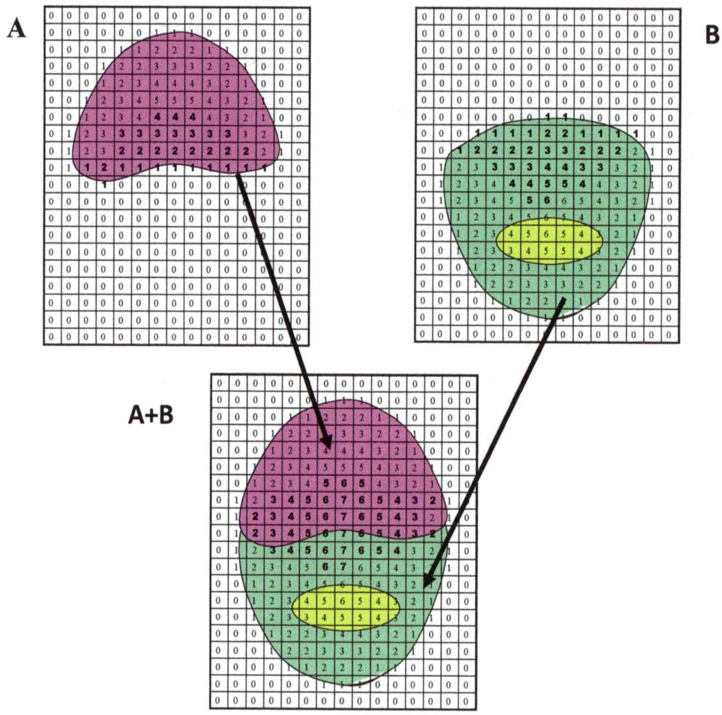


**Figure 3.** The core index distributions in schematic 2D maps of proteins A and B and their complex. Regions of protein density are colored red and green, respectively, and a region of protein B containing an inaccessible cavity as shown in light green. Regions outside of the protein are colored white. The numerical values of the core index for each grid point are labelled in each map. Bold numbers indicate the core indices of proteins A and B that change upon the formation of the AB complex.

These core-weighted correlation functions are designed to down-weight the regions overlapping with other components, while emphasizing the regions with no overlap. As explained above, the regions with significant overlap have small $f_m$ and large $f_n$, and thus a small weighting function. By down-weighting the overlapping regions, the core-weighted correlation functions can minimize the overlap effect in predicting the correct fit.

To perform rigid fitting, one needs to search a conformational space with three translational and three rotational degrees of freedom. The grid-threading Monte Carlo (GTMC) search is a combination of a grid search and Monte Carlo sampling [17]. The conformational space is split into searching grid points, which is not those defined for maps. Short Monte Carlo searches are performed starting from each of the searching grid points to identify local maxima close to the grid points. After searching out the local maxima, we can identify the global maximum among them. Figure 4 illustrates this procedure for a search in a 2D conformational space. The 2D conformational space is divided into a 3 × 3 grid. Among all the local maxima, the local maximum at (5.3) has the highest correlation and is identified as the global maximum. More details of the core-weighted fitting method can be found elsewhere [2].

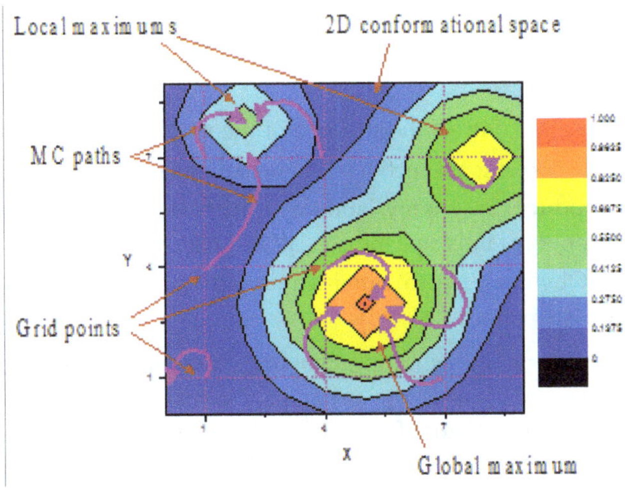


**Figure 4.** Illustration of the grid-threading Monte Carlo (GTMC) search in a 2D conformational space. The conformational space is split into a 3 × 3 grid. From each of the 9 grid points, a short MC search (shown as purple curves) is performed to locate a nearby local maximum. Among these local maxima, the global maximum is identified. Only conformations along the 9 Monte Carlo paths are explored during a GTMC search.

## 2.2. Flexible fitting

When EM/ET maps have enough resolution to show structural difference from X-ray or NMR structures, flexible fitting should be applied. Flexible fitting allows conformational changes to adapt to the EM/ET maps. Low-resolution EM/ET maps often do not have enough structural information to uniquely define thousands of atomic positions of molecular systems. To make up the informational gap, an atomic force field is used to determine most of the degrees of freedom. An atomic force field defines covalent bonded terms, nonbonded interactions, as well as solvation-free energies. Theoretically, an ideal atomic force field aided with sufficient molecular simulation can determine molecular structures without the need of experimental information. However, current force fields are not accurate enough to determine molecular structures from scratch. In addition, biological systems are often too complicated to be accurately described by overly simplified simulation systems. A middle approach is to utilize an atomic force field to supplement the missing information in low-resolution maps in determining the structure of molecular systems. In other words, low-resolution maps can provide structural information to compensate the inaccuracy in atomic force fields and X-ray or NMR structures provide reasonable initial conformations to simplify the conformational search in molecular simulation.

For a group of $N$ atoms, at the target conformation, these atoms should reproduce the target map at given experimental condition and these atoms must all sit at high-density positions. Because atomic masses are closely correlated to their numbers of electrons, we can define a map-restraint potential as the total products between the atomic mass, $m_a$ and the normalized density at the atom position, $\hat{\rho}(x_a, y_a, z_a)$ :

$$E_{\text{map}} = -c_{\text{map}} \sum_{a}^{N} m_a \hat{\rho}(x_a, y_a, z_a) \tag{5}$$

The restraint constant, $c_{\text{map}}$, controls the strength of the map restraint. The units of variables, $m_a$ and $c_{\text{map}}$, are g/mol and kcal/g, respectively. A normalized map, $\hat{\rho}(x_a, y_a, z_a)$, has $<\hat{\rho}>=0$ and $\delta\hat{\rho}=1$. Equation (5) produces an energy landscape in the shape of the density distribution $-c_{\text{map}}\hat{\rho}(x_a, y_a, z_a)$. This energy landscape induces atoms to move to positions of lower energies, or of higher densities. Obviously, eq. (5) mimics a simplified correlation between atom masses and the map density distribution.

It should be noted that this map-restraint potential only captures the low-resolution characteristics of molecular systems and is not designed to reproduce atomic structures by itself. Instead, when aided with an all-atom force field, which contains bonded interactions (bond lengths, bond angles, dihedral angles) and nonbonded interactions (van der Waals, electrostatic interactions, solvation), the map-restraint potential can help to stabilize conformations that agree with the restraint map. It is the combination of a force field and the map restraint that drives an atomic system to the target conformation. This map-restraint potential has an order of $O(N)$ and is very efficient to calculate as compared to other pair-wise nonbonded interactions, especially for large systems.

The map restraints drive atoms in a cooperative way so that atoms interacting through the force field contribute together to match the target map density distribution. The map restraints have the following convenient characteristics for a targeted conformational search:

1. The map-restraint energy surface is soft, which makes large-scale conformational transition feasible.
2. The map-restraint is atom-identity-blind, so the restraint energy calculation is of $O(N)$.
3. The map-restraint energy tolerates noises in target maps.
4. The map-restraint can be extended to maps of other properties such as partial charges, desolvation energies, and van der Waals interactions [18].

### 2.3. Targeted conformational search

In some cases, there are large-scale conformational changes between X-ray/NMR structures and the EM/ET experimental systems. The self-guided Langevin dynamics (SGLD) simulation method [19-22] is a method designed for an efficient search of conformational space. The equation of SGLD motion with map restraints has the following general form:

$$\dot{\mathbf{p}}_a = \mathbf{f}_a^{(\mathrm{ff})} + \mathbf{f}_a^{(\mathrm{map})} + \mathbf{g}_a - \gamma_a \mathbf{p}_a + \mathbf{R}_a \tag{6}$$

where $\dot{\mathbf{p}}_a$ is the time derivative of momentum; $\mathbf{f}_a^{(\mathrm{ff})}$ is the atomic force due to the force field; $\mathbf{f}_a^{(\mathrm{map})}$ is the map-restraint force from the restraint map; $\mathbf{R}_a$ is a random force, which is related to the mass, $m_a$, the collision frequency $\gamma_a$, and the simulation temperature $T$. There is a guiding force, $\mathbf{g}_a$, which is calculated based on the local average of momentum, $\mathbf{p}_a$, to accelerate conformational search through SGLD. Detailed description of the SGLD method and application can be found elsewhere [19, 21-23].

The method combining the map potential and the SGLD method is called MapSGLD [14]. This is a general method for targeted conformational search. Flexible fitting is a direct application of this method to efficiently identify conformations that match EM/ET maps.

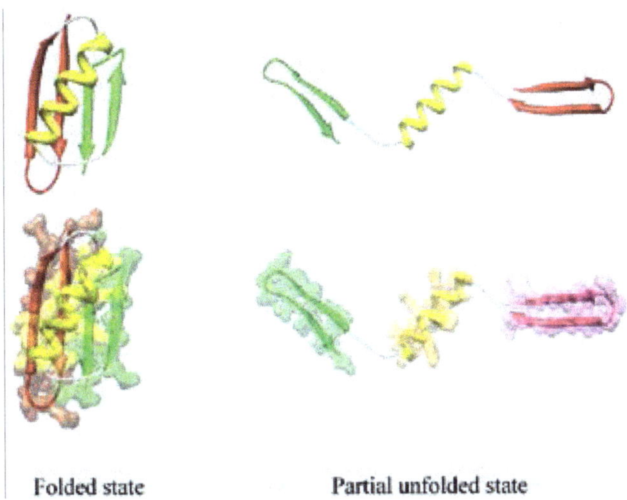


**Figure 5.** Use restraint maps to maintain structures of GB1 substructures during conformational search.

Targeted conformational search is a convenient way to study dynamic properties of macromolecular systems. In protein-folding studies, it has been observed that secondary structure elements fold first, followed by their arrangement to form tertiary structures. To study how the secondary structure elements fold into the tertiary structures, it requires these secondary structures to be maintained during simulations. Similarly, for protein assemblies, it is desired to simulate how individual proteins assemble to form the complex structure while these individual proteins remain folded. These are typical examples of targeted conformational search. Figure 5 shows a small protein, the B1 domain of streptococcal protein G, abbreviated here as GB1, in its folded and unfolded states under the map restraints to maintain its

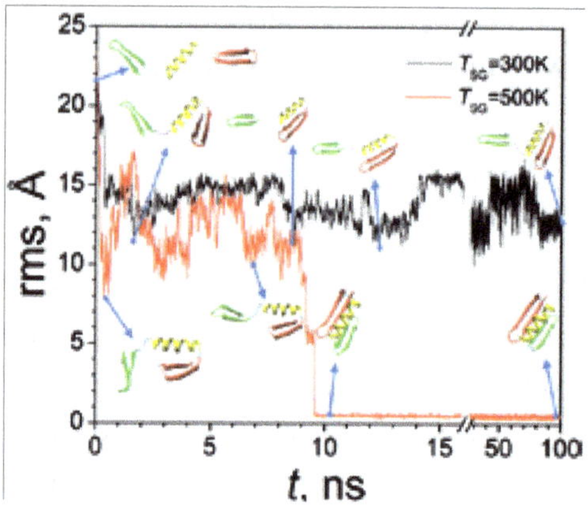


**Figure 6.** GB1 folding simulation by targeting the map of the folded state. The black line is for an LD simulation without enhanced sampling and the red line is for an SGLD simulation with an enhanced sampling.

secondary structures. GB1 has 56 residues with one $\alpha$-helix and one $\beta$-sheet. The $\beta$-sheet is made of two $\beta$-hairpins. Three restraint maps for the three secondary structure motifs were generated from the NMR structure: residues 1 to 19 for the N-terminal $\beta$-hairpin, residues 22 to 37 for the helix, and residues 42 to 56 for the C-terminal $\beta$-hairpin. Figure 6 shows two simulation results, one with $T_{SG}$ = 300 K and one with $T_{SG}$ =500 K. When $T_{SG}$ = 300 K = $T$, it was a normal conformational search and the SGLD simulation reduced to a regular Langevin dynamics (LD) simulation, where the simulation failed to reach the folded conformation in up to 100 ns. While in the case of $T_{SG}$ = 500 K, conformational motion was accelerated, which prompted the protein to reach the folded state in 9.5 ns. In other words, the map restraint itself is not enough to bring the protein to the folded state in 100 ns, but SGLD can significantly accelerate the search to find a target conformation.

## 3. Applications

### 3.1. Architecture of an icosahedral pyruvate dehydrogenase complex

The pyruvate dehydrogenase multienzyme complex couples the activity of three component enzymes (E1, E2, and E3) in the oxidative decarboxylation of pyruvate to generate acetyl-CoA, linking glycolysis and the tricarboxylic acid cycle. The *Bacillus stearothermophilus* PDH complex is assembled around a core of 60 dihydrolipoyl acetyltransferase (E2) chains arranged with icosahedral symmetry. Each E2 chain consists of three domains: (i) an N-terminal 9 kDa lipoyl domain, which visits the active sites of the pyruvate decarboxylase (E1) component and then those of E2 and dihydrolipoyl dehydrogenase (E3); (ii) a 4 kDa peripheral subunit-binding domain to which E1 and E3 bind tightly and mutually exclusively; and (iii) a C-terminal 28 kDa catalytic (acetyltransferase) domain, 60 copies of which assemble to form the icosahedral inner core. These domains are linked by stretches of extended, conformationally flexible polypeptide chain.

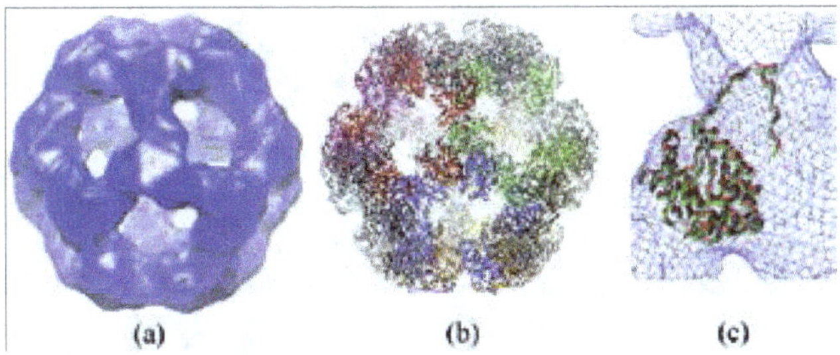

**Figure 7.** (a) The experimental map (at 14 Å resolution) of the icosahedral complex formed from 60 copies of the E2 catalytic domain of the pyruvate dehydrogenase. (b) The X-ray structure (PDB code: 1B5S) of the same complex. (c) Comparison of the location of the E2 catalytic domain obtained using a CW-GTMC search (green) with that from the X-ray structure (red) (rms = 2.13Å). The experimental EM map to fit is shown as blue wires.

Using the CW-GTMC method, we fit 60 copies of E2 into a 14 Å electron microscopic map of the icosahedral core of pyruvate dehydrogenase (Figure 7a), a 1.8 MDa complex comprised of 60 copies of the E2 catalytic domain, whose structure (Figure 7b) has been determined using X-ray crystallographic methods (PDB code: 1B5S). The CW-GTMC search with 5000 MC steps for each of 729 (3×3×3×3×3×3) grids identified all 60 global maximum positions [2]. Figure 7c compares the CW-GTMC fit and the corresponding X-ray structure position. The atomic structure of the E2 core from our rigid fit agrees well with the X-ray structure, validating the core-weighted GTMC method.

Since atomic coordinates of the *B. stearothermophilus* E1 enzyme are not available, the coordinates of the closely related E1 $\alpha2\beta2$ tetramer from *Pseudomonas putida* (PDB entry 1QS0) were used. From sequence alignment of the two enzymes, we found the presence of additional amino acid segments 42–51, 178–182, and 375–380 in the *P. putida* E1 $\alpha$-subunit and 187–192 in its $\beta$-subunit, which are not present in the *B. stearothermophilus* E1 enzyme. By omitting these residues from the *P. putida* structure, we obtained a model for the *B. stearothermophilus* E1 enzyme. Using the core-weighted grid-threading Monte Carlo (CW-GTMC) method, we fit the coordinates to a 28 Å electron cryo-microscopy map [1]. The transformation space was divided into 5 × 5 × 5 grids and orientation space was divided into 3 × 3 × 3 grids. At each grid point, a 1000-step Monte Carlo search was performed with an initial transformational step size of 15 Å and initial rotational step size of 30°.

The automatic rigid fitting procedure identified two types of best-fit conformations into which all starting positions converged. The spatial relationship between these positions is indicated in the plot shown in Figure 8. In the second-best fit identified by the automated procedure, the E1 tetramer is pushed toward the fivefold vertex, covering some regions of the density not sampled by the best fit. In this orientation, the long axis of E1 is still along the surface of the icosahedron, but it is translated toward the fivefold vertex relative to the best fit. Analysis shows that the path from the best to the second-best fit involves a swivelling motion of E1 around an axis perpendicular to the icosahedral surface, with very little movement of the location and orientation of the E1 active site relative to the inner core. It is interesting that, while neither the best nor the second-best fit appears to span the outer density fully, the fits in Figure 8 show that almost all regions of the outer density are included in one or other of the

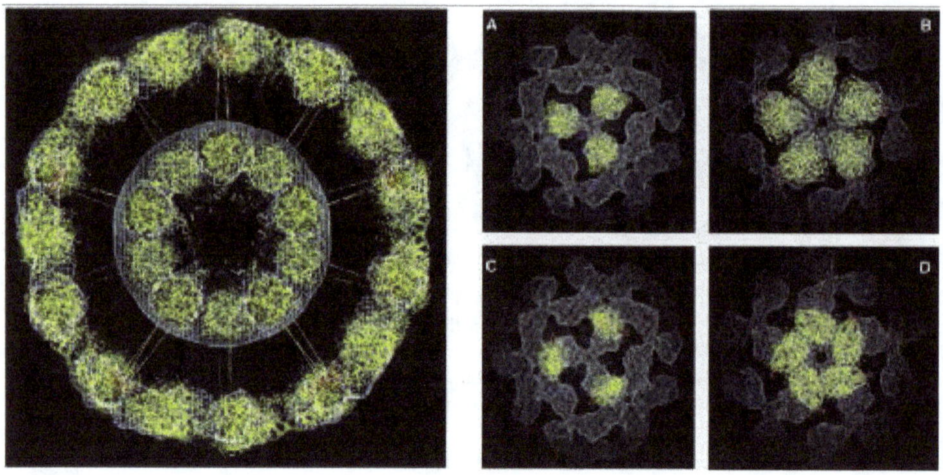


**Figure 8.** The E1E2 cryo-EM map at 28 Å resolution and the atomic structure obtained through CW-GTMC fitting. The left panel shows sectioned view of the density for the E1E2 complex with 60 docked copies of the atomic coordinates of the E1 $\alpha2\beta2$ tetramer, the E2 catalytic domain, and the E2 peripheral subunit-binding domain. To illustrate the probable location of the linker domain connecting the peripheral subunit-binding and catalytic domains of the E2 chain, 60 radial spokes have been included. Right panels show the views of the docked E1 molecules (in yellow) and E2 peripheral subunit-binding domains (in red) from threefold (A and C) and fivefold (B and D) vertices. The views in (A) and (B) are for the best fit obtained by the CW-GTMC, while the views in (C) and (D) are for the second-best fit.

two orientations of E1. Based on these findings our working hypothesis is that the two orientations reflect stable positions for the E1 tetramer in the E1E2 complex. Since each E1 molecule in the complex may sample both orientations over time, the smear of the outer density in our 3D reconstruction probably reflects the averaged contribution to the structure from these two populations.

### 3.2. Dynamics of a group II chaperonin

The advance in electron microscopy experiment technology results in high-resolution maps, to show dynamic details of macromolecular systems, such as domain motion, loop rearrangements, and side chain reorientations. Here, we describe flexible fitting of EM maps through the MapSGLD method to address the dynamics of macromolecular systems [14]. One advantage of using SGLD for the targeted conformational search is to promote large-scale conformational changes necessary for protein functions. We chose the open–close transitions of a group II chaperonin to demonstrate the application of this method. By flexible fitting a group II chaperonin, mn-cpn, using maps from the EM databank: EMD-5138 (close state) and EMD-5140 (open state) [24], we obtained dynamics of the chaperonin-folding chamber opening and closing, as well as the structures in these states.

Zhang and colleagues have modelled the structures of the closed state (PDB:3J03) [25] and the open state (PDB:3IYF) [24]. We used these model structures as starting conformations and use the maps of the opposite states as restraints to perform MapSGLD simulations to examine the capability of MapSGLD to identify target conformations.

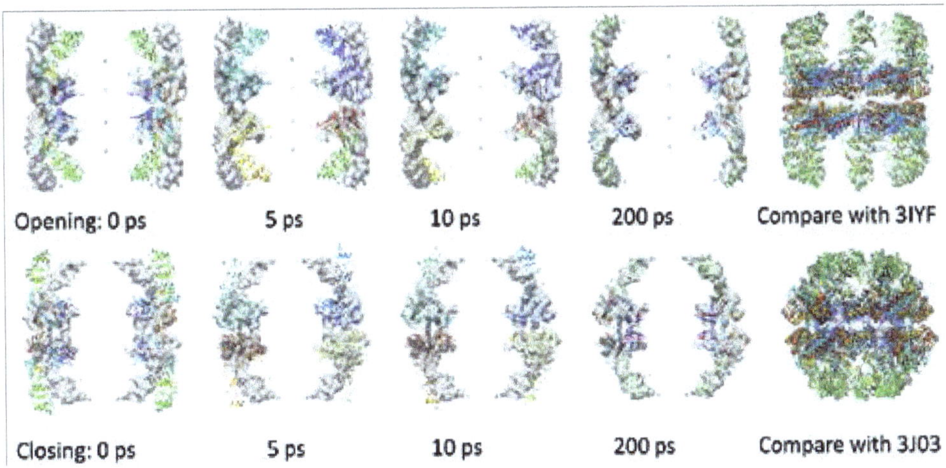


**Figure 9.** Opening (top) and closing (bottom) of the chaperonin-folding chamber during the MapSGLD simulations. The last column compares the final conformations with their corresponding PDB entries (in grey). The opening simulation started from the closed state (PDB: 3J03) and was restrained with the map at the open state: EMD-5140. The closing simulation started from the open state structure (PDB: 3IYF) and was restrained with the map at the closed state: EMD-5138.

Each monomer of the chaperonin assembly contains three domains, the apical domain (residues 205–334), the intermediate domain(residues 136–204, 335–371) and the equatorial domain (residues 1–135, 372–491). We used movable map restraints to maintain the structures of these domains. The movable map restraints were generated from the initial conformation with a resolution of 3 Å and a map-restraint constant of 0.1 kcal/g. The experimental EM maps were applied to the system as fixed map restraints with a restraint constant of 0.05 kcal/g. Figure 9 shows the conformations during the opening and closing of the folding chamber obtained from MapSGLD simulations. We can see the apparent difference in the starting conformation from the EM maps and the agreement of the final conformations with their restraint maps. The root-mean-square-deviation (rmsd) of the final conformation from the closing simulation was within 1.5 Å from the closed structure (PDB:3J03) and the rmsd of the final conformation from the opening simulation was within 2.0 Å from the opened structure (PDB:3IYF), demonstrating the reliability of the flexible fitting method. From the simulation trajectories we can extract atomic details of the folding chamber dynamics during opening and closing.

Figure 10 shows the energies and rmsd during these MapSGLD simulations. From the middle panel of Figure 10, we can see that the closing simulation reached within 1.51 Å from the closed structure in 20 ps and remained there with little change afterward. Also we can see that the opening simulation reached within 2.01 Å from the opened structure in 20 ps and remained there with little change afterward.

Examining the energy profiles shown in the bottom panel of Figure 10, we can see that in order for the closed assembly to open up, it first went through an energy barrier. It reached a peak at about 15ps, and began an energy decrease throughout the rest of the simulation. Throughout the closing simulation, there was no energy barrier. It was the energy barrier overcoming ability of SGLD made it efficient to overcome of the energy barrier during the opening process.

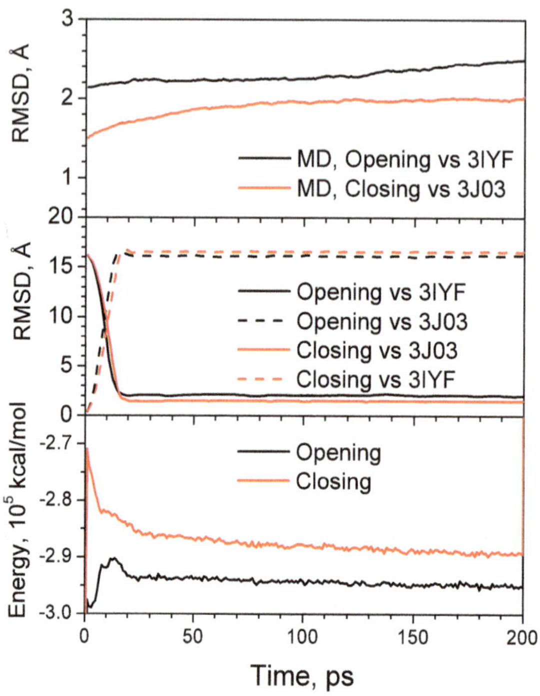


**Figure 10.** The opening and closing simulation profiles. Bottom panel: the molecular potential energy profiles during the MapSGLD simulations. Middle panel: the rmsd profiles during the MapSGLD simulations. Top panel: the rmsd profiles during the MD simulations continuing from the final conformations of the MapSGLD simulations.

Driven by the map-restraint potential, MapSGLD simulations search for conformations matching the EM maps. It is the map-restraint potential that makes the targeted conformation a global free-energy minimum state. Without the map-restraint potential, experimental structures may not be the global free-energy minimum states at this simplified simulation conditions. There are many factors that affect the global minimum, such as inaccuracies in the force field, overly simplified setup of the simulated system, or inadequate description of the effects of the solvent. To illustrate this point, we performed conventional molecular dynamics simulations without the map restraint from the final conformations of the MapSGLD simulations, and the rmsd profiles are shown in the top panel of Figure 10. We can see that in the MD simulations, the systems slowly drifted away from the corresponding states, which indicates that the simplification in simulation conditions can cause conformational deviations from experimental observations and the map restraint can help overcome the effect of the simplification in simulation conditions.

### 3.3. High-resolution structure determination from EM maps

A milestone in high-resolution structure determination using cryo-electron microscopy (cryo-EM) has been reached recently [26] through the structure of a complex between *Escherichia coli* β-galactosidase and the cell-permeant inhibitor phenylethyl β-D-thiogalactopyranoside (PETG), determined by cryo-EM at an average resolution of ~2.2 angstroms (Å). Besides the PETG ligand, ~800 water molecules and for magnesium and sodium ions are identified in the map. To achieve resolutions close to 2 Å using single-particle cryo-EM, preparation of specimens of adequate quality and intrinsic protein flexibility, rather than imaging or image-processing technologies, are the major bottlenecks. Figure 11 shows the EM map and the protein structure fit into the map. The right panel shows side chains and their map density. At this resolution, the EM density of heavy atoms is recognizable.

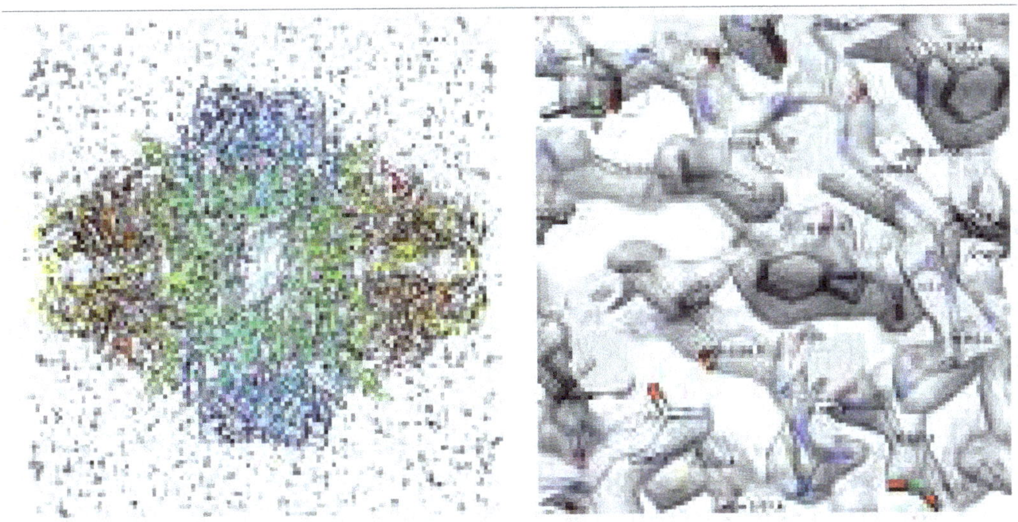

**Figure 11.** The 2.2Å resolution EM density map of a complex between *Escherichia coli* β-galactosidase and the cell-permeant inhibitor phenylethyl β-D-thiogalactopyranoside (PETG) and the flexible fit structure of the complex. The left panel shows the whole complex (as ribbon) and the right panel shows residues around residue A355 shown (sticks).

### 3.4. Kinesin walking on microtube [27]

To understand kinesin walking mechanism, it is desirable first to see how kinesin complexes with microtubule. Due to the heterogeneity of the kinesin–microtubule complex, it is difficult to determine the complex structure through X-ray crystallography. Instead, the complex models are often built from low-resolution electron microscopy maps. The kinesin–microtubule complex system was built by fitting the kinesin motor domains (1BG2 or 2NCD) and tubulin (1JFF) into the complex EM map (EMD-5011). Figure 12 illustrates the structure obtained by the CW-GTMC rigid fitting. Rigid fitting provides a reasonable first guess of the complex structure at the limit of the map resolution. Their surface elements, such as loops and

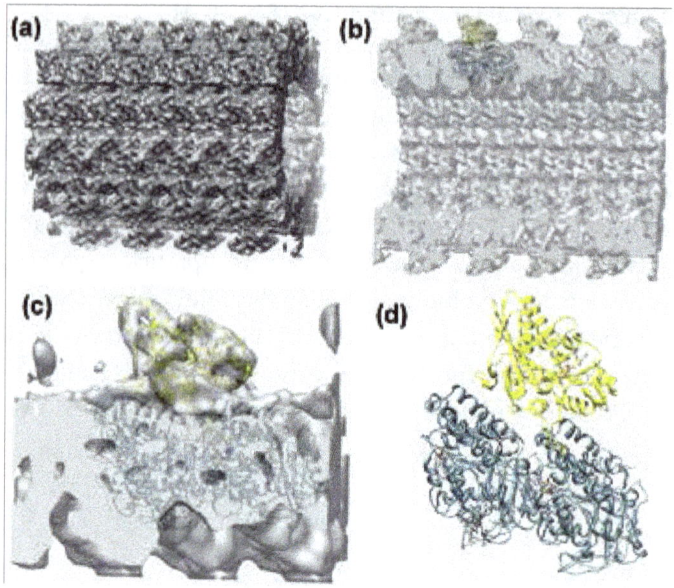


**Figure 12.** Fitting X-ray structures of kinesin (PDB: 1BG2 or 2NCD) (yellow) and microtube (PDB: 1JFF) (cyan) into the 8 Å resolution EM map of a kinesin–microtube complex to obtain initial conformations of the complex. (a) The EM map EMD-5011. (b) A cross-section view of the fitting result. (c) An enlarged view of the fitting. (d) The complex structure from the fitting.

side chains, must adjust to suit their environment. Energy minimizations were performed on these fitting results to derive energetically favorable structures.

Based on the EM fit complex structures, SGLD simulations were performed to study the dynamics of these complexes at different nucleotide states: 1BG2/ADP, 1BG2/ATP, 2NCD/ADP, and 2NCD/ATP. SGLD simulations showed significant conformational change and the final conformations are compared with the initial fit structures in Figure 13. As can be seen clearly, the motor domains rotated certain angles in related to their initial positions. After the SGLD simulations, 1BG2 in the ADP-binding state rotated about -40° and in the ATP-binding state rotated about 30°. 2NCD in the ADP-binding state rotated approximately -20° and in the ATP-binding state rotated approximately -60°. In other words, 1BG2 rotated count clockwise for about 70°, while 2NCD rotated clockwise for about 40°, when changing from the ADP-binding state to the ATP-binding state. This directional difference in rotation clearly corresponds to the directional walking of the two types of kinesin. Because the coiled-coil linker to cargo extends near the C-terminal of the β6 strand, a counter clockwise rotation will result in a plus-end-directed movement of the linker, and a clockwise rotation will result in a minus-end-directed movement of the linker. This kinesin-specific directional rotation provides clues to understanding the mechanism of kinesin walking directions.

These application examples illustrate the atomic structure derivation procedures through rigid fitting and flexible fitting. These derived atomic structures open a gateway for further understanding of dynamics and functions of molecular systems.

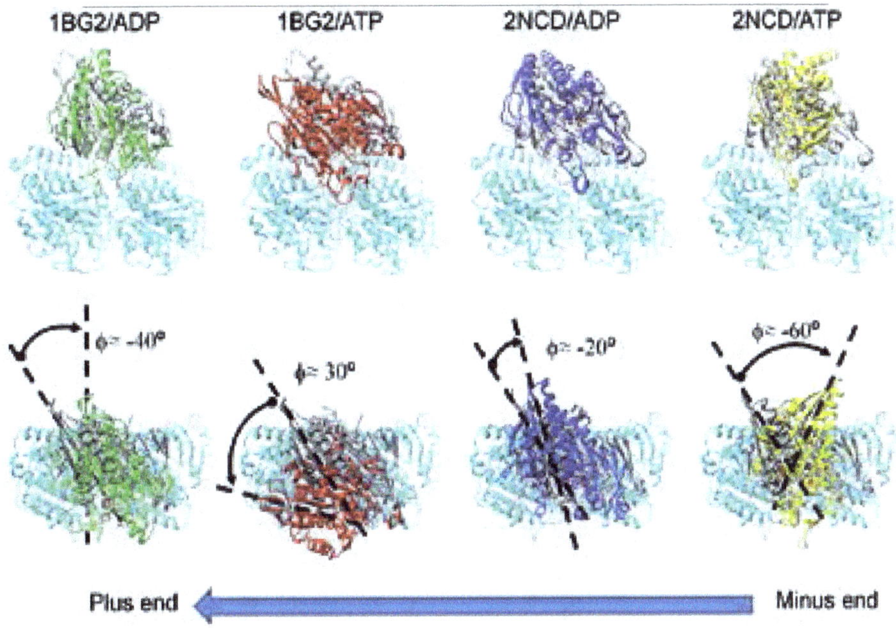


**Figure 13.** The kinesin–microtubule complexes before and after the SGLD simulations. The initial kinesin structures are shown in grey and microtubules are colored cyan. 1BG2/ADP, 1BG2/ATP, 2NCD/ADP, and 2NCD/ATP are colored green, red, blue, and yellow, respectively. The upper row shows their side views and the bottom row shows their top views. Rotation angles are calculated with the β6 strand in the kinesin motor domains against the initial conformation.

## Acknowledgements

This research was supported by the Intramural Research Programs of National Heart, Lung, and Blood Institute (Z01 HL001027-30). Katherine Wu of Massachusetts Institute of Technology performed the kinesin-microtube simulation.

## Author details

Xiongwu Wu* and Bernard R. Brooks

*Address all correspondence to: wuxw@nhlbi.nih.gov

Laboratory of Computational Biology, National Heart, Lung, and Blood Institute, National Institutes of Health, Bethesda, USA

## References

[1] Milne, J. L., Shi, D., Rosenthal, P. B., Sunshine, J. S., Domingo, G. J., Wu, X., Brooks, B. R., Perham, R. N., Henderson, R., and Subramaniam, S. Molecular architecture and

mechanism of an icosahedral pyruvate dehydrogenase complex: a multifunctional catalytic machine, *Embo J* (2002) *21*, 5587-5598.

[2] Wu, X., Milne, J. L., Borgnia, M. J., Rostapshov, A. V., Subramaniam, S., and Brooks, B. R. A core-weighted fitting method for docking atomic structures into low-resolution maps: application to cryo-electron microscopy, *J Struct Biol* (2003) *141*, 63-76.

[3] Roseman, A. M. Docking structures of domains into maps from cryo-electron microscopy using local correlation, *Acta Crystallogr D Biol Crystallogr* (2000) *56*, 1332-1340.

[4] Spahn, C. M., Penczek, P. A., Leith, A., and Frank, J. A method for differentiating proteins from nucleic acids in intermediate-resolution density maps: cryo-electron microscopy defines the quaternary structure of the Escherichia coli 70S ribosome, *Structure* (2000) *8*, 937-948.

[5] Antzutkin, O. N., Leapman, R. D., Balbach, J. J., and Tycko, R. Supramolecular structural constraints on Alzheimer's beta-amyloid fibrils from electron microscopy and solid-state nuclear magnetic resonance, *Biochemistry* (2002) *41*, 15436-15450.

[6] Wriggers, W., and Birmanns, S. Using situs for flexible and rigid-body fitting of multiresolution single-molecule data, *J Struct Biol* (2001) *133*, 193-202.

[7] Wriggers, W., Milligan, R. A., and McCammon, J. A. Situs: A package for docking crystal structures into low-resolution maps from electron microscopy, *J Struct Biol* (1999) *125*, 185-195.

[8] Milne, J. L., Wu, X., Borgnia, M. J., Lengyel, J. S., Brooks, B. R., Shi, D., Perham, R. N., and Subramaniam, S. Molecular structure of a 9-MDa icosahedral pyruvate dehydrogenase subcomplex containing the E2 and E3 enzymes using cryoelectron microscopy, *J Biol Chem* (2006) *281*, 4364-4370.

[9] Tama, F., Miyashita, O., and Brooks, C. L., 3rd. Normal mode based flexible fitting of high-resolution structure into low-resolution experimental data from cryo-EM, *J Struct Biol* (2004) *147*, 315-326.

[10] DiMaio, F., Tyka, M. D., Baker, M. L., Chiu, W., and Baker, D. Refinement of protein structures into low-resolution density maps using rosetta, *J Mol Biol* (2009) *392*, 181-190.

[11] Bradley, P., Misura, K. M. S., and Baker, D. Toward high-resolution de Novo structure prediction for small proteins, *Science* (2005) *309*, 1868-1871.

[12] Trabuco, L. G., Villa, E., Mitra, K., Frank, J., and Schulten, K. Flexible fitting of atomic structures into electron microscopy maps using molecular dynamics, *Structure* (2008) *16*, 673-683.

[13] Trabuco, L. G., Villa, E., Schreiner, E., Harrison, C. B., and Schulten, K. Molecular dynamics flexible fitting: a practical guide to combine cryo-electron microscopy and X-ray crystallography, *Methods* (2009) *49*, 174-180.

[14] Wu, X., Subramaniam, S., Case, D. A., Wu, K. W., and Brooks, B. R. Targeted conformational search with map-restrained self-guided Langevin dynamics: Application to flexible fitting into electron microscopic density maps, *J Struct Biol* (2013) *183*, 429-440.

[15] Case, D. A., Babin, V., Berryman, J. T., Betz, R. M., Cai, Q., Cerutti, D. S., Cheatham Iii, T. E., Darden, T. A., Duke, R., Gohlke, H., Gusarov, S., Homeyer, N., Janowski, P., Kaus, J., Kolossvary, I., Kovalenko, A., Lee, T. S., LeGrand, S., Luchko, T., Luo, R., Madej, B., Merz, K. M., Paesani, F., Roe, D. R., Roitberg, A., Sagui, C., Salomon-Ferrer, R., Seabra, G., Simmerling, C. L., Smith, W., Swails, J., Walker, R. C., Wan, J., Wolf, R. M., Wu, X., and Kollman, P. A. AMBER 14, University of California, San Francisco (2014).

[16] Brooks, B. R., Brooks, C. L., 3rd, Mackerell, A. D., Jr., Nilsson, L., Petrella, R. J., Roux, B., Won, Y., Archontis, G., Bartels, C., Boresch, S., Caflisch, A., Caves, L., Cui, Q., Dinner, A. R., Feig, M., Fischer, S., Gao, J., Hodoscek, M., Im, W., Kuczera, K., Lazaridis, T., Ma, J., Ovchinnikov, V., Paci, E., Pastor, R. W., Post, C. B., Pu, J. Z., Schaefer, M., Tidor, B., Venable, R. M., Woodcock, H. L., Wu, X., Yang, W., York, D. M., and Karplus, M. CHARMM: the biomolecular simulation program, *J Comput Chem* (2009) *30*, 1545-1614.

[17] Allen, M. P., and Tildesley, D. J. *Computer Simulations of Liquids*, Clarendon Press, Oxford (1987).

[18] Wu, X., and Brooks, B. R. Modeling of Macromolecular assemblies with map objects., *Proceedings of the 2007 international conference on bioinformatics & computational biology.* (2007) *II*, 411-417.

[19] Wu, X., and Brooks, B. R. Self-guided Langevin dynamics simulation method, *Chem Phys Lett* (2003) *381*, 512-518.

[20] Wu, X., and Brooks, B. R. Force-momentum-based self-guided Langevin dynamics: a rapid sampling method that approaches the canonical ensemble, *J Chem Phys* (2011) *135*, 204101.

[21] Wu, X., and Brooks, B. R. Toward canonical ensemble distribution from self-guided Langevin dynamics simulation, *J Chem Phys* (2011) *134*, 134108.

[22] Wu, X., Damjanovic, A., and Brooks, B. R. Efficient and Unbiased Sampling of Biomolecular Systems in the Canonical Ensemble: A Review of Self-Guided Langevin Dynamics, In *Advances in Chemical Physics* (Rice, S. A., and Dinner, A. R., Eds.), pp. 255-326, John Wiley & Sons, Inc., Hoboken (2012).

[23] Wu, X., and Wang, S. Self-guided molecular dynamics simulation for efficient conformational search, *J Phys Chem B* (1998) *102*, 7238-7250.

[24] Zhang, J., Baker, M. L., Schroder, G. F., Douglas, N. R., Reissmann, S., Jakana, J., Dougherty, M., Fu, C. J., Levitt, M., Ludtke, S. J., Frydman, J., and Chiu, W. Mechanism of folding chamber closure in a group II chaperonin, *Nature* (2010) *463*, 379-383.

[25] Zhang, J., Ma, B., DiMaio, F., Douglas, Nicholai R., Joachimiak, Lukasz A., Baker, D., Frydman, J., Levitt, M., and Chiu, W. Cryo-EM structure of a Group II chaperonin in the prehydrolysis ATP-bound state leading to lid closure, *Structure* (2011) *19*, 633-639.

[26] Bartesaghi, A., Merk, A., Banerjee, S., Matthies, D., Wu, X., Milne, J. L. S., and Subramaniam, S. 2.2 Å resolution cryo-EM structure of β-galactosidase in complex with a cell-permeant inhibitor, *Science* (2015) *348*, 1147-1151.

[27] Wu, K. W. The hydrolysis engine concept for motor proteins, In *Society for Science and The Public*, p. 20, Washington, DC (2013).

# 11

# Development of Myeloid Dendritic Cells under the Influence of Sexual Hormones Visualized using Scanning and Transmission Electron Microscopy

Josef Neumüller, Sylvia Emanuela Neumüller-Guber, Johannes Huber, Adolf Ellinger and Thomas Wagner

**Abstract**

Dendritic cells (DCs) are antigen-presenting cells, which are mediated by MHC-class II molecules reacting with T-helper cells, eliciting a broad spectrum of immune reactions at cellular and humoral levels depending on their subtypes. DCs are also able to cross-present peptides from intracellular proteins as well as from intracellular pathogens via MHC-class I molecules by inducing MHC-class I–restricted cytotoxic T cells, which are also able to destroy cells undergoing malignant transformation. DCs originate from $CD34^+$ hematopoietic stem cells but can also develop from monocytes. The local or systemic milieu of cytokines and steroid hormones significantly influences the generation of particular DC subtypes such as the classical myeloid DCs such as cDC1 and cDC2 as well as the plasmacytoid DCs. These subtypes are able to induce specific Th1- and Th17-dependent, Th2-dependent, or regulatory immune responses, respectively. Immature DCs take up extracellular pathogens that are presented by MHC molecules that are upregulated during maturation. Immature and mature DCs can be characterized by morphological and biochemical features that are outlined in this article. In addition, DCs are under control of sexual hormones. Estrogen receptor ligands are potent modulators of hemopoiesis and immune function in health and disease, influencing key cytokines promoting the maturation of DCs. DC differentiation is mainly regulated by binding of estradiol to ER$\alpha$. Estrogen promotes the differentiation of immature DC subsets derived from bone marrow precursors or from myeloid progenitors. In contrast to estrogen, progesterone inhibits DC maturation, causing a decreased immunity in pregnancy or in postmenopausal women, where elevated levels of progesterone result in the production of Th2 cytokines. The influence of estrogen and progesterone on DC maturation has been demonstrated in own *in vitro* experiments using fluorescence microscopy and cell sorting and, above all, by visualization using SEM and TEM. At the end of this article, pits and falls concerning the treatment of malignancies with living DC vaccines are discussed.

**Keywords:** Dendritic cells, MHC antigens, antigen presentation, cross-presentation, cytokines, transcription factors, steroid hormones, anti-tumor vaccine

## 1. Introduction

Dendritic cells (DCs) are antigen-presenting cells (APCs) interacting with CD4-positive T-Helper (Th) cells via peptide-presenting MHC (HLA) class II molecules and the $\alpha/\beta$ or $\gamma/\delta$ T-cell receptor and via costimulatory molecules. These interactions provide initiation or modulation of specific kinds of immune reactions triggering effector cells via direct cell contacts as well as by intercellular communications involving a complex cytokine network.

In the first part, the differentiation of myeloid DCs, lymphoid DCs as well as of Langerhans cells (LCs) are explained, including their morphological and immunological characteristics based on their local or systemic sites of activity. Local areas mainly concern the secondary lymphoid organs such as lymph nodes and spleen as well as the mucosa-associated lymphoid tissues (MALT), including the Waldeyer's tonsillar ring and the mucosa lining the respiratory and the gastrointestinal tract. Although the objective of this article is rather morphological, we think that an immunological overview is helpful for understanding the functional interrelations of DCs not only with responding immune cells but also with other cells in the vicinity of locally acting DCs. This will be concisely discussed in the light of our present scientific knowledge.

In the second part, we show own results concerning ultrastructural details from the maturation of myeloid DCs under *in vitro* conditions using TEM and SEM. We started the differentiation of myeloid DCs outgoing from $CD34^+$ hematopoietic stem cells (HSCs) of human umbilical cord blood. In the course of differentiation in a conditioned cell culture medium, the HSCs differentiated into immature DCs and finally into mature myeloid DCs.

In the third part, the maturation of myeloid DCs under the influence of sexual hormones is demonstrated using scanning electron microscopy (SEM), showing images taken from our own investigations. We visualized the mDC maturation under the influence of β-estrogen and progesterone by SEM after flow cytometry and cell sorting.

In the fourth part, the impact of mDCs for anti-tumor therapies is discussed. In this respect, different DC isolation protocols (from HSCs or from monocytes) as well as DC anti-tumor targeting methods are presented. Several efforts to produce an efficacious anti-tumor vaccine using primed autologous DCs were undertaken and clinical trials were carried out with varying success. Some clinical trials were promising, but other clinical applications were disappointing since recurrence of the respective tumor could not be prevented.

## 2. Functional aspects of DCs in the immune system

During the past years, the view about the differentiation lines of DCs has been changed significantly [1]. In this respect, different efforts were made to find an appropriate nomenclature. Some researchers use a development-orientated nomenclature concerning the development dependency from flt-3L or from macrophage colony–stimulating factor (M-CSF). Another feature of DCs concerns their maturation either from HSCs or from monocytes. Above all, specialized DCs are located in the intestine and other MALTs or in the integument. The DCs of the skin can be divided into epidermal Langerhans cells (LCs) characterized by cytoplasmic Birbeck granules containing Langerin and dermal Langerin-positive as well as Langerin-negative DCs [2]. Langerin (CD207) is a transmembrane protein belonging to the family of C-type II $Ca^{2+}$-dependent lectins with an extracellular carbohydrate-binding domain specific for mannosyl residues and an intracellular domain with a proline-rich motif. It is taken up from the cell surface by clathrin-dependent receptor-mediated endocytosis and traffics via early endosomes (EEA1) to the endosomal recycling compartment (ERC) but does not reach lysosomes for degradation. Langerin is also localized at the inner band of Birbeck granules [3]. These organelles have a unique ultrastructural morphology and consist of superimposed and zippered membrane components [3]. A second type of nomenclature distinguishes classical or conventional DCs (myeloid DCs) from plasmacytoid DCs based on the distinct morphology, markers, and gene expression profiles. Furthermore, one can distinguish between resident DCs moving directly to lymphatic compartments and migratory DCs entering to a respective tissue and traveling via lymphatic vessels to lymph nodes. Both cell types originate from blood cell precursors. More recently, newer criteria including not only developmental but also functional characteristics refer to regulatory mechanisms such as the initiation of immune tolerance, the role of hematopoietins such as G-CSF, M-CSF, flt-3L, GM-CSF and their receptors responsible for the generation of monocyte-dependent or independent DCs, as well as the significance of various transcription factors, for example, PU.1, mafs, ID2, IRFs, E2-2, and NF-κB, that are responsible for DC development according transcriptional programs [1, 4].

### 2.1. Antigen presentation

The most important DC function is the uptake of antigenic material via phagocytosis or pinocytosis by immature DCs and its processing into peptides and presentation via the peptide binding groove of MHC (in human HLA) molecules by mature DCs. Uptake of antigenic material is mediated by four alternative pathways: endocytosis, pinocytosis, phagocytosis and macropinocytosis. How the antigenic material is taken up by DCs depends on its features such as the form of the antigen, its solubility, or whether it is part of an immune complex or still associated with a pathogen [5]. During maturation, DCs not only change their dendritic morphology and their ability to take up antigenic material but also upregulate the expression of MHC-class I and MHC-class II molecules. In addition, the mature phenotype is characterized by specific differentiation markers [6]. Antigen presentation is initiated by direct contact with respective subtypes of T-Helper cells via binding to the T-cell receptor (TCR) complex, the CD4 molecule, as well as to additionally required costimulatory molecules expressed at the cell surface of both involved cells [7]. Antigen processing involves an interaction of the endocytic and lysosomal pathway in the MHC-class II compartment (MIIC) and the presentation by MHC-class II molecules. This process requires an activation of the lysosomal machinery [8] and a stepwise association of the degradation products with the MHC-class II binding groove. This complex process has been summarized in an excellent review by Seliger et al. [9]. Exogenous antigens are internalized via the endocytic pathway. HLA class II heterodimers assemble in the endoplasmic reticulum (ER) with the invariant chains (Ii) to form nonameric a/b–Ii complexes [(ab)3Ii3], targeted to MIIC. The invariant chain Ii protects the binding groove against association with intracellular degradation products shredded by proteasomes. The HLA-class II–associated Ii is degraded in distinct steps, leaving class II–associated Ii peptide (CLIP) within the HLA-class II binding groove. CLIP can then be exchanged for antigenic peptides, a process catalyzed by HLA-DM molecules. The HLA-DM–dependent peptide loading is regulated by HLA-DO molecules. Peptide-loaded HLA-class II molecules are then transported to the cell surface for presentation to $CD4^{+}$ T cells.

At the EM level, multivesicular and multilamellar bodies are characteristic organelles related to the MCII compartment [10–15]. Although MHC-class II molecules are abundantly expressed at the plasma membrane of DCs, they are also present in immature DCs where they have only a short half-life since they are directed in a recycling pathway. They are actively sorted to luminal vesicles of multivesicular bodies after ubiquitination and subsequently transferred to lysosomes. An increase of MHC-class II molecules at the plasma membrane could be explained by a recruitment of antigen-loaded MHC-class II molecules to the cell surface during maturation by allowing intraluminal vesicles to fuse back with the delimiting membrane of multivesicular bodies. However, there is strong evidence that the enhanced cell surface expression of antigen-presenting MHC-class II molecules is due to their increased synthesis but to a reduced lysosomal degradation [16]. Nevertheless, the stable expression of peptide-loaded MHC-class II molecules requires the intersection of the endocytic and the lysosomal pathway as explained above.

### 2.2. Cross-presentation

As most of nucleated cells, DCs also express classical MHC-class I molecules presenting intracellular peptides prepared by proteasomes and cytosolic proteases. The peptides are transported by the TAP transporters (heterodimeric multimembrane-spanning polypeptides belonging to the ABC transporter family) from the cytosol into the ER, where their assembly with MHC-class I molecules takes place, assisted by various chaperones, such as calnexin, calreticulin, ER60, and tapasin [9].

In DCs, two alternative pathways for cross-presentation have been reported: the "cytosolic" and the "vacuolar" pathway. In the cytosolic pathway, phagocytosed proteins are transported to proteasomes, where they are degraded and bound to MHC-class I molecules in the ER and directed to the plasma membrane for antigen presentation. This retrotranslocation machinery involves the ER membrane proteins sec61 and p97 ATPase [17]. Alternatively, after degradation by proteasomes, ingested proteins can be re-transported to phagosomes, where they are associated with MHC-class I molecules. In the vacuolar pathway, exogenous proteins are transported to phagosomes, where degradation, alternative to the proteasomes, takes place.

Consecutively, appropriate peptides are bound to the antigen-binding groove of MHC-class I molecules [18, 19].

Bacterial and cellular antigens are also efficiently cross-presented in association with heat-shock proteins such as hsp 70, hsc 70, and hsp 90 and the ER chaperones grp94/gp96 and their bacterial homologues. Peptide complexes are ingested after binding to specific receptors such as CD91 (the $\alpha_2$-macroglobulin receptor) and the scavenger receptors LOX-1 and SR-A [5, 20–23].

Only the MHC-class I–restricted CD8$^+$ cytotoxic T-cell (CTL) response can destroy intracellular pathogens such as viruses and intracellular parasitic bacteria as well as cells undergoing malignant transformation. Only professional APCs such as DCs and some macrophages express an efficient combination of co-receptors and MHC-class I molecules to stimulate naive CD8$^+$ T cells [24]. However, cross-presentation of intracellular antigens after degradation of intracellular peptides belonging to internal cell structures and compartments is important for the maintenance of peripheral tolerance to self-antigens (cross-tolerance) [5, 25]. Cross-presentation is also important for vaccination against tumors, which is discussed at the end of this article.

### 2.3. Types of DCs and their characteristics

The enormous progress in DC research has been achieved mainly in the mouse system. Although many homologies exist between the human and murine immune system, there are also significant differences in the expression of differentiation markers, transcription signatures, and cytokine release. In the mouse system, DCs can be subdivided into classical (cDCs) and plasmacytoid (pDCs) DCs. However, under inflammatory conditions, a new subtype of so-called inflammatory DCs appears. DCs acting in the murine immune system, cDCs, can be further subdivided into "CD8-like" DCs with a Clec9A$^+$/CD141$^+$ phenotype and CD11b-like" DCs with a CD1c$^+$ phenotype. Both of these subsets include resident DCs that are located in secondary lymphoid organs such as lymph nodes, tonsils, and spleen as well as migratory DCs, which are present in peripheral tissues as well as in non-lymphoid organs (skin, liver, lung, kidney, intestine, and other organs). After antigen contact, cells of this subtype migrate via lymphatic vessels into local lymph nodes [19]. In addition, special DC types are present in the skin: the epidermal LCs and dermal DCs, as mentioned above.

*In vitro*, human DCs can be either derived from CD34$^+$ HSCs, or isolated from bone marrow, umbilical cord blood, or purified monocytes (Fig. 1a and b). In this respect, monocytes represent an immature cell type with a potential to differentiate either into macrophages or into DCs. *In vivo*, similar differentiation lines can be demonstrated, whereas the decision to follow a distinct differentiation line is made in the bone marrow. Surface markers, transcription signatures, as well as growth factors driving differentiation have been summarized in several reviews [6, 26–30].

Depending on their stimulation, DCs are able to promote different T-cell responses. In contrast to the murine immune system, in human system, pDCs (bearing MHC-class II antigens, CD4, CD123, CD80, and CD86) can provoke a Th2 and Th0/Tr1 response, whereas mDCs (bearing

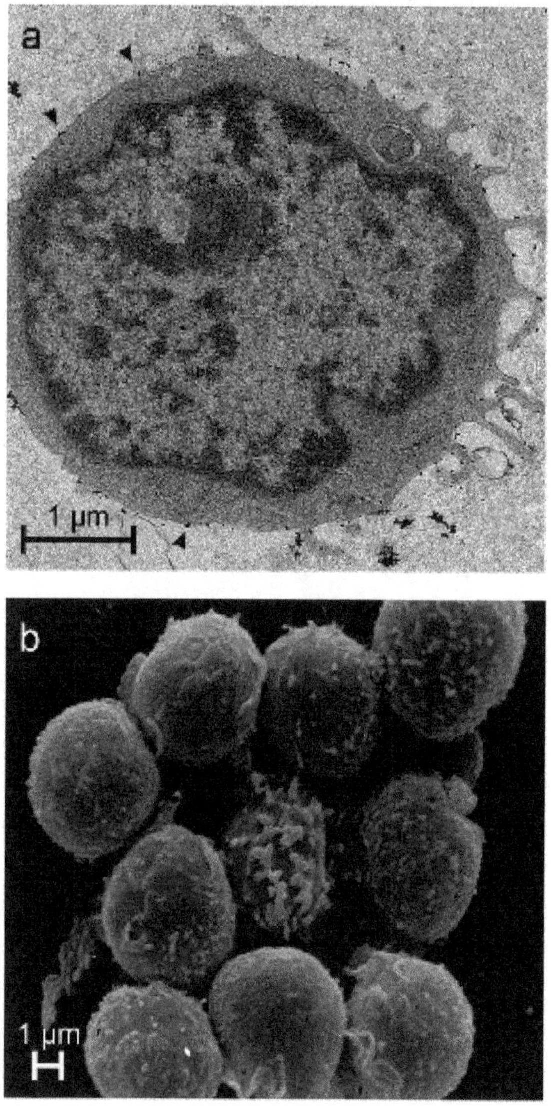


**Figure 1.** CD34$^+$ HSCs, isolated with paramagnetic particles, coated with an anti-CD34 monoclonal antibody shown by TEM (a) and SEM (b). In Figure 1a, arrowheads mark binding of anti-CD34 antibody-coated amorphous paramagnetic particles (average size: 50 nm).

MHC-class II antigens CD11c, CD80, and CD86) can provoke a Th1 response [6]. Naive CD4$^+$ T cells can differentiate into Th1, Th2, Th17, and regulatory T cells (Tregs). While Th1 cells secrete mainly IFN-γ to eliminate intracellular antigens, Th2 cells produce IL-4, IL-5, and IL-13 to eliminate helminthes and extracellular pathogens [31]. The IL-17-synthesizing Th17 cells are functional in the clearance of extracellular bacteria and fungi [32]. Tregs secrete anti-inflammatory cytokines such as IFN-β and IL-10 and are essential for the maintenance of self-tolerance and for the prevention of autoimmunity [4, 33]. After activation of the TCR, the signals for CD4⁺ T-cell differentiation are mediated by binding of specific cytokines to their cognitive receptors that are associated with Janus kinases (Jaks). After phosphorylation of the intracellular domains of the cytokine receptor and activation of Jaks, STATs are also phosphorylated, inducing the expression of genes responsible for the initiation of differentiation by the action of master regulators responsible for the synthesis of Th subtype–specific cytokines [4, 34]. STATs are signal transducers and activators of transcription pathways initiating the expression of master regulator transcription factors belonging to a complex network of other transcription factors. Different STATs are required for Th-specific differentiation lines: STAT4 and 1 for Th1 cells, STAT6 and 5 for Th2 cells, STAT3 for Th17 cells, and STAT5 for Tregs [4].

According to a recent excellent and comprehensive review [35], we can summarize that DCs belong to the entire phagocytic system including monocytes/macrophages in different compartments such as lymphoid and non-lymphoid tissue and blood. In this nomenclature, DCs are subdivided into classical $CD141^+$ myeloid DCs (cDC1), $CD1^+$ myeloid DCs (cDC2), and plasmacytoide DCs (pDCs). DCs can also differentiate from monocytes (moDCs). In the murine system, different development pathways for moDCs and DCs derived from HSCs have been demonstrated. However, because of the research at the transcriptional level, it became evident that these restrictions are not so strong in both humans and mice as previously suggested.

**Human DCs can be classified according to Haniffa et al.** [35].

### *2.3.1. $CD141^+$ myeloid DCs (cDC1)*

CD141+ myeloid DCs (cDC1)secrete TNFα, CXCL10, and IFNγ, but little IL-12p70. They cross-present necrotic cell–derived and soluble antigens via CLEC9A, but cross-presentation is not totally restricted to this cell type. CD141 (BDCA-3, thrombomodulin), an integral membrane protein, is expressed not only at the cell surface on DCs but also on endothelial and mesothelial cells as well as on monocytes. By binding to thrombin, it serves as its cofactor and reduces blood coagulation by converting thrombin to an anticoagulant enzyme and participates in inhibition of fibrinolysis. In addition, it regulates C3b inactivation by coagulation factor 1 [36].

### *2.3.2. $CD1c^+$ myeloid DCs (cDC2)*

CD1c+ myeloid DCs (cDC2) are characterized by a high expression of the toll-like receptors (TLRs) 1, 2, 4, 5, and 8. They are able to present glycolipid antigens and react with mycobacteria via CD1a and c. They also exhibit reactivity against fungi by involving the receptors Dectin-1 (CLEC7A) and Dectin-2 (CLEC6A), the receptors DEC205 (CD205; CLEC13B), and the macrophage mannose receptor (CD206; CLEC13D). Upon stimulation, these cells secrete IL-1β, TNFα, IL-8, and IL-10. In blood, cDC2 cells can abundantly secrete the enzyme retinaldehyde dehydrogenase (RALDH)2 dependent on the level of vitamin D3. They are able to initiate Th2, Th1, and also Th17 response dependent on stimulation by respective antigens.

CD1c (BDCA-1) acts as antigen-presenting protein binding self- and non-self–lipid and glycolipid antigens and presenting them to T-cell receptors on natural killer T cells [37].

#### *2.3.3. Plasmacytoid DCs (pDCs)*

Plasmacytoid DCs (pDCs) strongly express TLR7 and TLR9, enabling the recognition of autologous nucleic acids and viral antigens. Challenging pDCs with viral antigens results in a potent secretion of class-I interferons, mainly IFN$\alpha$. In response to inflammation, pDCs are recruited to tissues and lymph nodes. They display two main functions: they can polarize Th1 and Th2 response and play an important role in immune tolerance but also in autoimmune disease by initiation of a Treg response. The functional capacity of pDCs increases gradually during fetal life. In this respect, it could be demonstrated that preterm neonates possess an impaired BDCA4 expression and produce lower levels of IFN$\alpha$. Ultrastructurally, they show an immature morphology [38]. In addition, the *in vitro* maturation of blood DC depends on the upregulation of BDCA3 on pDCs [39].

## 3. Maturation of myeloid DCs (mDCs)

When challenged by immune stimulatory and pathogenic antigens, mDCs mature into regulatory or stimulatory DCs. The development of regulatory DCs requires tolerogenic stimuli via anti-inflammatory cytokines and mediators such as TGF-β, IL-10, and $PGE_2$. Regulatory DCs provide induction of Tregs, inhibiting the proliferation of $CD4^+$ $CD8^+$ T-cells and producing anti-inflammatory cytokines such as IL-10 and TGF-β, as well as immune-regulatory molecules such as indoleamine 2,3-dioxygenase (IDO) and programmed cell death protein 1 (PD-1), which suppress activation and proliferation of effector T cells and efficiently induce Tregs [36]. In contrast, stimulatory DCs develop under the influence of proinflammatory signals: inflammatory cytokines, chemokines, pathogen-associated molecular patterns (PAMPs) such as lipopolysaccharide (LPS), flagellin, lipoteichoic acid, peptidoglycan, and viral double-stranded RNA (dsRNA) [40–42]. Additional stimulators are damage-associated molecular patterns (DAMPs) representing host molecules that can initiate and perpetuate inflammatory response that are not caused by infectious agents. DAMPs can be proteins like heat shock proteins or breakdown products of the extracellular matrix such as hyaluronan fragments or other molecules such as ATP, uric acid, heparin, sulfate, or DNA [43–46].

### 3.1. Immature mDCs

Immature mDCs circulate either in the peripheral blood or migrate into infected tissues where they take up pathogens or cell debris. This uptake is mediated by type C lectin receptors such as CD206, DEC205, or CLEC4A or in the case of PAMPs by toll-like receptors. Antigenic material is ingested by phagocytosis or pinocytosis and directed to the intracellular degradation machinery [36].

In our own experiments, we differentiated mDCs from $CD34^+$ HSCs of human cord blood: Fig. 1a and b. We followed a cultivation protocol established by Strunk et al. [47]. Shortly, mononuclear cells were obtained by centrifugation over the flotation medium Ficoll®-Paque Premium (Sigma-Aldrich, Vienna, Austria). DCs were incubated with paramagnetic particles coated with a monoclonal antibody against CD34, isolated using a magnetic cell separator (MACS®, Miltenyi Biotech, Bergisch Gladbach, Germany), and cultivated in a conditioned RPMI-1640-medium containing 20% FCS, 100 ng/ml GM-CSF, 20 ng/ml SCF, 20 ng/ml TNF-$\alpha$, and 0,5ng/ml TGF-$\beta$1. CD34$^+$ stem cells differentiated into immature mDCs showed an LC-like phenotype (HLA-DR$^{dim}$, CD1a$^+$, and Langerin$^+$) and grew in clusters. Cultures containing clusters were layered over 7.5% bovine serum albumin solution in phosphate-buffered saline (BSA/PBS, Sigma, St. Louis, MO), where they could be collected and transferred to fibronectin-coated chambered coverslips. Further differentiation to mature DCs has been achieved by addition of 10 µg/ml LPS + 50 ng/ml TNF-$\alpha$ to the culture medium or by supplementing the medium with steroid hormones as outlined later on.

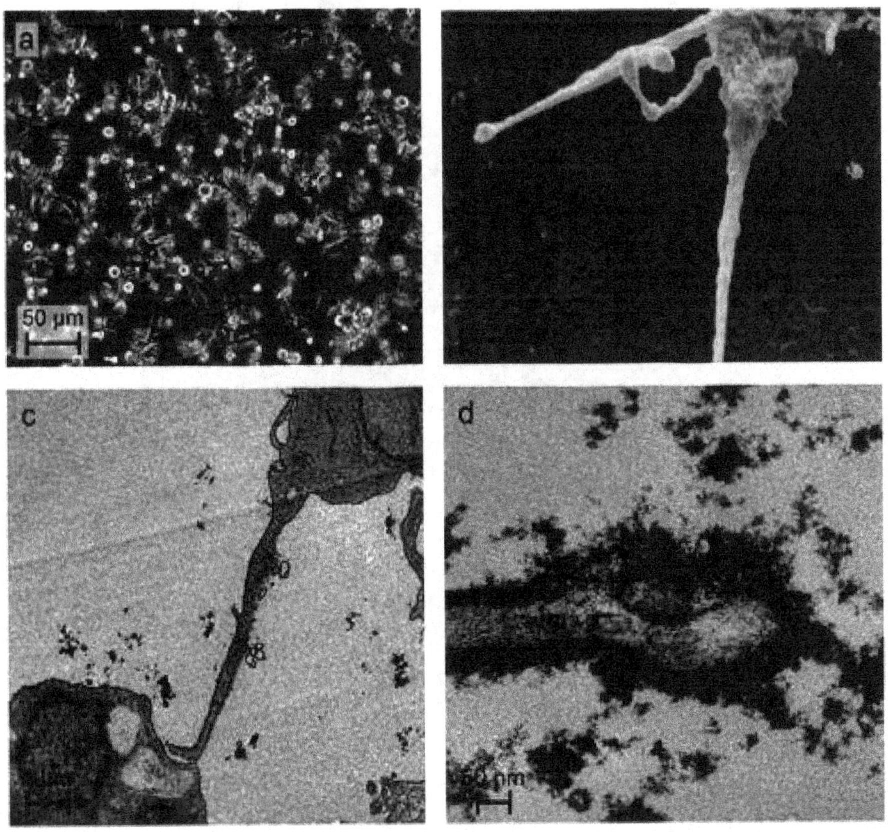


**Figure 2.** Immature dendritic cells growing in conditioned cell culture medium on the surface of fibronectin-coated chambered coverslips. Fig. 2a shows the appearance of the DC morphology in phase contrast. Fig. 2b and c show extended cell projections with knob-like structure in SEM (Fig. 2b) and TEM (Fig. 2c and at higher magnification in Fig. 2d). Note the prominent coat of the glycocalyx in Fig. 2d. In Fig. 2c paramagnetic particles are visible, which are still present after cell separation but partially detached from the cell surface.

Immature mDCs preferentially extended large dendritic cell processes (Fig. 2 ac) which a knob-like structure as shown in Fig. 2d where we suppose a concentration of the mentioned receptors. In this respect, DCs, located in the lamina propria of the small intestine, extending long cell processes that survey antigens in the lumen of the gut have been demonstrated [48–50].

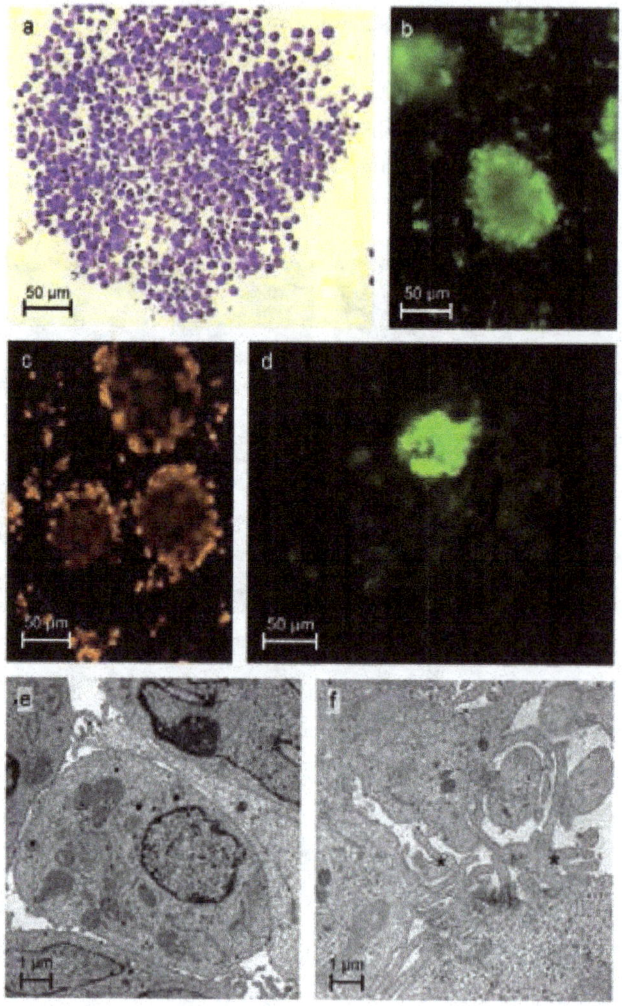


**Figure 3.** Cultivation and characterization of immature mDC clusters using light microscopy (Fig. 3a), fluorescence microscopy (Fig. 3b, c, and d), and TEM (Fig. 3e and f). Fig. 3a shows a low-magnification image of an adhered cell cluster taken from a 1 μm semithin section stained with toluidine blue. Fig. 3b shows staining with an FITC-conjugated monoclonal antibody against CD1a, and Fig. 3c, a cell surface staining with a PE-conjugated monoclonal anti-HLA DR antibody. In Fig. 3d, only the center of the cluster is positive for E-cadherin cell surface expression using an FITC-conjugated monoclonal antibody against E-cadherin. TEM images show cells with relatively smooth surface in the center of a cluster (Fig. 3e) as well as cells at the periphery of clusters forming interdigitating cell projections, which are marked by stars (Fig. 3f).

After collection of clusters and transfer to fibronectin-coated polystyrol dishes, they adhered again to the growing surface (Fig. 3a). At this stage, they displayed still an immature phenotype till day 7 of cultivation. Under the influence of maturation signals, they differentiated starting from the margins of clusters (Fig. 3). Using fluorescence microscopy, it could be shown that CD1a staining was present at the surface of all mDCs (Fig. 3b), while HLA-DR expression could be found only at the periphery of clusters (Fig. 3c). E-cadherin could also be demonstrated but only in the center of the clusters (Fig. 3d). TEM imaging of the clusters revealed that in their center the surface of cells remained relatively smooth (Fig. 3e), while at the periphery, many interdigitating cell projections were formed (Fig. 3f).

In TEM, immature mDCs display characteristics of high activity as they develop an abundant rough ER and a prominent Golgi apparatus as well as an extended trans-Golgi network (TGN) (Fig. 4a and b). The MIIC compartment is characterized by the presence of abundant lysosomes and multilamellar bodies (Fig. 4c and d). Also autophagosomes engulfing several multilamellar bodies could be demonstrated (Fig. 4e). Interestingly particular endosomes aligned in a pearl-like structure could be shown (Fig. 4f). We regard these structures as special organelles of the endosomal pathway. In a previous publication we could identify similar structures in endothelial progenitor cells also derived from $CD34^{+}$ HSCs of human cord blood [51, 52]. In this respect, immature mDCs are highly active in antigen uptake and processing.

Regarding *in vitro* conditions, above all under the influence of TGF-β1, mDCs develop a characteristic LC-like ultrastructural morphology by displaying Birbeck granules and Birbeck-like organelles already described in early publications [47, 53, 54], which we could also verify in our experiments (Fig. 5). Birbeck granules store langerin, a mannose-binding lectin.

### 3.2. Mature mDCs

The maturation of mDCs is accompanied by morphological and functional changes: Mature mDCs lose their capacity to adhere to plastic surfaces. The few long cell processes of immature DCs are replaced by numerous short filopodia (Figs. 6a and b) and veil-like membrane projections while the uptake of antigenic materials is downregulated in favor of antigen presentation as well as of cross-presentation [55]. In this respect, only mature DCs are entirely able to cooperate with lymphocytes (Fig. 6b) by presenting antigens via MHC-class I and II molecules [56]. Maturation of mDCs requires the expression of costimulatory molecules such as CD80, and CD86 and the integrin receptor LFA-1 (CD11a). In addition, the cytokine tumor necrosis factor alpha (TNF-α), LPS, and the CD40 ligand (CD40L) have been found to induce DC-maturation [56]. The degeneration of aged mature mDCs is shown in Figs. 6c and d.

It has been demonstrated that CD83 is expressed only in mature DCs [55]. CD83 is a 45-kD, type-1 membrane glycoprotein belonging to the Ig superfamily. CD83 is released from activated cells, and the soluble form of CD83 has a strong immunosuppressive effect. The immunoregulatory function of CD83 implicates its use for therapy of cancer and autoimmune diseases [57].

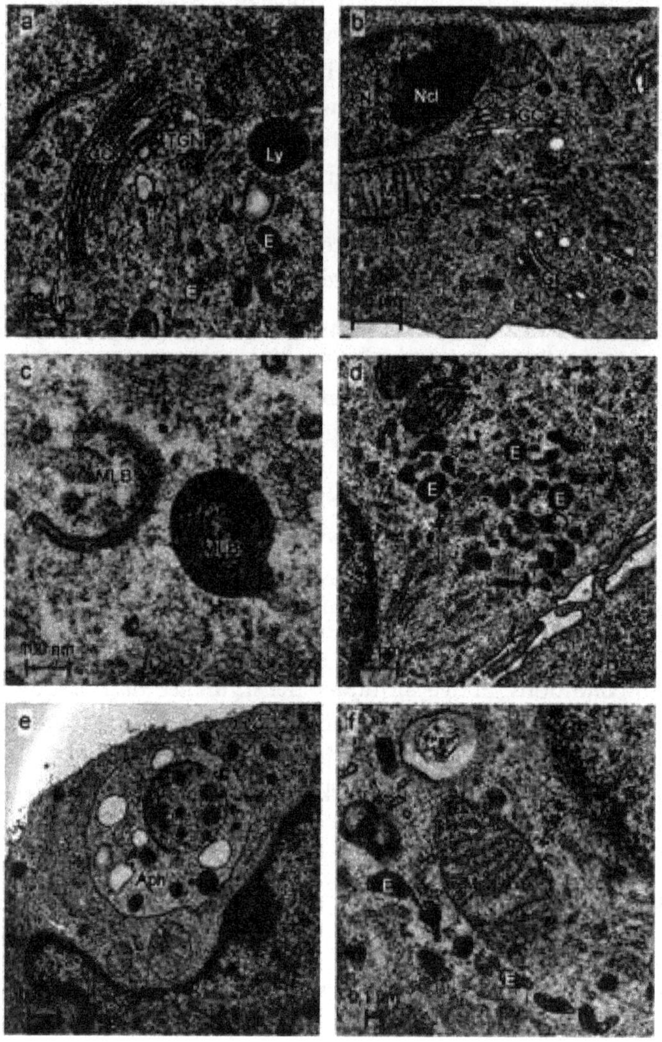


**Figure 4.** Particular ultrastructural characteristics of immature mDCs visualized by TEM: Well-expressed Golgi complexes (GC) and trans-Golgi networks (TGN) as well as endosomes (E), lysosomes (Ly), mitochondria (M), and an active nucleus (N) with a distinct nucleolus (Ncl) are visible in Figs. 4a, b, d, and e. Multilamellar bodies (MLB) shown in Fig. 4c are part of the MIIC compartment. An autophagosomes (APh) with degraded material is shown in Fig. 4 e. In Fig 4f, a pearl-like structure of aligned endosomes (E) is visible.

# 4. How sexual hormones influence the development and maturation of mDCs

*In vitro*, mDCs can be derived from $CD34^+$ HSCs from bone marrow but also from umbilical cord blood. Triggered by the cytokines GM-CSF and TNF$\alpha$, HSCs differentiate into inter-

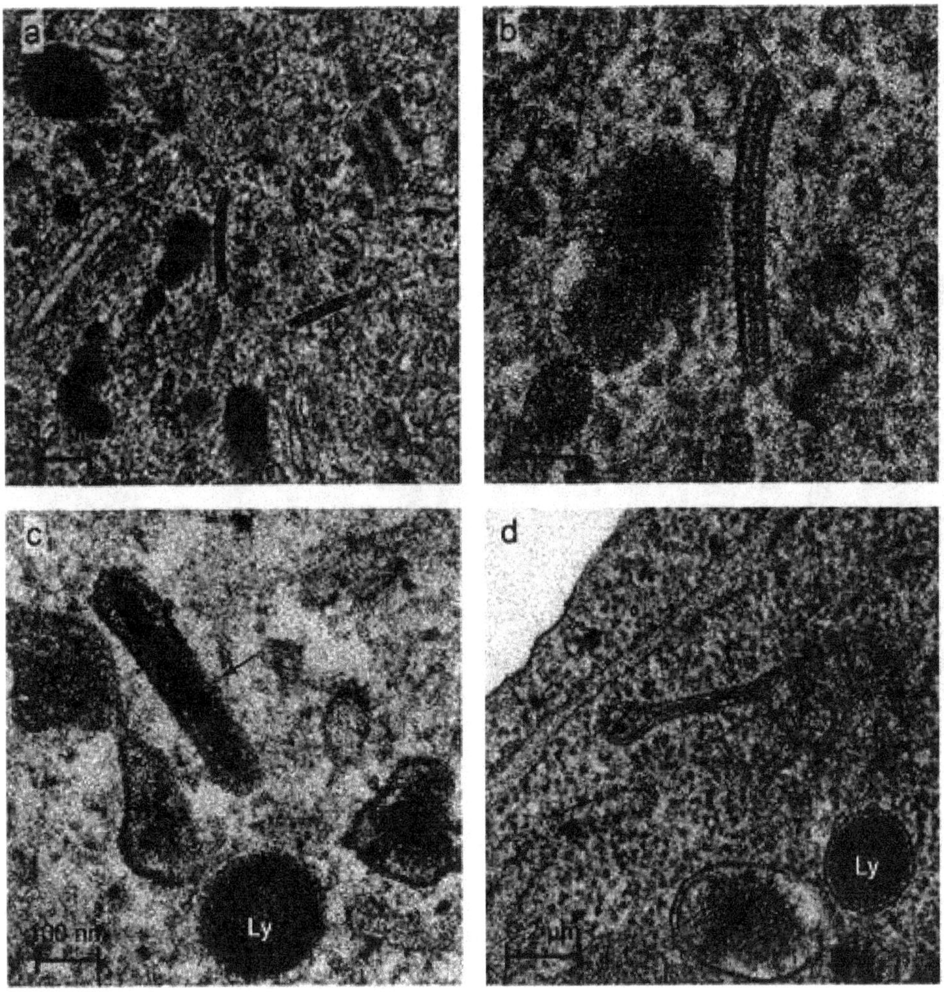


**Figure 5.** Formation of Birbeck granules and related structures visualized by TEM: Classical Birbeck granules are shown in Figs. 5a and b (arrowheads). They are rod-shaped (Fig. 5a and b) or drumstick-like (Fig. 5d; arrowhead).organelles with a central linear density and a striated appearance. Occasionally, organelles resembling endothelial Weibel-Palade bodies with central microtubular structures (arrow) occur (Fig. 5c). In addition, lysosomes (Ly) are visible.

mediate $CD14^+/CD1a^-$ precursors, which develop consecutively into mature DCs but also to LC-like cells. Bone marrow-derived $CD34^+/CD10^+/Lin^-$ cells mature directly into DCs under the influence of a cytokine cocktail containing IL-1, IL-7, TNF$\alpha$, GM-CSF, SCF, and Flt3-ligand [30]. ER ligands are potent modulators of hemopoiesis and immune function in health and disease influencing key cytokines such as GM-CSF and Flt3 ligand that promote the maturation of mDCs. However, myeloid progenitors differentiate variably under the influence of 17$\beta$ estradiol with each of the two cytokines: While GM-CSF promotes the development of LC-like mDCs, Flt3 ligand decreases the number of plasmacytoid, lymphoid, and myeloid DCs in a dose-dependent manner. DC differentiation is mainly regulated by binding of estradiol to

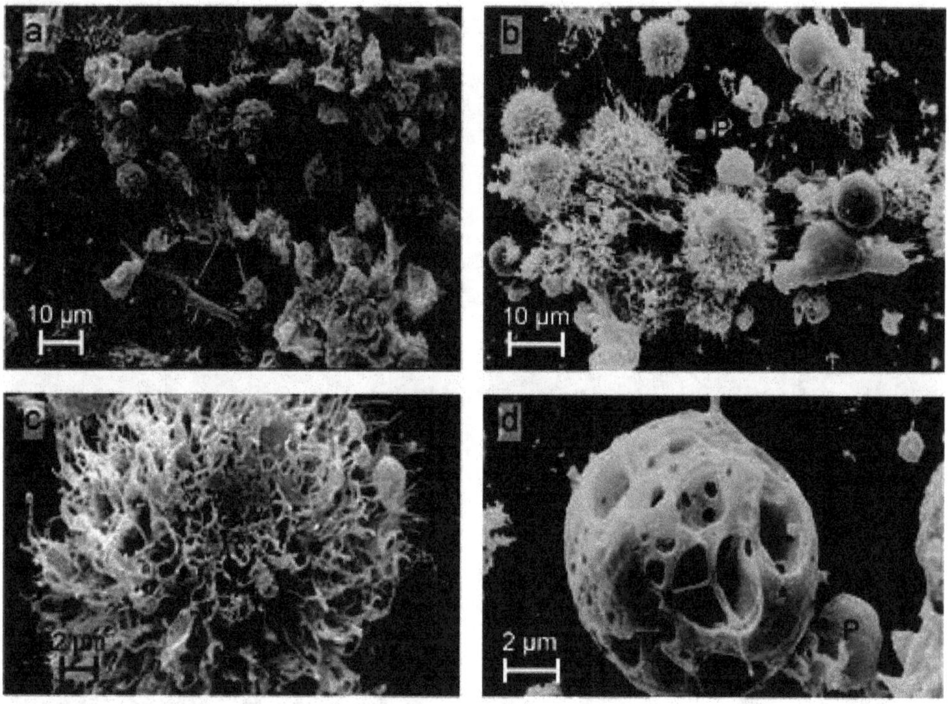


**Figure 6.** Ultrastructural characteristics of mature and apoptotic/necrotic myeloid DCs investigated by SEM. Figs. 6a and b show clusters of mature DCs with numerous microvillus-like cell projections. Adhesion of lymphocytes (L) and platelets (P) is visible in Fig. 6b. In Fig. 6a, still some immature DCs with long cell projections are visible. Fig. 6c shows abundant cell projections as a sign of beginning degeneration, distinctly pronounced in Fig, 6d.

ER$\alpha$. Therefore, endogenous or ER ligands produced by pharmaceutical companies may differentially affect DC development in health and disease concerning the DC-mediated immunity [58]. ER$\alpha$ and $\beta$ belong to the nuclear receptor superfamily encoded by the ESR1 and ESR2 genes, respectively. Different types of ER receptor isoforms are generated by homeotypic or heterotypic dimerization of both chains [59, 60]. Activation of ERs exhibits an epigenetic modification of DNA by chromatin-modifying co-regulators and other transcription factors that initiate DNA transcription [61]. It has been shown that estrogen preferentially promotes the differentiation of immature CD11c$^{+}$ CD11b$^{int}$ DCs from bone marrow precursors [62] as well as the GM-CSF-mediated differentiation of a CD11b$^{+}$ subset from myeloid progenitors via the interferon regulatory factor (IRF4). This transcription factor is induced by GM-CSF [63].

In contrast to ER activation, therapeutic selective ER modulators such as tamoxifen and raloxifene have been shown to impair DC differentiation and activation. As competing with estrogen, both drugs possess antagonistic effects on ER function. They were not able to provide DC differentiation and have, therefore, no effect as ER agonists. Tamoxifen has beneficial effects in respect to the treatment of breast cancer, while it increases the risk for endometrial cancer, whereas raloxifene can be successfully used against osteoporosis since it exerts similar effects than estrogen. However, tamoxifen and raloxifene medication can lead to a partial immune suppression at the level of antigen presentation and T-cell activation by downregulating the expression of MHC-class II and costimulatory molecules. Unfortunately, this downregulation can interfere with cellular immune therapy against cancer such as vaccination with DCs. [64].

In pregnancy, an immune-modulatory function of progesterone has been discussed [65]. It could be demonstrated that pregnancy-associated elevated levels of progesterone induced the production of IL-10 by mature DCs derived from human blood monocytes, leading to a Th2 immune response. In this respect, the regulation of the Th1- and Th2- related cytokines is important for a successful pregnancy [66]. Impaired DC function has also been reported in the mouse immune system. In this respect, progesterone-treated DCs were characterized by a decreased expression of Ia molecules (MHC-class II), CD80, and CD86 as well as by an increased production of IL-10, and a decreased secretion of IL-12. In addition, an impaired stimulatory capacity for $CD4^+$ helper cells was demonstrated [67].

*In vitro* experiments with metroxyprogesterone acetate, an anti-contraceptive drug, could impressively demonstrate that TLR3 agonist stimulation decreased the expression of CD40 and CD80 and impaired the initiation of T-cell proliferation [68].

In general, most of functions of the innate and adaptive immune system decline with aging. Therefore, vaccination response against pathogens is also frequently impaired. Postmenopausal women produce higher levels of proinflammatory cytokines such as MCP1, TNF$\alpha$, and IL-6, which has been implicated in the development of several diseases such as osteoporosis (by activation of osteoclasts) but also in diabetes, atherosclerosis, and cardiovascular diseases. Estrogen ($E_2$) and progesterone are the master cytokines of the immune system in the female reproductive tract (FRT). Although there are still gaps in our knowledge of the immune system of the FRT, above all concerning the mucosal parts of FRT, there is good evidence for successful treatment of postmenopausal women using an appropriate hormone therapy [69, 70].

### 4.1. Investigations performed in our laboratory

We visualized the maturation of mDCs cultivated using the differentiation protocol as described above under the influence of $10^{-5}$M β-estradiol or alternatively to $10^{-5}$M progesterone. After 1 week of cultivation, clusters of immature DCs were transferred to new chambered coverslips containing the same conditioned medium containing the respective hormones and cultivated for another week. Using flow cytometry and cell sorting (Fig. 7a and b), immature (Fig. 7c) and mature mDCs (Fig. 7d) were collected and prepared for SEM. In addition, the percentage of mDCs positive for CD1a, CD14, CD83, CD1a/CD83, and CD14/CD83 was determined.

#### *4.1.1. Immunological characterization with fluoro-labelled antibodies*

Immunological characterization with fluoro-labelled antibodies was performed by staining of cells using a fluorescein-isothiocyanate (FITC)-conjugated monoclonal antibody against CD1a (BD Pharmingen™, Vienna, Austria), clone HI149, mouse IgG1, a phycoerythrin (PE)-

conjugated monoclonal anti-HLA DR antibody (BD Biosciences, Vienna, Austria), clone L243, mouse $IgG_{2a}$, and a FITC-conjugated monoclonal antibody against E-cadherin from BD Biosciences, clone 36/E cadherin, mouse $IgG_{2a}$. Fluorescence microscopy was performed using an inverted Nikon Eclipse TE-300 Microscope (Nikon Coop., Tokyo, Japan) with conventional filter packs for blue and green light excitation. Micrographs were taken with a Nikon Coolpix 5000 digital microscope camera.

#### *4.1.2. For TEM investigations*

For TEM investigations cells were attached to poly L-lysine coated coverslips, washed twice with $Ca^{++}$- $Mg^{++}$-free PBS, and fixed in 2.5% glutaraldehyde in 0.1 M cacodylate buffer at pH 7.2 for 60 minutes at 4°C. After washing with cacodylate buffer, cells were postfixed with 1% aqueous $OsO_4$ for 90 minutes at 4°C. Cells were dehydrated stepwise with ethanol and embedded in Epon. After polymerization, 70 nm ultrathin sections parallel to the growing plane were performed using a diamond knife. Acquisition of digital images was performed using a bottom-mounted CCD camera (Gatan Inc., Munich, Germany) in a Tecnai 20™ transmission electron microscope (FEI company, Eindhoven, The Netherlands).

#### *4.1.3. For SEM investigations*

For SEM investigations, cells were treated according to the TEM preparation protocol till fixation and then subjected to a critical point drying procedure and sputtered with gold. Micrographs were taken using a Nikon F3 reflex camera (Nikon Coop.) mounted on a Stereoscan S90 scanning electron microscope (Cambridge Instrument Company, Cambridge, UK).

#### *4.1.4. Flow cytometry and cell sorting*

Flow cytometry and cell sorting was performed using a FACSort machine (Becton Dickinson) which utilizes a "catcher tube", a mechanical sorting device, located in the upper portion of the flow cell. By moving in and out of the sample stream, it allows to collect a population of gated cells. In our investigations, sorted $CD1a^{bright}/CD83^{dim}$ immature mDCs (Fig. 7c) and sorted $CD1a^{dim}/CD83^{brigh}$ mature mDCs (Fig. 7d) were shown in SEM. Fig. 7a shows the gating process of mature DCs by forward and side scatter and Fig. 7b, the gating of $CD1a^{dim}/CD83^{brigh}$ mDCs at the fluorescence dot blot (FL1: labeling with a FITC-conjugated anti-CD1a monoclonal antibody; FL2: labeling with a PE-conjugated monoclonal antibody against CD83) using the Attractor software (Becton Dickinson). In addition, the software allows neglecting cell debris and, therefore, cell quantification in terms of absolute counts using reference beads. For demonstration of mature mDCs, a monoclonal PE-conjugated antibody against CD83 (Clone HB15e (RUO) from mouse, IgG1 κ chain) has been used.

#### *4.1.5. Results from our own studies*

It could be demonstrated that already at day 7 of cultivation above all the percentage of CD1a and CD14 positive DCs decreased under the influence of progesterone, while it increased under the influence of β-estradiol (Table 1a). At day 14 of cultivation, a similar situation occurred but CD83 in combination with CD1a and CD14 was markedly reduced due to progesterone treatment but distinctly increased by β-estradiol (Table 1b). In addition, the maturation process concerning the formation of abundant cell projections was demonstrated by SEM analysis, and was still present at day 7 (Table 1a), but more pronounced on day 14 (Table 1b).

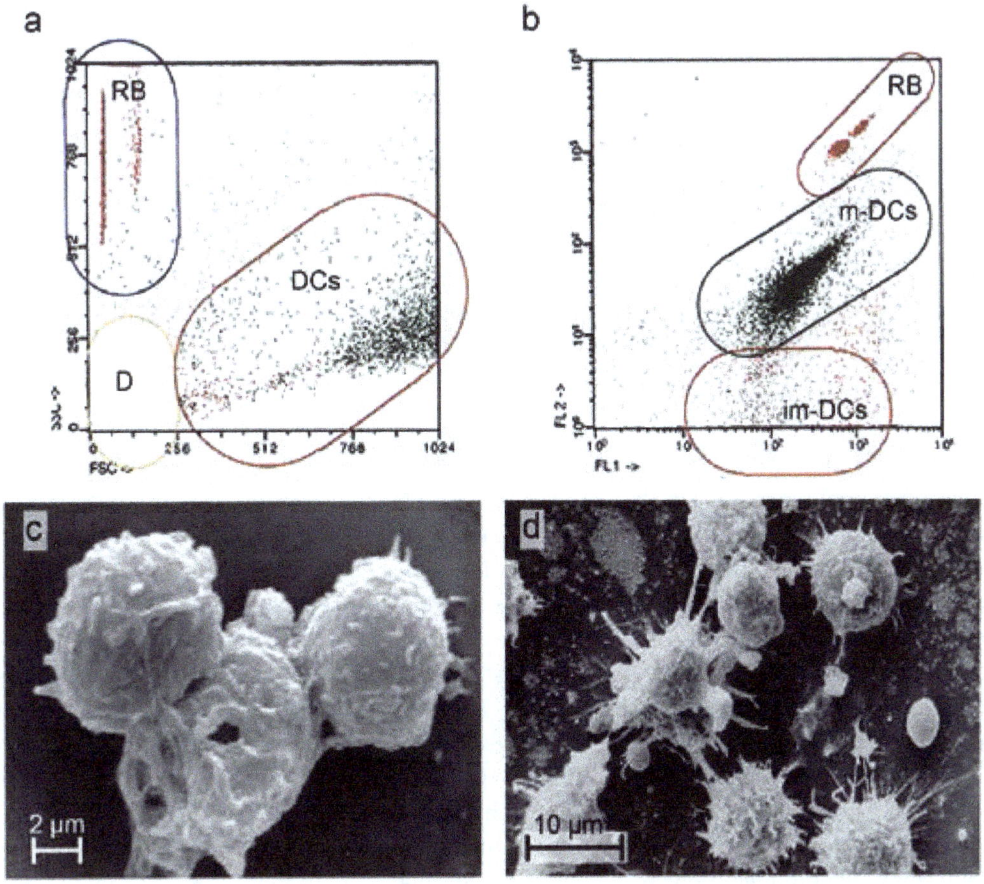


**Figure 7.** Visualization of immature mDCs by SEM after flow cytometric cell sorting by gating of CD1a$^{bright}$/CD83$^{dim}$ mDCs (Fig. 7c) and of mature mDCs by gating of CD1a$^{dim}$/CD83$^{brigh}$ mDCs (Fig. 7d). Fig. 7a shows the gating process of mature DCs using a forward and side scatter dot plot, Fig. 7b the gating of CD1a$^{dim}$/CD83$^{brigh}$ mDCs at the fluorescence dot plot (FL1, determination of cell labeling with an FITC-conjugated anti-CD1a monoclonal antibody; FL2, cell labeling with a PE-conjugated monoclonal antibody against CD83 using the Attractor software, which allows neglecting cell debris and therefore also cell quantification in terms of absolute counts by using reference beads).

| Hormone | Flow cytometry and SEM on day 7 | | | | | |
|---|---|---|---|---|---|---|
| DC markers | CD1a$^+$ | CD14$^+$ | CD83$^+$ | CD1a$^+$/CD83$^+$ | CD14$^+$/CD83$^+$ | SEM: mature DCs |
| Control (only medium) | 16 | 23 | 18 | 6 | 4 | 16 |
| Progesterone | 7 | 5 | 17 | 4 | 3 | 14 |
| β-estradiol | 17 | 39 | 11 | 5 | 6 | 26 |

(a)

| Hormone | Flow cytometry and SEM on day 14 | | | | | |
|---|---|---|---|---|---|---|
| DC markers | CD1a$^+$ | CD14$^+$ | CD83$^+$ | CD1a$^+$/CD83$^+$ | CD14$^+$/CD83$^+$ | SEM: mature DCs |
| Control (only medium) | 57 | 30 | 64 | 45 | 24 | 29 |
| Progesterone | 26 | 57 | 56 | 24 | 10 | 15 |
| β-estradiol | 51 | 25 | 74 | 45 | 23 | 31 |

(b)

**Table 1.** a. Percentage of DC marker expression measured by flow cytometry and visualization of mature mDCs by SEM on day 7 of cultivation under the influence of steroid hormones. b. Percentage of DC marker expression measured by flow cytometry and visualization of mature mDCs by SEM on day 14 of cultivation under the influence of steroid hormones.

## 5. Generation of DC anti-tumor vaccines

Since two decades, mDCs, in their immature as well as mature state, have been widely used for experimental as well as for clinical purposes. The Nobel Prize Laureate in physiology or medicine Ralph Steinman was a pioneer in DC research. Sorry to say that Steinman never got to know the decision of the Nobel Prize committee since he died on October 3, 2011 [71]. He postulated that DCs could provide an immune attack on cancer, broad enough to encompass multiple targets, including mutant proteins expressed by the cancer. In addition, DCs could be able to activate and expand different arms of cell-mediated immunity such as NK cells, γδ- and αβ-T cells that can recognize different alterations in cancer cells [72]. Still in 1989 an institute for tumor therapy using DC vaccines was founded in Duderstadt near Göttingen (Germany) based on early investigations [73, 74] of the working group of Peters with the focus on gynecological-oncological diseases. Early therapeutic approaches included the following steps: (1) the freezing and storage of tumor material after operation; (2) blood drawing from the tumor patient immediately after operation; (3) isolation of mononuclear cells by gradient centrifugation; cultivation of cells in conditioned medium containing IL4 and GM-CSF in order to obtain immature mDCs; (4) confrontation of mDCs with thawed, irradiated, and homogenized tumor material; and (5) administration of mDCs by creating an intradermal blister. However, early clinical trials have frequently shown that after short recurrence of the disease a relapse occurred which was difficult to treat. The main question in this context is why the immune system fails to attack the tumor. The answer consists in the downregulation of antigen-presentation and T-cell activation of DCs in the tumor environment. The administration of primed DCs fails also because of the immunosuppressive action of tumor cells [75]. The reason for these pitfalls has been outlined in a recent review by Tran Janco et al. [76]. In tumors, tumor-infiltrating DCs (TIDCs) are functional but several receptors and cytokines of tumor cells compromise them leading to an environment that favors tumor progression. The expression of CD11c allows distinguishing three DC subsets: CD11c$^{bright}$, CD11c$^{intermediate}$, and CD11c$^{dim}$; CD11c$^{intermediate}$ expressing DCs are the predominant group of TIDCs. They fail to express enough quantities of costimulatory molecule, resulting in a hampered T-cell activation. In addition, they produce high levels of the immune-inhibitory molecule, programmed death 1 ligand (PD-L1). PD1 (CD279), a cell surface receptor belonging to the immunoglobulin molecule superfamily, and its ligands PD-L1 and PD-L2, impair the effector phase of immune cells. TIDCs also comprise a high number of pDCs, which promote the expansion of regulatory T cells (Tregs) characterized by high expression of the forkhead box P3 (FOXP3), a master regulator in the development of these cells. The consequence is an immune-tolerance preventing the elimination of the tumor. A further problem, concerning tumor expansion, is the secretion of the cytokine vascular endothelial growth factor (VEGF) and the chemokines CCL2, CXCL1, and CXCL5, leading to a hypervascularization of the tumor and intravasation of metastatic tumor cells. Tumor-induced transcription factors such as STAT3 induce S100A9 protein, preventing full maturation of DCs and consecutively block their responsiveness to local danger signals [77]. In this respect, the high concentration of IL-10 in the tumor cell microenvironment can also lead to STAT3 activation [76].

How could an anti-cancer therapy prevent the immune-compromising function of TIDCs? To influence the function of TIDCs, the PD1 and ligands can be inhibited by antagonist antibodies to PD1 and PD-L1 [78]. Strategies to promote DC maturation or to introduce the delivery of oligonucleotide-coated nanoparticles by DCs in order to influence immune functions at the epigenetic level by modulating the activity of miRNAs have been elaborated [79].

As mentioned above, *ex vivo* expansion of DCs originating from different sources has been used in clinical and pre-clinical trials [80]. The most commonly used approach is the differentiation of DCs from peripheral blood mononuclear cells (PBMCs) obtained via leukapheresis. After cultivation in conditioned medium containing GM-CSF and IL-4 for several days, they differentiate into immature CD14$^-$/CD83$^-$ DCs. Targeting these cells with maturation stimuli leads to further differentiation into mature CD14$^-$/CD83$^+$ DCs, that express high quantities of MHC-Class I and II as well as costimulatory molecules. A faster method includes the isolation of CD14$^+$ cells by immunomagnetic isolation and cultivation for 2 days in GM-CSF and IL-4 conditioned medium. Subsequently, the addition of proinflammatory molecules such as TNF-$\alpha$, IL-1$\beta$, IL-6, and $PGE_2$ for further 24 hours induces maturation. Generation of DCs from CD34$^+$ HSCs requires mobilization of these cells into the peripheral blood of the patient prior to leukapheresis. Cultivation and expansion of these cells involve the cytokines GM-CSF, Flt3L, and TNF-$\alpha$ and results in mDCs with the typical LC-like morphology as described above. These cells are highly potent to initiate T-cell response. The most important step before vaccination is the complete maturation of DCs. In this respect, proinflammatory cytokines and $PGE_2$ are able to enhance the expression of costimulatory molecules, CD40L and CD70. In addition, TLR agonists can optimize DC function [80]. Loading of anti-tumor peptides to immature or mature

DCs result in an effective antigen presentation. It can be performed with homogenized and irradiated tumor material but also with known tumor-associated antigens. Loading of DCs with tumor antigens can be also carried out using recombinant bacterial or viral vectors or by RNA transfection of DCs. [81, 82]. Finally, DC transfection with mRNAs related to tumor-associated antigens can bypass the use of bacterial or viral vectors [69].

Nevertheless, an efficient DC vaccination requires strategies to overcome the immune modulating tumor environment as outlined above.

## 6. Conclusion

DCs are powerful mediators of innate and adaptive cellular immune response. There are different DC subtypes in respect to their localization and their functions in immune regulation or antigen presentation. The function of mDCs is highly controlled by steroid hormones. In this respect, β-estradiol supports differentiation and maturation of DCs, whereas progesterone has inhibitory effects. In the FRT, progesterone has an immune-suppressive effect during pregnancy. The estrogen deficiency after menopause also decreases the overall immune responsiveness. Also tumors of the female reproductive system are dependent on steroid hormones with implications for tumor prevention and tumor therapy. Finally, DCs can be propagated *ex vivo,* primed with tumor material, and used for vaccination against tumors. Nevertheless, the immunosuppressive tumor environment has been taken into consideration for an effective therapy.

## 7. Abbreviations

BDCAs, blood DC-specific antigens; BSA, bovine serum albumin; CD, cluster of differentiation; cDC1s, classical $CD1^+$ DCs; Clec9A, group V C-type lectin-like cell surface receptor mediating endocytosis but not phagocytosis; CXCL10, chemokine CXCL10; DCs, dendritic cells; E2-2, basic helix-loop-helix transcription factor (E protein); EEA1, early endosome antigen 1; ER$\alpha$, estrogen receptor $\alpha$; ERC, endosomal recycling compartment; EM, electron microscopy; ER, endoplasmic reticulum; ER60, chaperon thioreductase ER60; flt-3L, fms-like transcript 3 Ligand; FRT, female reproductive tract; G-CSF, granulocyte colony-stimulating factor; GM-CSF, granulocyte macrophage colony-stimulating factor; HLA, human leukocyte antigens; ID2, transcriptional regulator belonging to the inhibitor of DNA binding (ID) family; IDO, indoleamine 2,3-dioxygenase; IFN-$\alpha$, interferon alpha; IFN-$\beta$, Interferon beta; IFN-$\gamma$, interferon gamma; Ig, immunoglobulin; IL-10, Interleukin 10; IRFs, interferon regulatory transcription factors; LC, epidermal Langerhans cells; LPS, lipopolysaccharide; MACS, magnetic cell separator; maf, maf transcription factor family; MALT, mucosa-associated lymphoid tissue; M-CSF, macrophage colony-stimulating factor; MIIC, MHC-class II compartment; mDCs, myeloid dendritic cells; MHC, major histocompatibility complex; NF-κB, nuclear factor kappa-light-chain-enhancer of activated B cells; PAMPs, pathogen-associated molecular patterns; PBS, phosphate-buffered saline; pDCs, plasmacytoid dendritic cells; $PGE_2$, prostaglandin $E_2$; PU.1, hematopoietic transcription factor PU.1; SCF, stem cell factor;

SEM, scanning electron microscopy; STATs, signal transducers and activators of transcription; TAPs, transporter proteins associated with antigen processing; TCR, T-cell receptor; TEM, transmission electron microscopy; Th0, Th1, Th2, and Th17, T-helper cells 0, 1, 2 or 17; TLRs, toll-like receptors; TGF-β1, transforming growth factor β1; Treg, regulatory T-cells

## Acknowledgements

The authors gratefully acknowledge Mrs. Federenko Ivanna, Mrs. Beatrix Mallinger, and Mrs. Regina Wegscheider for their skillful and excellent technical assistance, and thank Mr. Ulrich Kaindl and Mr. Thomas Nardelli for their valuable help with the artwork.

## Author details

Josef Neumüller[1,2], Sylvia Emanuela Neumüller-Guber[3], Johannes Huber[4], Adolf Ellinger[1] and Thomas Wagner[5*]

*Address all correspondence to: thomas.wagner@medunigraz.at

1 Medical University of Vienna, Center for Anatomy and Cell Biology, Department of Cell Biology and Ultrastructure Research, Vienna, Austria

2 Blood Donation Center of the Austrian Red Cross for Vienna, Lower Austria and Burgenland, Vienna, Austria

3 Medical University of Innsbruck, Sektion for Virologie, Department for Hygiene, Microbiology, Social Medicine, Innsbruck, Austria

4 Medical University of Vienna, Department of Gynecologic Endocrinology and Reproductive Medicine, Vienna, Austria

5 Medical University of Graz, Department of Blood Group Serology and Transfusion Medicine, Graz, Austria

## References

[1] Steinman RM, Idoyaga J. Features of the dendritic cell lineage. Immunological Reviews. 2010;234(1):5-17. DOI: 10.1111/j.0105-2896.2009.00888.x

[2] Romani N, Clausen BE, Stoitzner P. Langerhans cells and more: Langerin-expressing dendritic cell subsets in the skin. Immunological Reviews. 2010; 234(1):120-141.DOI: 10.1111/j.0105-2896.2009.00886.x

[3] Mc Dermott R, Ziylan U, Spehner D, et al. Birbeck granules are subdomains of endosomal recycling compartment in human epidermal Langerhans cells, which form where Langerin accumulates. Molecular Biology of the Cell. 2002; 13(1):317-335. DOI: 10.1091/mbc.01-06-0300

[4] Christie D, Zhu J. Transcriptional regulatory networks for CD4 T cell differentiation. Current Topics in Microbiology and Immunology. 2014; 381:125-172. DOI: 10.1007/82_2014_372

[5] Brode S, Macary PA. Cross-presentation: Dendritic cells and macrophages bite off more than they can chew! Immunology. 2004; 112(3):345-351. DOI: 10.1111/j.1365-2567.2004.01920.x

[6] Grabbe S, Kampgen E, Schuler G. Dendritic cells: Multi-lineal and multi-functional. Immunology Today. 2000; 21(9):431-433. URL: http://www.ncbi.nlm.nih.gov/pubmed/11012244

[7] Kared H, Camous X, Larbi A. T cells and their cytokines in persistent stimulation of the immune system. Current Opinion in Immunology. 2014; 29:79-85. DOI: 10.1016/j.coi.2014.05.003

[8] Trombetta ES, Ebersold M, Garrett W, Pypaert M, Mellman I. Activation of lysosomal function during dendritic cell maturation. Science. 2003; 299(5611):1400-1403. DOI: 10.1126/science.1080106

[9] Seliger B, Maeurer MJ, Ferrone S. Antigen-processing machinery breakdown and tumor growth. Immunology Today. 2000; 21(9):455-464. URL: http://www.ncbi.nlm.nih.gov/pubmed/10953098

[10] Bykovskaia SN, Shurin GV, Graner S, et al. Differentiation of immunostimulatory stem-cell- and monocyte-derived dendritic cells involves maturation of intracellular compartments responsible for antigen presentation and secretion. Stem Cells 2002; 20(5):380-393. DOI: 10.1634/stemcells.20-5-380

[11] Kaneko T, Okiji T, Kan L, Suda H, Takagi M. An immunoelectron-microscopic study of class II major histocompatibility complex molecule-expressing macrophages and dendritic cells in experimental rat periapical lesions. Archives of Oral Biology. 2001; 46(8):713-720. URL: http://www.ncbi.nlm.nih.gov/pubmed/11389863

[12] Paillot R, Laval F, Audonnet JC, Andreoni C, Juillard V. Functional and phenotypic characterization of distinct porcine dendritic cells derived from peripheral blood monocytes. Immunology. 2001; 102(4):396-404. URL: http://www.ncbi.nlm.nih.gov/pubmed/11328373

[13] Geuze HJ. The role of endosomes and lysosomes in MHC class II functioning. Immunology Today. 1998; 19(6):282-287. URL: http://www.ncbi.nlm.nih.gov/pubmed/9639994

[14] Smyth CM, Logan G, Boadle R, Rowe PB, Smythe JA, Alexander IE. Differential subcellular localization of CD86 in human PBMC-derived macrophages and DCs, and ultrastructural characterization by immuno-electron microscopy. International Immunology. 2005; 17(2):123-132. DOI: 10.1093/intimm/dxh193

[15] Platt CD, Ma JK, Chalouni C, Ebersold M, Bou-Reslan H, Carano RA, Mellman I, Delamarre L. Mature dendritic cells use endocytic receptors to capture and present antigens. Proceedings of the National Academy of Sciences of the United States of America. 2010; 107(9):4287-4292. DOI: 10.1073/pnas.0910609107

[16] ten Broeke T, van Niel G, Wauben MH, Wubbolts R, Stoorvogel W. Endosomally stored MHC class II does not contribute to antigen presentation by dendritic cells at inflammatory conditions. Traffic. 2011; 12(8):1025-1036. DOI: 10.1111/j.1600-0854.2011.01212.x

[17] Ackerman AL, Giodini A, Cresswell P. A role for the endoplasmic reticulum protein retrotranslocation machinery during crosspresentation by dendritic cells. Immunity. 2006; 25(4):607-617. DOI: 10.1016/j.immuni.2006.08.017

[18] Joffre OP, Segura E, Savina A, Amigorena S. Cross-presentation by dendritic cells. Nature Reviews Immunology. 2012; 12(8):557-569. DOI: 10.1038/nri3254

[19] Segura E, Amigorena S. Cross-presentation by human dendritic cell subsets. Immunology Letters. 2014; 158(1-2):73-78. DOI: 10.1016/j.imlet.2013.12.001

[20] Basu S, Binder RJ, Ramalingam T, Srivastava PK. CD91 is a common receptor for heat shock proteins gp96, hsp90, hsp70, and calreticulin. Immunity. 2001; 14(3):303-313. URL: http://www.ncbi.nlm.nih.gov/pubmed/11290339

[21] Berwin B, Hart JP, Rice S, Gass C, Pizzo SV, Post SR, Nicchitta CV. Scavenger receptor-A mediates gp96/GRP94 and calreticulin internalization by antigen-presenting cells. EMBO Journal. 2003; 22(22):6127-6136. DOI: 10.1093/emboj/cdg572

[22] Binder RJ, Han DK, Srivastava PK. CD91: A receptor for heat shock protein gp96. Nature Immunology. 2000; 1(2):151-155. DOI: 10.1038/77835

[23] Delneste Y, Magistrelli G, Gauchat J, et al. Involvement of LOX-1 in dendritic cell-mediated antigen cross-presentation. Immunity. 2002; 17(3):353-362. URL: http://www.ncbi.nlm.nih.gov/pubmed/12354387

[24] Guermonprez P, Valladeau J, Zitvogel L, Thery C, Amigorena S. Antigen presentation and T cell stimulation by dendritic cells. Annual Review of Immunology. 2002; 20:621-667. DOI: 10.1146/annurev.immunol.20.100301.064828

[25] Carbone FR, Kurts C, Bennett SR, Miller JF, Heath WR. Cross-presentation: A general mechanism for CTL immunity and tolerance. Immunology Today 1998; 19(8): 368-373. URL: http://www.ncbi.nlm.nih.gov/pubmed/9709505

[26] Liu K, Nussenzweig MC. Origin and development of dendritic cells. Immunological Reviews. 2010; 234(1):45-54. DOI: 10.1111/j.0105-2896.2009.00879.x

[27] Makala LH, Nagasawa H. Dendritic cells: A specialized complex system of antigen presenting cells. Journal of Veterinary Medical Science/Japanese Society of Veterinary Science. 2002; 64(3):181-193. URL: http://www.ncbi.nlm.nih.gov/pubmed/11999435

[28] Mackensen A, Herbst B, Kohler G, et al. Delineation of the dendritic cell lineage by generating large numbers of Birbeck granule-positive Langerhans cells from human peripheral blood progenitor cells in vitro. Blood. 1995; 86(7):2699-2707. URL: http://www.ncbi.nlm.nih.gov/pubmed/7545468

[29] Manz MG, Traver D, Miyamoto T, Weissman IL, Akashi K. Dendritic cell potentials of early lymphoid and myeloid progenitors. Blood. 2001; 97(11):3333-3341. URL: http://www.ncbi.nlm.nih.gov/pubmed/11369621

[30] Ardavin C, Martinez del Hoyo G, Martin P, et al. Origin and differentiation of dendritic cells. Trends in Immunology. 2001; 22(12):691-700. URL: http://www.ncbi.nlm.nih.gov/pubmed/11739000

[31] Zhou L, Chong MM, Littman DR. Plasticity of CD4+ T cell lineage differentiation. Immunity. 2009; 30(5):646-655. DOI: 10.1016/j.immuni.2009.05.001

[32] Korn T, Bettelli E, Oukka M, Kuchroo VK. IL-17 and Th17 Cells. Annual Review of Immunology. 2009; 27:485-517. DOI: 10.1146/annurev.immunol.021908.132710

[33] Josefowicz SZ, Lu LF, Rudensky AY. Regulatory T cells: Mechanisms of differentiation and function. Annual Review of Immunology. 2012; 30:531-564. DOI: 10.1146/annurev.immunol.25.022106.141623

[34] O'Shea JJ, Plenge R. JAK and STAT signaling molecules in immunoregulation and immune-mediated disease. Immunity. 2012; 36(4):542-550. DOI: 10.1016/j.immuni.2012.03.014

[35] Haniffa M, Bigley V, Collin M. Human mononuclear phagocyte system reunited. Seminars in Cell & Developmental Biology. 2015; 41:59-69. DOI: 10.1016/j.semcdb.2015.05.004

[36] Chistiakov DA, Sobenin IA, Orekhov AN, Bobryshev YV. Myeloid dendritic cells: Development, functions, and role in atherosclerotic inflammation. Immunobiology. 2015; 220(6):833-844. DOI: 10.1016/j.imbio.2014.12.010

[37] Scharf L, Li NS, Hawk AJ, et al. The 2.5 A structure of CD1c in complex with a mycobacterial lipid reveals an open groove ideally suited for diverse antigen presentation. Immunity. 2010; 33(6):853-862. DOI: 10.1016/j.immuni.2010.11.026

[38] Schüller SS, Sadeghi K, Wisgrill L, et al. Preterm neonates display altered plasmacytoid dendritic cell function and morphology. Journal of Leukocyte Biology. 2013; 93(5):781-788. DOI: 10.1189/jlb.1011525

[39] van der Aa E, van Montfoort N, Woltman AM. BDCA3(+)CLEC9A(+) human dendritic cell function and development. Seminars in Cell & Developmental Biology. 2015; 41:39-48. DOI: 10.1016/j.semcdb.2014.05.016

[40] Arnold-Schrauf C, Berod L, Sparwasser T. Dendritic cell specific targeting of MyD88 signalling pathways in vivo. European Journal of Immunology. 2015; 45(1):32-39. DOI: 10.1002/eji.201444747

[41] Maverakis E, Kim K, Shimoda M, et al. Glycans in the immune system and The Altered Glycan Theory of Autoimmunity: A critical review. Journal of Autoimmunity. 2015; 57:1-13. DOI: 10.1016/j.jaut.2014.12.002

[42] Trda L, Boutrot F, Claverie J, Brule D, Dorey S, Poinssot B. Perception of pathogenic or beneficial bacteria and their evasion of host immunity: Pattern recognition receptors in the frontline. Frontiers in Plant Science. 2015; 6:219. DOI: 10.3389/fpls.2015.00219

[43] Land WG. The role of damage-associated molecular patterns in human diseases: Part I – Promoting inflammation and immunity. Sultan Qaboos University Medical Journal. 2015; 15(1):e9-e21. URL: http://www.ncbi.nlm.nih.gov/pubmed/25685392

[44] Land WG. The role of damage-associated molecular patterns (DAMPs) in human diseases: Part II: DAMPs as diagnostics, prognostics and therapeutics in clinical medicine. Sultan Qaboos University Medical Journal. 2015; 15(2):e157-170. URL http://www.ncbi.nlm.nih.gov/pubmed/26052447

[45] Lotze MT, Deisseroth A, Rubartelli A. Damage associated molecular pattern molecules. Clinical Immunology. 2007; 124(1):1-4. DOI: 10.1016/j.clim.2007.02.006

[46] Venereau E, Ceriotti C, Bianchi ME. DAMPs from cell death to new life. Frontiers in Immunology. 2015; 6:422. DOI: 10.3389/fimmu.2015.00422

[47] Strunk D, Rappersberger K, Egger C, Strobl H, Kromer E, Elbe A, Maurer D, Stingl G. Generation of human dendritic cells/Langerhans cells from circulating CD34+ hematopoietic progenitor cells. Blood. 1996; 87(4):1292-1302. URL: http://www.ncbi.nlm.nih.gov/pubmed/8608217

[48] Chang SY, Song JH, Guleng B, et al. Circulatory antigen processing by mucosal dendritic cells controls CD8(+) T cell activation. Immunity. 2013; 38(1):153-165. DOI: 10.1016/j.immuni.2012.09.018

[49] Niess JH, Brand S, Gu X, et al. CX3CR1-mediated dendritic cell access to the intestinal lumen and bacterial clearance. Science. 2005; 307(5707):254-258. DOI: 10.1126/science.1102901

[50] Rescigno M, Rotta G, Valzasina B, Ricciardi-Castagnoli P. Dendritic cells shuttle microbes across gut epithelial monolayers. Immunobiology. 2001; 204(5):572-581. DOI: 10.1078/0171-2985-00094

[51] Neumüller J, Neumüller-Guber SE, Lipovac M, Mosgoeller W, Vetterlein M, Pavelka M, Huber J. Immunological and ultrastructural characterization of endothelial cell cultures differentiated from human cord blood derived endothelial progenitor cells. Histochemistry and Cell Biology. 2006; 126(6):649-664. DOI: 10.1007/s00418-006-0201-6

[52] Srisen K, Rohrl C, Meisslitzer-Ruppitsch C, Ranftler C, Ellinger A, Pavelka M, Neumüller J. Human endothelial progenitor cells internalize high-density lipoprotein. PloS One. 2013; 8(12):e83189. DOI: 10.1371/journal.pone.0083189

[53] Strobl H, Knapp W. TGF-beta1 regulation of dendritic cells. Microbes and Infection/Institut Pasteur. 1999; 1(15):1283-1290. URL: http://www.ncbi.nlm.nih.gov/pubmed/10611756

[54] Strobl H, Riedl E, Scheinecker C, Bello-Fernandez C, Pickl WF, Rappersberger K, Majdic O, Knapp W. TGF-beta 1 promotes in vitro development of dendritic cells from CD34+ hemopoietic progenitors. Journal of Immunology. 1996; 157(4):1499-1507. URL: http://www.ncbi.nlm.nih.gov/pubmed/8759731

[55] Romani N, Reider D, Heuer M, Ebner S, Kampgen E, Eibl B, Niederwieser D, Schuler G. Generation of mature dendritic cells from human blood. An improved method with special regard to clinical applicability. Journal of Immunological Methods. 1996; 196(2):137-151. URL: http://www.ncbi.nlm.nih.gov/pubmed/8841452

[56] Reis e Sousa C. Dendritic cells in a mature age. Nature Reviews Immunology. 2006; 6(6):476-483. DOI: 10.1038/nri1845

[57] Fujimoto Y, Tedder TF. CD83: A regulatory molecule of the immune system with great potential for therapeutic application. Journal of Medical and Dental Sciences. 2006; 53(2):85-91. URL: http://www.ncbi.nlm.nih.gov/pubmed/16913569

[58] Carreras E, Turner S, Paharkova-Vatchkova V, Mao A, Dascher C, Kovats S. Estradiol acts directly on bone marrow myeloid progenitors to differentially regulate GM-CSF or Flt3 ligand-mediated dendritic cell differentiation. Journal of Immunology. 2008; 180(2):727-738. URL: http://www.ncbi.nlm.nih.gov/pubmed/18178810

[59] Heldring N, Pike A, Andersson S, et al. Estrogen receptors: How do they signal and what are their targets. Physiological Reviews. 2007; 87(3):905-931. DOI: 10.1152/physrev.00026.2006

[60] Kovats S. Estrogen receptors regulate innate immune cells and signaling pathways. Cellular Immunology. 2015; 294(2):63-69. DOI: 10.1016/j.cellimm.2015.01.018

[61] Mann M, Cortez V, Vadlamudi RK. Epigenetics of estrogen receptor signaling: role in hormonal cancer progression and therapy. Cancers. 2011; 3(3):1691-1707. DOI: 10.3390/cancers3021691

[62] Paharkova-Vatchkova V, Maldonado R, Kovats S. Estrogen preferentially promotes the differentiation of CD11c+ CD11b(intermediate) dendritic cells from bone marrow precursors. Journal of Immunology. 2004; 172(3):1426-1436. URL: http://www.ncbi.nlm.nih.gov/pubmed/14734718

[63] Carreras E, Turner S, Frank MB, et al. Estrogen receptor signaling promotes dendritic cell differentiation by increasing expression of the transcription factor IRF4. Blood. 2010; 115(2):238-246. DOI: 10.1182/blood-2009-08-236935

[64] Nalbandian G, Paharkova-Vatchkova V, Mao A, Nale S, Kovats S. The selective estrogen receptor modulators, tamoxifen and raloxifene, impair dendritic cell differentiation and activation. Journal of Immunology. 2005; 175(4):2666-2675. URL: http://www.ncbi.nlm.nih.gov/pubmed/16081843

[65] Huck B, Steck T, Habersack M, Dietl J, Kammerer U. Pregnancy associated hormones modulate the cytokine production but not the phenotype of PBMC-derived human dendritic cells. European Journal of Obstetrics, Gynecology, and Reproductive Biology. 2005; 122(1):85-94. DOI: 10.1016/j.ejogrb.2005.02.017

[66] Piccinni MP, Scaletti C, Maggi E, Romagnani S. Role of hormone-controlled Th1- and Th2-type cytokines in successful pregnancy. Journal of Neuroimmunology. 2000; 109(1):30-33. URL: http://www.ncbi.nlm.nih.gov/pubmed/10969178

[67] Xu Y, He H, Li C, Shi Y, Wang Q, Li W, Song W. Immunosuppressive effect of progesterone on dendritic cells in mice. Journal of Reproductive Immunology. 2011; 91(1-2):17-23. DOI: 10.1016/j.jri.2011.06.101

[68] Quispe Calla NE, Ghonime MG, Cherpes TL, Vicetti Miguel RD. Medroxyprogesterone acetate impairs human dendritic cell activation and function. Human Reproduction. 2015; 30(5):1169-1177. DOI: 10.1093/humrep/dev035

[69] Ghosh M, Rodriguez-Garcia M, Wira CR. The immune system in menopause: Pros and cons of hormone therapy. Journal of Steroid Biochemistry and Molecular Biology. 2014; 142:171-175. DOI: 10.1016/j.jsbmb.2013.09.003

[70] Gameiro C, Romao F. Changes in the immune system during menopause and aging. Frontiers in Bioscience. 2010; 2:1299-1303. URL: http://www.ncbi.nlm.nih.gov/pubmed/20515802

[71] Ljunggren HG. Dendritic cells, dendritic cell-based vaccines and Ralph Steinman. Journal of Internal Medicine. 2012; 271(2):174-176. DOI: 10.1111/j.1365-2796.2011.02495.x

[72] Steinman RM, Banchereau J. Taking dendritic cells into medicine. Nature. 2007; 449(7161):419-426. DOI: 10.1038/nature06175

[73] Gieseler R, Heise D, Soruri A, Schwartz P, Peters JH. In-vitro differentiation of mature dendritic cells from human blood monocytes. Developmental Immunology. 1998; 6(1-2):25-39. URL: http://www.ncbi.nlm.nih.gov/pubmed/9716903

[74] Lenzner S, Scholtes U, Peters JH. Focussing human B cell specificity by immunoselection via antigen-presenting cells in vitro. Immunobiology. 1998; 198(5):539-551. DOI: 10.1016/S0171-2985(98)80077-6

[75] Soruri A, Zwirner J. Dendritic cells: Limited potential in immunotherapy. International Journal of Biochemistry & Cell Biology. 2005; 37(2):241-245. DOI: 10.1016/j.biocel.2004.07.003

[76] Tran Janco JM, Lamichhane P, Karyampudi L, Knutson KL. Tumor-infiltrating dendritic cells in cancer pathogenesis. Journal of Immunology. 2015; 194(7):2985-2991. DOI: 10.4049/jimmunol.1403134

[77] Cheng P, Corzo CA, Luetteke N, et al. Inhibition of dendritic cell differentiation and accumulation of myeloid-derived suppressor cells in cancer is regulated by S100A9 protein. Journal of Experimental Medicine. 2008; 205(10):2235-2249. DOI: 10.1084/jem.20080132

[78] Sznol M, Chen L. Antagonist antibodies to PD-1 and B7-H1 (PD-L1) in the treatment of advanced human cancer. Clinical Cancer Research: An Official Journal of the American Association for Cancer Research. 2013; 19(5):1021-1034. DOI: 10.1158/1078-0432.CCR-12-2063

[79] Cubillos-Ruiz JR, Baird JR, Tesone AJ, et al. Reprogramming tumor-associated dendritic cells in vivo using miRNA mimetics triggers protective immunity against ovarian cancer. Cancer Research. 2012; 72(7):1683-1693. DOI: 10.1158/0008-5472.CAN-11-3160

[80] Sabado RL, Bhardwaj N. Directing dendritic cell immunotherapy towards successful cancer treatment. Immunotherapy. 2010; 2(1):37-56. DOI: 10.2217/imt.09.43

[81] Brockstedt DG, Dubensky TW. Promises and challenges for the development of Listeria monocytogenes-based immunotherapies. Expert Review of Vaccines. 2008; 7(7):1069-1084. DOI: 10.1586/14760584.7.7.1069

[82] Jenne L, Schuler G, Steinkasserer A. Viral vectors for dendritic cell-based immunotherapy. Trends in Immunology. 2001; 22(2):102-107. URL: http://www.ncbi.nlm.nih.gov/pubmed/11286712

# Permissions

All chapters in this book were first published by InTech Open; hereby published with permission under the Creative Commons Attribution License or equivalent. Every chapter published in this book has been scrutinized by our experts. Their significance has been extensively debated. The topics covered herein carry significant findings which will fuel the growth of the discipline. They may even be implemented as practical applications or may be referred to as a beginning point for another development.

The contributors of this book come from diverse backgrounds, making this book a truly international effort. This book will bring forth new frontiers with its revolutionizing research information and detailed analysis of the nascent developments around the world.

We would like to thank all the contributing authors for lending their expertise to make the book truly unique. They have played a crucial role in the development of this book. Without their invaluable contributions this book wouldn't have been possible. They have made vital efforts to compile up to date information on the varied aspects of this subject to make this book a valuable addition to the collection of many professionals and students.

This book was conceptualized with the vision of imparting up-to-date information and advanced data in this field. To ensure the same, a matchless editorial board was set up. Every individual on the board went through rigorous rounds of assessment to prove their worth. After which they invested a large part of their time researching and compiling the most relevant data for our readers.

The editorial board has been involved in producing this book since its inception. They have spent rigorous hours researching and exploring the diverse topics which have resulted in the successful publishing of this book. They have passed on their knowledge of decades through this book. To expedite this challenging task, the publisher supported the team at every step. A small team of assistant editors was also appointed to further simplify the editing procedure and attain best results for the readers.

Apart from the editorial board, the designing team has also invested a significant amount of their time in understanding the subject and creating the most relevant covers. They scrutinized every image to scout for the most suitable representation of the subject and create an appropriate cover for the book.

The publishing team has been an ardent support to the editorial, designing and production team. Their endless efforts to recruit the best for this project, has resulted in the accomplishment of this book. They are a veteran in the field of academics and their pool of knowledge is as vast as their experience in printing. Their expertise and guidance has proved useful at every step. Their uncompromising quality standards have made this book an exceptional effort. Their encouragement from time to time has been an inspiration for everyone.

The publisher and the editorial board hope that this book will prove to be a valuable piece of knowledge for researchers, students, practitioners and scholars across the globe.

# List of Contributors

**Luděk Frank**
Institute of Scientific Instruments, Czech Academy of Sciences, Brno, Czech Republic

**Mohsen Asadi Asadabad and Mohammad Jafari Eskandari**
Materials Research School, NSTRI, Isfahan, Iran

**Abed Malti**
INRIA, France

**Meltem Sezen**
Sabanci University Nanotechnology Research and Application Center (SUNUM), Orhanlı-Tuzla, Istanbul, Turkey

**Jitka Stráská, Josef Stráský, Peter Minárik, Miloš Janeček and Robert Král**
Charles University in Prague, Department of Physics of Materials, Prague, Czech Republic

**Hideki Masuda**
National Institute for Materials Science, Tsukuba, Japan

**Diogo Benchimol de Souza and Bianca Martins Gregório**
Urogenital Research Unit, Department of Anatomy, State University of Rio de Janeiro, Rio de Janeiro, Brazil

**Marlene Benchimol**
Unigranrio University, Rio de Janeiro, Brazil

**Fernanda Amorim de Morais Nascimento**
Urogenital Research Unit, Department of Anatomy, State University of Rio de Janeiro, Rio de Janeiro, Brazil
Basic Nutrition and Dietetics Nucleus, Federal University of Rio de Janeiro, Macaé, Brazil

**Xiongwu Wu and Bernard R. Brooks**
Laboratory of Computational Biology, National Heart, Lung, and Blood Institute, National Institutes of Health, Bethesda, USA

**Mogana Das Murtey**
Cluster of Integrative Medicine, Advanced Medical and Dental Institute (AMDI), Universiti Sains Malaysia , Bertam Kepala Batas, Penang, Malaysia

**Patchamuthu Ramasamy**
Quest International University Perak (QIUP), Ipoh, Perak Darul Ridzuan, Malaysia

**Adolf Ellinger**
Medical University of Vienna, Center for Anatomy and Cell Biology, Department of Cell Biology and Ultrastructure Research, Vienna, Austria

**Josef Neumüller**
Medical University of Vienna, Center for Anatomy and Cell Biology, Department of Cell Biology and Ultrastructure Research, Vienna, Austria
Blood Donation Center of the Austrian Red Cross for Vienna, Lower Austria and Burgenland, Vienna, Austria

**Sylvia Emanuela Neumüller-Guber**
Medical University of Innsbruck, Sektion for Virologie, Department for Hygiene, Microbiology, Social Medicine, Innsbruck, Austria

**Johannes Huber**
Medical University of Vienna, Department of Gynecologic Endocrinology and Reproductive Medicine, Vienna, Austria

**Thomas Wagner**
Medical University of Graz, Department of Blood Group Serology and Transfusion Medicine, Graz, Austria

# Index

www.ingramcontent.com/pod-product-compliance
Lightning Source LLC
LaVergne TN
LVHW080807220926
848485LV00011B/286
* 9 7 8 1 6 4 7 2 8 4 5 3 4 *